Green Energy and Technology

Ayhan Demirbas

Biohydrogen

For Future Engine Fuel Demands

Springer

Ayhan Demirbas, Professor of Energy Technology
Sila Science and Energy
Trabzon
Turkey

ISSN 1865-3529
ISBN 978-1-4471-2286-9 e-ISBN 978-1-84882-511-6
DOI 10.1007/978-1-84882-511-6
Springer Dordrecht Heidelberg London New York

British Library Cataloguing in Publication Data
A catalogue record for this book is available from the British Library

Cover design: WMXDesign, Heidelberg, Germany

Printed on acid-free paper

Springer is part of Springer Science+Business Media (www.springer.com)

Preface

Today's world faces three critical problems: (1) high fuel prices, (2) climatic changes, and (3) air pollution. Experts suggest that current oil and gas reserves are only sufficient to last a few more decades. It is well-known that transport is almost totally dependent on fossil fuels, particularly petroleum-based fuels such as gasoline, diesel fuel, liquefied petroleum gas, and compressed natural gas. Petroleum-based fuels are well-established products that have served industry and consumers for more than one hundred years, and for the foreseeable future, automotive fuels will still be largely based on liquid hydrocarbons. However, time is running out and petroleum, once considered inexhaustible, is now being depleted at a rapid rate. As the amount of available petroleum decreases, the need for alternate technologies to produce liquid fuels that could potentially help prolong the liquid fuels culture and mitigate the forthcoming effects of the shortage of transportation fuels increases. There are several reasons for biofuels to be considered as relevant technologies by both developing and industrialized countries. They include energy security reasons, environmental concerns, foreign exchange savings, and socioeconomic issues related to the rural sectors of all countries in the world.

This book discusses the production, storage, transportation, usage, economy, policy, and environmental impacts of biohydrogen. Biohydrogen is a replacement for fossil and biorenewable liquid fuels. In this book, the modern biomass-based transportation fuels bioethanol, biomethanol, biodiesel, bio-oil, biogas, and biohydrogen are briefly reviewed. Biomass conversion technologies are important for obtaining biofuels such as bioethanol, biodiesel, bio-oil, and biohydrogen.

A comprehensive definition of biohydrogen is hydrogen produced chemically, thermochemically, biologically, biochemically, and biophotolytically from all biomass materials. Biohydrogen is a renewable biofuel produced from biorenewable feedstocks by chemical, thermochemical, biological, biochemical, and biophotolytical methods. Biohydrogen is an environmentally friendly alternative automotive fuel that can be used in an internal combustion engine.

Hydrogen can be produced from biorenewable feedstocks via thermochemical conversion processes such as pyrolysis, gasification, steam gasification, steam reforming of bio-oils, and supercritical water gasification (SWG) of biomass. Hydrogen can also be produced by renewable biological systems. For example, there are three types of microorganisms that generate hydrogen: cyanobacteria, anaerobic bacteria, and fermentative bacteria.

The hydrogen economy is a vision for a future in which hydrogen replaces fossil fuels. Economically, the wasteful hydrogen process translates to electricity from hydrogen and fuel cells costing at least four times as much as electricity from the grid. In fact, electricity would be used much more efficiently if it were sent directly to the appliances instead. The transition to a hydrogen economy would require a huge investment in new infrastructure to produce, store and deliver hydrogen to end-users, to establish hydrogen stationary systems, as well as to develop and manufacture fuel cells. The transition to the hydrogen-powered system could take several decades due to the slow turnover of the existing stock of capital. The transition to a hydrogen economy is likely to begin later in most developing economies than in industrialized countries.

Hydrogen is a synthetic energy carrier. The synthesis of hydrogen requires energy. Since production, packaging, storage, transfer and delivery of hydrogen gas, in essence all key components of an economy, are extremely energy consuming, alternatives should be considered. The production technology would be site-specific and include steam reforming of methane and electrolysis in hydropower-rich countries. Conventionally produced hydrogen gas costs about twice that of natural gas or oil and about three times that of coal. At present only the space industry seems to be willing to pay the high cost of hydrogen energy.

Hydrogen can be transported by two systems: (1) a road delivery system (cryogenic liquid trucks, compressed tube trailers), and (2) a pipeline delivery system. Placing the pipelines in sewers, securing utility status, or converting existing natural gas pipelines to carry a mixture of hydrogen and natural gas could reduce hydrogen pipeline costs.

Hydrogen could be a peaceful energy carrier for all countries. Hydrogen might be the next great fuel, as it is available worldwide and water is its only byproduct. Hydrogen has received increased attention as a renewable and environmentally friendly option to help meet today's energy needs.

Policy makers will need to pay more attention to the implications for the transition to a hydrogen economy. A major dilemma now faced by developing countries is how to invest in hydrogen research and development for the transition to a hydrogen economy. Hydrogen's share in the energy market is increasing with the implementation of fuel cell systems for sustainable energy supply. The concept of sustainable development embodies the idea of the interlinking and balance between economic, social and environmental concerns.

Hydrogen is currently more expensive than conventional energy sources. There are different technologies presently being practiced to produce hydrogen economically from biomass. Biohydrogen technology will play a major role in the future since it is able to utilize renewable sources of energy.

The use of biohydrogen for transport in order to reduce greenhouse gas emissions and the environmental impact of transport has a strategic importance. Since hydrogen can be burned in such a way that it produces no harmful emissions, this makes it an attractive alternative to fossil fuels. But hydrogen is only as clean as the technologies used to produce and use it. If hydrogen is produced without emitting any carbon dioxide or other climate-destabilizing greenhouse gases, it could form the basis of a truly sustainable energy system.

Vehicles and stationary power generation fueled by hydrogen are zero-emission devices at the point of use, which benefits local air quality. Hydrogen-powered fuel cells could contribute to reducing or eliminating emissions of carbon dioxide and other greenhouse gases from road transportation vehicles. The production of hydrogen electrolytically, using clean solar power or other forms of renewable energy is essentially pollution-free. The feedstock, water, is composed of hydrogen and oxygen. Hydrogen production or distribution would produce no CO2. An internal combustion engine fueled by hydrogen can be adjusted so that the emission of NOx is 200 times less than in present vehicles. Emissions of NOx increase with the combustion temperature, the length of the high-temperature combustion period, and the availability of hydrogen, up to a point.

This book on biohydrogen attempts to address the needs of energy researchers, chemical engineers, energy specialists, engineers, agriculturists, crop cultivators, fuel processors, policy makers, environmentalists, environmental engineers, automobile engineers, college students, research faculty and others interested in a practical tool for pursuing their interests in relation to bioenergy. Each chapter in the book begins with fundamental explanations for general readers and ends with in-depth scientific details suitable for expert readers. The book may even be adopted as a textbook for college courses that deal with biohydrogen.

Trabzon, Turkey, November 2008 *Ayhan Demirbas*

Contents

Chapter 1
Introduction

1.1 Introduction

At the start of the twenty-first century, we are facing significant energy challenges. The world's energy requirements are currently satisfied by fossil fuels, which serve as the primary energy source. Consequently, overwhelming scientific evidence concludes that this unfettered use of fossil fuels has caused the world's climate to change, with potentially disastrous effect. Moreover, oil crises, which surfaced in 1973, provided a reminder that breaking an energy paradigm based on fossil fuel dependency would lead to economic and environmental advantages. In response, a sustained program of research and development into many areas of hydrogen as an energy carrier started in 1977, with the initiation of the International Energy Agency (Vijayaraghavan and Soom, 2006).

Hydrogen is seen as the energy carrier of the future. Hydrogen is storable and transferable, has high heat value per mass unit, and its sources are globally distributed. Hydrogen is strategically important as it has low emission, is environmentally benign, and represents a cleaner and more sustainable energy system. Hydrogen seems to be the future energy carrier by virtue of being renewable. All primary energy sources can be used in the hydrogen-producing process (Veziroglu, 1975). Hydrogen produced from renewable resources such as water, organic waste or biomass, either biologically or photobiologically, is termed as "biohydrogen." Biohydrogen can be generated from renewable sources, namely biomass, and more particularly by the photobiological method and dark fermentation process. Biohydrogen production has several advantages when compared to photoelectrochemical or thermochemical processes due to the low energy requirement and investment cost. Most of the biologically produced hydrogen in the biosphere is derived from microbial fermentation processes. These organisms decompose organic matter into carbon dioxide and hydrogen (Gorman, 2002).

Hydrogen production from fossil fuels is based on steam reforming of natural gas, thermal cracking of natural gas, partial oxidation of heavier than naphtha hydrocarbons, coal gasification and pyrolysis or gasification. Among these methods, the steam reforming process alone produces nearly 90% of hydrogen. The biological methods of hydrogen production are based on biophotolysis of water by algae and cyanobacteria, photodecomposition of organic compounds by photosynthetic bacteria, fermentative hydrogen production from organic compounds and hybrid systems using photosynthetic and fermentative bacteria. Hydrogen production from water is based on electrolysis, photolysis, thermochemical process and direct thermal decomposition or thermolysis (Das and Veziroglu, 2001).

1.1.1 Global Energy Sources and the Present Energy Situation

Energy affects all aspects of modern life. Energy plays a vital role in our everyday lives. Energy is one of the vital inputs to the socioeconomic development of any country. The demand for energy is increasing at an exponential rate due to the exponential growth of world population. An energy system includes an energy supply sector and energy end-use technologies. The objective of an energy system is to deliver to consumers the benefits that energy use offers, including illumination, cooked food, comfortable indoor temperatures, refrigeration, telecommunications, education, and transportation. Energy end-use equipment such as stoves, light bulbs, vehicles, and machinery convert the final energy into useful energy, which provides the desired benefits: the energy services (UNDP, 2004). Heat, electricity, and vehicle fuel are the main forms of energy that people use every day.

Advanced energy-efficient technologies reduce the energy needed to provide energy services, thereby reducing environmental and national security costs of using energy and potentially increasing its reliability. Energy resources are not used for their own sake, but for the services they provide. These energy services are fundamental to our modern economy; they heat, cool, and light our buildings, power our industrial processes, process our food, fuel our transport, and energize our communications and information technologies.

Energy sources are classified into three groups: fossil, fissile and renewable. The term fossil refers to an earlier geologic age. Fossil fuels were formed a great many years ago and are not renewable. The fossil energy sources are petroleum, coal, bitumens, natural gas, oil shales, and tar sands. The main fissile energy sources are uranium and thorium. Figure 1.1 shows fossil, nuclear and renewable energy consumptions of the world in 2005.

Oil is the largest single source of energy consumed by the world's population. Unfortunately petroleum oil is in danger of becoming short in supply. Hence, the future trend is towards using alternate energy sources. Fortunately, technological development is making this transition possible. Figure 1.2 shows global oil production scenarios based on today's production. A peak in global oil production may occur between 2015 and 2030.

Roughly 80% of the world's energy comes from fossil fuels, where oil is the world's leading energy resource. The oil resources are concentrated in a few regions.

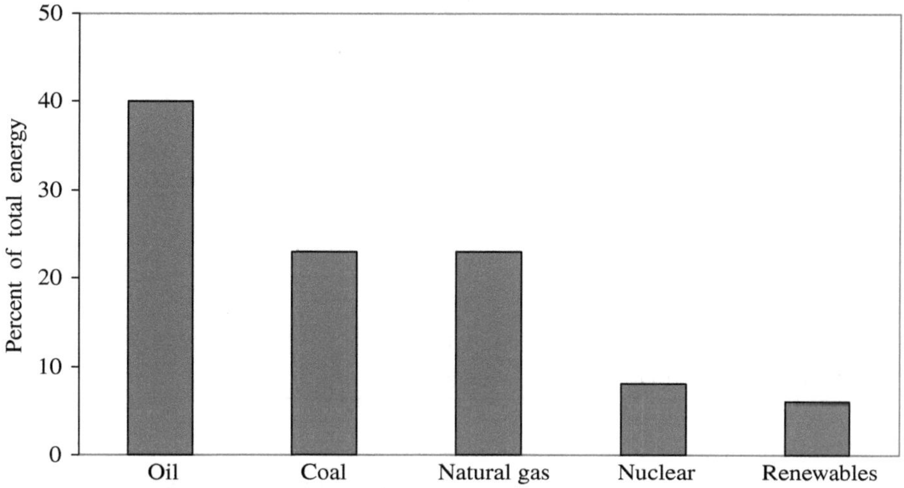

Fig. 1.1 Global energy consumption in 2005

Countries in the Middle East and Russia hold 70% of the world's dwindling reserves of oil and gas. Coal is the most abundant and widely distributed fossil fuel. Global natural gas reserves are large and currently yield a reserve/production ratio of 50 to 60 years. Nuclear power's share of worldwide electricity supplies has been steady at 16–17% for many years, but reactor safety and waste disposal problems are still matters of concern.

The renewable energy sources such as biomass, hydropower, wind, solar (thermal and photovoltaic), geothermal, marine, and hydrogen will play an important role in the world's future. Approximately, half of the global energy supply will be derived from renewables in 2040. It is also projected that electricity generation from renewables will contribute more than 80% to the total global electricity supply by the year 2040 (EWEA, 2005; EREC, 2006).

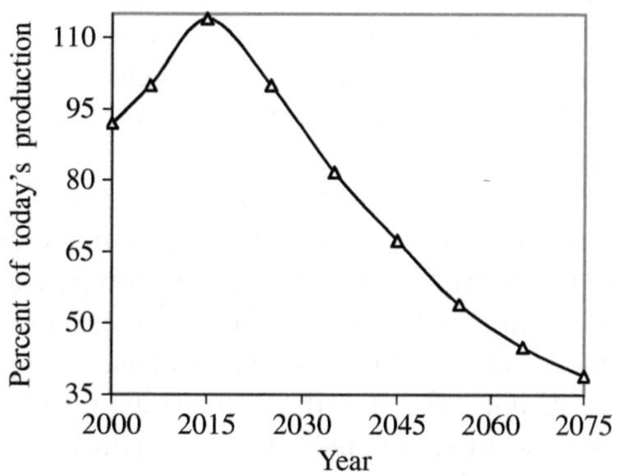

Fig. 1.2 Global oil production scenarios based on today's production. Adapted from Demirbas (2008)

Table 1.1 Global primary energy consumption, by energy source for 2002 (shares in 10.2 Gtoe)

Energy source	Percentage of total
Oil	35.1
Coal	22.6
Natural gas	21.7
Nuclear	6.9
Traditional biomass	9.3
Hydropower	2.3
Modern biomass	1.4
Geothermal, solar, wind, etc.	0.8

Renewable energy sources such as wind, solar, biomass and hydropower are important today, and they will be important in the future. Table 1.1 shows the global primary energy consumption as shares of 10.2 gigatons of oil equivalent (Gtoe) by energy source in 2002 (UNDP, 2004).

Table 1.2 shows the global electricity production as shares of 15.48 terawatt hour (TWh), per energy source for the year 2002. Hydropower is the largest and most important renewable resource and generates about 17% of the world's electricity. Non-hydro renewables are expected to make a growing contribution to global power generation. Biomass has the potential to become the world's largest and most sustainable renewable energy source. However, to move forward from this potential stage, both production and end-use technologies must be modernized. Wind is often considered to be the most advanced of the renewables, after hydropower. Geothermal is an important renewable resource and it can be utilized for base-load electricity production. The best geothermal fields are located within well-defined belts of geologic activity. Solar radiation, the earth's primary source of energy, is being increasingly utilized, while photovoltaic (PV) power generation is still the most expensive solar technology (WEC, 2004). Other energy resources are peat, oil shale, tar sand, landfill, wave, tidal, OTEC (ocean thermal energy conversion), natural bitumen and extra-heavy oil and wood.

Considering observations from the last 30 years, oil has been losing share, gas has been growing, coal has been coming down, while nuclear and hydropower are both constant at about 6%. Renewables still make a minor contribution to world's energy market.

The present world energy supply system is facing three basic problems: (1) limitation of fossil fuel resources, (2) climate change by carbon dioxide emission, and (3) insecurity by nuclear weapon competence and radioactive materials. The strategic goal therefore should be to transition to unlimited resources, use zero-emission fuels and accept no options for abuse.

About 98% of carbon emissions result from fossil fuel combustion. Reducing the use of fossil fuels would considerably reduce the amount of carbon dioxide

Table 1.2 World electricity production per energy source for 2002 (shares of 15.476 TWh). Adapted from UNDP (2004)

Energy source	Percentage of total
Coal	38.3
Natural gas	18.1
Nuclear	17.0
Hydropower	16.5
Petroleum products	7.5
Combustible renewables and waste	1.1
Small hydropower	1.0
Geothermal, solar, wind, etc.	0.6

and other pollutants produced. This can be achieved by either using less energy altogether or by replacing fossil fuels by renewable fuels. Therefore, current efforts focus on advancing technologies that emit less carbon (e.g., high-efficiency combustion) or no carbon at all, such as nuclear, hydrogen, solar, wind, geothermal, or on using energy more efficiently and on developing sequestration of carbon dioxide emitted during fossil fuel combustion.

Another problem with petroleum fuels is their uneven distribution in the world; for example, the Middle East has 63% of the global reserves and is the dominant supplier of petroleum. This energy system is unsustainable because of equity issues as well as environmental, economic, and geopolitical concerns that have far-reaching implications. Interestingly, renewable energy resources are more evenly distributed than fossil or nuclear resources. Also, the energy flows from renewable resources are more than three orders of magnitude higher than current global energy needs. Thus, sustainable renewable energy sources such as biomass, hydropower, wind, solar (both thermal and photovoltaic), geothermal, and marine will play an important role in the world's future energy supply.

1.2 Conventional Fossil Fuel Sources

Conventional fuel sources are petroleum and coal (including lignite). Our modern way of life is intimately dependent upon fossil fuels or mineral fuels. Conventional energy sources based on petroleum and coal have proven to be highly effective drivers of economic progress, but at the same time damaging to the environment and to human health (Akella et al., 2009).

One of our most important sources of energy today is oil and coal. Fossil fuels are found deposited in rock formations. Fossils are non-renewable and relatively rare resources. More importantly, the major energy demand is fulfilled by fossil fuels. Fossil fuels are fossil source fuels, that is, hydrocarbons found within the top layer of the Earth's crust. It is generally accepted that they formed from the fossilized remains of dead plants and animals by exposure to heat and pressure in the Earth's crust over hundreds of millions of years.

1.2.1 Petroleum and Heavy Oil Refining

Petroleum (word derived from the Greek *petra* meaning *rock*, and *elaion* meaning *oil*, or Latin *oleum* meaning *oil*) or crude oil, sometimes colloquially called "black gold," is a thick, dark brown or greenish liquid. It is used to describe a broad range of hydrocarbons that are found as gases, liquids, or solids, occurring in nature. Petroleum consists of a complex mixture of various hydrocarbons, largely of the alkane and aromatic compounds, but may vary much in appearance and composition. The physical properties of petroleum vary greatly. The color ranges from pale yellow through red and brown to black or greenish. The two most common forms are crude oil and natural gas. Petroleum is a fossil fuel because it was formed from the remains of tiny sea plants and animals that died millions of years ago, and sank to the bottom of the oceans. This organic mixture was subjected to enormous hydraulic pressure and geothermal heat. Over time, the mixture changed, breaking down into compounds made of hydrocarbons by reduction reactions, resulting in the formation of oil-saturated rocks. The oil rose and became trapped under non-porous rocks, which are sealed with salt or clay layers.

According to the well-accepted biogenic theory, crude oil, like coal and natural gas, is the product of compression and heating of ancient vegetation and animal remains over geological time scales. According to this theory, organic matter is formed from the decayed remains of prehistoric marine animals and terrestrial plants. Over many centuries this organic matter, mixed with mud, was buried under thick sedimentary layers. The resulting high pressure and heat caused the remains to transform, first into a waxy material known as kerogen, and then into liquid and gaseous hydrocarbons by the catagenesis process. The fluids then migrated through adjacent rock layers until they became trapped underground in porous rocks known as reservoirs, forming an oil field, from which the liquid could then be removed by drilling and pumping. These reservoirs are at different depths in different parts of the world, but typically lie between 4 and 5 km underground. The thickness of the oil layer is about 150 m, known generally as the "oil window." Three important elements of an oil reservoir are: a rich source rock, a migration conduit, and a trap (seal) that forms the reservoir.

The first oil wells were drilled in China in the fourth century, or perhaps earlier. The wells, as deep as 243 m, were drilled using bits attached to bamboo poles. The oil was burned to produce heat needed in the production of salt from brine evaporation. By the tenth century, extensive bamboo pipelines connected oil wells with salt springs. Ancient Persian tablets also indicate the medicinal and lighting uses of petroleum in the upper echelons of their society.

In the eighth century, the streets of the newly constructed Baghdad were paved with tar, derived from easily accessible petroleum from natural fields in the region.

In the ninth century, oil fields were exploited to produce naphtha in Baku, Azerbaijan. These fields, described by the geographer Masudi in the tenth century, had increased output to hundreds of shiploads in the thirteenth century as described by Marco Polo.

The modern history of petroleum began in 1846, with the discovery of refining kerosene from coal by Atlantic Canada's Abraham Pineo Gesner. Poland's Ignacy Łukasiewicz discovered a means of refining kerosene from the more readily available "rock oil" *(petr-oleum)* in 1852, and in the following year the first rock oil mine was built in Bobrka, near Krosno in southern Poland. The discovery rapidly spread around the world, and Meerzoeff built the first Russian refinery in the more established oil fields at Baku in 1861, which produced about 90% of the world's oil. In fact, the battle of Stalingrad was fought over Baku (now the capital of the Azerbaijan Republic).

James Miller Williams drilled the first commercial oil well in North America in 1858 in Oil Springs, Ontario, Canada. In the following year, Edwin Drake discovered oil near Titusville, Pennsylvania. There he pioneered a new method for producing oil from the ground, in which he drilled using piping to prevent borehole collapse, allowing for the drill to penetrate deeper into the ground. Previous methods for collecting oil had been limited. For example, ground collection of oil consisted of gathering it from where it occurred naturally, such as from oil seeps or shallow holes dug into the ground. The methods of digging large shafts into the ground also failed, as collapse from water seepage almost always occurred. The significant advancement that Drake made was to drive a 10-m iron pipe through the ground into the bedrock below. This allowed Drake to drill inside the pipe, without the hole collapsing from the water seepage. The principle behind this idea is still employed today by many companies for petroleum drilling. Drake's well was 23 m deep, which is very shallow compared to today's well depth of 1000–4000 m. Although technology has improved the odds since the time of Edwin Drake, petroleum exploration today is still a gamble. For example, only about 33 in every 100 exploratory wells have oil, and the remaining 67 come up "dry."

For about 10 years Pennsylvania was the one great oil producer of the world, however since 1870, the industry has spread all over the globe. From the time of the completion on the Baku field of the first flowing well, Russia has ranked second in the list of oil-producing countries, while the region of Galicia in Central Europe and Romania became prominent in 1878 and 1880, respectively. In time, Sumatra, Java, Burma and Borneo, where active development began in 1883, 1886, 1890 and 1896, respectively, will perhaps rank among the chief sources of oil supplies of the world.

The Middle East produces 32% of the world's oil (Table 1.3), but more importantly it has 64% of the total proven oil reserves in the world (Table 1.4). Also,

Table 1.3 Percentage of petroleum production by region

Middle East	Latin America	Eastern Europe	North America	Asia and Pacific	Africa	Western Europe
32	14	13	11	11	10	9

Table 1.4 Percentage of total proven reserves by region

Middle East	Latin America	Eastern Europe	North America	Asia and Pacific	Africa	Western Europe
64	12	6	3	4	9	2

its reserves are depleting at a slower rate than any other region in the world. The Middle East provides more than half of OPEC's total oil exports and has a major influence on the worldwide crude oil prices, despite the fact that OPEC produces less than half the oil in the world.

Crude oil is a complex mixture that is between 50 and 95% hydrocarbon by weight. Table 1.5 shows the average elemental composition of crude oil. The oil industry classifies "crude" by its production location (e.g., "West Texas Intermediate, WTI" or "Brent"), relative density (API gravity), viscosity ("light," "intermediate," or "heavy"), and sulfur content ("sweet" for low sulfur, and "sour" for high sulfur). Additional classification is due to conventional and non-conventional oil as shown in Table 1.6.

The petroleum industry can be divided into two subgroups, upstream producers (exploration, development and production of crude oil or natural gas) and downstream transporters (tanker, pipeline transport), and also includes refiners, retailers, and consumers.

Table 1.5 Average elemental composition of crude oil

Element	Percent composition
Carbon	83.0–87.0
Hydrogen	10.0–14.0
Nitrogen	0.1–2.0
Sulfur	0.05–6.0
Oxygen	0.05–1.5

Table 1.6 Classification of crude oils

Class	Viscosity of oil (measured in °API)
Light crude	> 31.1
Medium oil	22.3–31.1
Heavy oil	< 22.3
Deep sea oil above 500 m water depth	–
Extra heavy oil below (including tar sands)	< 10
Oil shale	–
Bitumen from tar sands	–

Table 1.7 Main crude oil fractions

Fraction	Boiling range (K)	Number of carbon atoms
Natural gas	< 293	C_1 to C_4
Petroleum ether	293–333	C_5 to C_6
Ligroin (light naphtha)	333–373	C_6 to C_7
Gasoline	313–478	C_5 to C_{12}, and cycloalkanes
Jet fuel	378–538	C_8 to C_{14}, and aromatics
Kerosene	423–588	C_{10} to C_{16}, and aromatics
No. 1 diesel fuel	443–595	C_9 to C_{18}, and aromatics
No. 2 diesel fuel	448–638	C_{12} to C_{24}, and aromatics
No. 3 diesel fuel	458–663	C_{10} to C_{20}, and aromatics
Gas oil (No. 4 and 5 fuel oils)	> 548	C_{12} to C_{70}, and aromatics
Gas oil (No. 6 fuel oils)	> 548	C_{12} to C_{70}, and aromatics
Lubricating oils	> 673	> C_{20}
Asphalt or petroleum coke	Non-volatile esidue	Polycyclic structures

The first step in refining crude oil involves separating the oil into different hydrocarbon fractions by distillation. Main fractions of crude oil are given in Table 1.7. Since there are a number of factors that influence the boiling point of a hydrocarbon, these petroleum fractions are complex mixtures. An oil refinery cleans and separates the crude oil into various fuels and byproducts, including gasoline, diesel fuel, heating oil, and jet fuel. Since various components boil at different temperatures, refineries use a heating process called distillation to separate the components. For example, gasoline has a lower boiling point than kerosene, allowing the two to be separated by heating to different temperatures. Another important job of the refineries is to remove contaminants from the oil, for example, sulfur from gasoline or diesel to reduce air pollution from automobile exhausts. After processing at the refinery, gasoline and other liquid products are usually shipped out through pipelines, which are the safest and cheapest way to move large quantities of petroleum across land.

1.2.2 Petroleum Products and Fuels

An important non-fuel use of petroleum is to produce chemical raw materials. The two main classes of petrochemical raw materials are olefins (including ethylene and propylene) and aromatics (including benzene and xylene isomers), both of which are produced in large quantities. A very important aspect of petrochemicals is their extremely large scale. The olefins are produced by chemical cracking by using steam or catalysts, and the aromatics are produced by catalytic reforming. These two basic building blocks serve as feedstock to produce a wide range of chemicals and materials including monomers, solvents, and adhesives. From the monomers, polymers or oligomers are produced for use as plastics, resins, fibers, elastomers, certain lubricants, and gels.

Gasoline is a blend of hydrocarbons with some contaminants, including sulfur, nitrogen, oxygen, and certain metals. The four major constituent groups of gasoline are olefins, aromatics, paraffins, and naphthenes. The important characteristics of gasoline are density, vapor pressure, distillation range, octane, and chemical composition.

Diesel fuel is produced by distilling raw oil, which is extracted from bedrock. Diesel fuel is a fossil fuel. Diesel fuel consists of hydrocarbons with between 9 and 27 carbon atoms in a chain as well as a smaller amount of sulfur, nitrogen, oxygen and metal compounds. It is a general property of hydrocarbons that the autoignition temperature is higher for more volatile hydrocarbons. The hydrocarbons present in the diesel fuels include alkanes, naphthenes, olefins and aromatics. In addition other substances are added to improve the characteristics of diesel fuel. Its boiling point is between 445 and 640 K. A good diesel fuel is characterized by low sulfur and aromatic content, good ignition quality, the right cold weather properties, low content of pollutants and also the right density, viscosity and boiling point. No. 2 diesel fuel is commonly used in diesel engines. Its boiling point is between 458 and 663 K and it includes C_{12}–C_{24} hydrocarbons and aromatics compounds.

1.2.3 Coal

The first known and the oldest fossil fuel is coal. Among all fossil energy resources, the coal reserve of the world is the largest. Coal is basically carbon, left over from bacterial action upon decaying plant matter in the absence of oxygen, usually under silt and water. The first step in coal formation yields peat, compressed plant matter, which still contains twigs and leaves. The second step is the formation of brown coal or lignite. Lignite has already lost most of the original moisture, oxygen, and nitrogen. It is widely used as a heating fuel but is of little chemical interest. In the third stage, the lignite is successively changed to subbituminous, bituminous, and anthracite coal. Not all of the deposits were subjected to the same degree of conversion. The bituminous coal is the most abundant form of coal and is the source of coke for smelting, coal tar, and many forms of chemically modified fuels. The chemical properties of typical coal samples are given in Table 1.8.

Depending on the geological age, the organic material will transform into the following members of the coal family:

1. *Lignite*. This contains considerable amount of water and volatiles and some mineral matters. It is youngest in geological age.
2. *Bituminous*. Moderate amount of moisture, volatiles, and inorganic materials are in bituminous coal.
3. *Anthracite*. This contains very little water or moisture but a large amount of carbon. It is the oldest in geological age.

Table 1.8 Chemical properties of typical coal samples

	Low-rank coal	High-volatile coal	High-rank coal
Carbon, %	75.2	82.5	90.5
Hydrogen, %	6.0	5.5	4.5
Oxygen, %	17.0	9.6	2.6
Nitrogen, %	1.2	1.7	1.9
Sulfur, %	0.6	0.7	0.5
Moisture, %	10.8	7.8	6.5
Calorific value, MJ/kg	31.4	35.0	36.0

The worldwide coal production is roughly equal to the gas production and only second to that of oil. Coal is produced in deep mines (hard coal) and in surface mines (lignite). Coal has played a key role as a primary source of organic chemicals as well as a primary energy source. Coal may become more important both as an energy source and as the source of carbon-based materials, especially aromatic chemicals in the twenty-first century (Schobert and Song, 2002). Coal accounted for 26% of the world's primary energy consumption and 37% of the energy consumed worldwide for electricity generation (Demirbas, 2006a).

The worldwide coal production and consumption in 1998 were 5043 and 5014 million short tons, respectively. The known world recoverable coal reserves in 1999 were 1087 billion short tons (AER, 1999; IEA, 2000). Coal reserves are rather evenly spread around the globe: 25% are in the USA, 16% in Russia, and 11.5% in China. Although coal is much more abundant than oil and gas on a global scale, coalfields can be depleted in a region. The known world recoverable coal reserves in 1999 were 989 billion tons.

The coal mining industry has exacted a heavy price from past generations. Despite environmental issues and competitive pressure from other fuels, coal is expected to maintain a major share of the world's future energy use. Large increases in coal use are projected for the developing countries of Asia. Coal's role in energy use worldwide has shifted substantially over the decades, from a fuel used extensively in all sectors of the economy to one that is now used primarily for electricity generation and in a few key industrial sectors, such as steel, cement, and chemicals. Although coal has lost market share to petroleum products and natural gas, it continues to be a key source of energy because of the dominant role it has maintained in its core markets and its success in penetrating markets in emerging economies. For coal to remain competitive with other sources of energy in the industrialized countries of the world, continuing technological improvements in all aspects of coal extraction are necessary.

Nearly all the different forms of coal are used in some way or other. For instance, peat has been used for burning in furnaces, lignite is used in power station and home stoves, whereas bituminous coal is used extensively for the generation of electricity. Coke is used primarily in the steel industry, where high temperatures are required.

More than 45% of the world's electricity is generated from coal and it is the major fuel for generating electricity worldwide. The known coal reserves in the world are enough for consumption for over 215 years, while the known oil reserves are only about 39 times the world's consumption and the known natural gas reserves are about 63 times the world's consumption level in 1998. Coal is not only plentiful, but also economical. Advanced coal-fired power generation technologies should be developed worldwide to generate at minimum economic coal, improve thermal efficiency and meet environmental requirements.

Sustainability goals indicate the importance of evolving fossil energy technologies towards the long-term goal of near-zero air pollutant and greenhouse gas emissions, without complicated end-of-pipe control technologies. Near-term technologies and strategies should support this long-term goal. The technological revolution under way in power generation, where advanced systems are replacing steam turbine technologies, supports this long-term goal.

Coal, as the most abundant and widely distributed fossil fuel, can be mined with modern techniques, transported and stored efficiently and cost-effectively. International coal trade is growing steadily and there is vigorous competition on supply and price. But coal's future commercial development depends critically on its environmental acceptability and in particular on the success of the power generation industry in reducing sulfurous and other polluting emissions.

The use of low-sulfur coals and the introduction of more efficient flue gas desulfurization facilities in conventional power stations have contributed to a significant lowering of emissions, but further improvement by these means is limited.

Coal may become more imported both as an energy source and as the source of organic chemicals in this century (Demirbas and Güllü, 1998). Coal science is an applied discipline, using methods of science and engineering for research into genesis, characterization, conversion and utilization of coal and related solid fuels (Demirbas, 1999).

Energy production from fossil fuels, including coal, results in high greenhouse gas emissions. Global utilization of fossil fuels for energy needs is rapidly resulting in critical environmental problems throughout the world. We now know that petroleum's success created problems with global warming, energy security, and environmental conservation.

Coal has been utilized as a primary source of energy, as well as a primary source of organic chemicals. It is a complex, heterogeneous combustible material, made up of portions that are either useful (carbon and hydrogen), or useless (diluents such as moisture, ash and oxygen, or contaminants such as sulfur and heavy metals). Coal can be defined as a sedimentary rock that burns, and as it was formed by the decomposition of plant matter, can be found in many forms. There are four classes of coal: anthracite, bituminous, subbituminous, and lignite. Elemental analysis gives empirical formulas such as $C_{137}H_{97}O_9NS$ for bituminous coal and $C_{240}H_{90}O_4NS$ for high-grade anthracite.

As early as 1800, coal gas or town gas was made by heating coal in the absence of air. Coal gas is rich in CH_4 and gives off up to 20.5 kJ per liter of gas burned. Coal gas became so popular that most major cities and many small towns had a local gas house in which it was generated, and gas burners were adjusted to burn a fuel. Coal can be converted into water gas with steam by the following (Demirbas, 2007):

$$C + H_2O \rightarrow CO + H_2 \tag{1.1}$$

Water gas burns to give CO_2 and H_2O, releasing roughly 11.2 kJ per liter of gas consumed.

Water gas, formed by the reaction of coal with oxygen and steam, is a mixture of CO, CO_2, and H_2. The ratio of H_2 to CO can be increased by adding water to this mixture, taking advantage of a reaction known as the water–gas shift reaction (Demirbas, 2007):

$$CO + H_2O \rightarrow CO_2 + H_2 \tag{1.2}$$

The concentration of CO_2 can be decreased by the reaction of CO_2 with coal at high temperatures to form CO:

$$C + CO_2 \rightarrow 2CO \tag{1.3}$$

Water gas from which the CO_2 has been removed is called synthesis gas because it can be used as a starting material for a variety of organic and inorganic compounds. For example, it can be used as the source of H_2 for the synthesis of methanol:

$$CO + 2H_2 \rightarrow CH_3OH \tag{1.4}$$

Methanol can then be used as a starting material for the synthesis of alkenes, aromatic compounds, acetic acid, formaldehyde, and ethyl alcohol (ethanol). Synthesis gas can also be used to produce methane, or synthetic natural gas (SNG) (Demirbas, 2007):

$$CO + 3H_2 \rightarrow CH_4 + H_2O \tag{1.5}$$

$$2CO + 2H_2 \rightarrow CH_4 + CO_2 \tag{1.6}$$

The first step toward making liquid fuels from coal involves the manufacture of synthesis gas (CO and H_2) from coal. In 1925, Franz Fischer and Hans Tropsch developed a catalyst that converted CO and H_2 at 1 atm and 250 to 300°C into liquid hydrocarbons. By 1941, Fischer–Tropsch plants produced 740 000 tons of petroleum products per year in Germany (Dry, 1999). Fischer–Tropsch technology is based on a complex series of reactions that use H_2 to reduce CO to CH_2 groups linked to form long-chain hydrocarbons (Schulz, 1999):

$$nCO + (n + m/2)\,H_2 \rightarrow C_nH_m + nH_2O \tag{1.7}$$

At the end of World War II, Fischer–Tropsch technology was under study in most industrial nations. Coal can be gasified to produce synthesis gas (syngas), which can be converted to paraffinic liquid fuels and chemicals by the Fischer–Tropsch synthesis. Liquid product mainly contains benzene, toluene, xylene (BTX), phenols, alkylphenols and cresol. The low cost and high availability of crude oil, however, led to a decline in interest in liquid fuels made from coal.

Another approach to liquid fuels is based on the reaction between CO and H_2 to form methanol, CH_3OH:

$$CO + 2H_2 \rightarrow CH_3OH \tag{1.8}$$

Methanol can be used directly as a transportation fuel, or it can be converted into gasoline with catalysts such as the ZSM-5 zeolite catalyst.

Methanol can also be produced from syngas with hydrogen and carbon monoxide in a 2:1 ratio. Coal-derived methanol typically has low sulfur and other impurities. Syngas from coal can be reformed by reacting with water to produce hydrogen. Ammonium sulfate from coal tar by pyrolysis can be converted to ammonia. The humus substances can be recovered from brown coal by alkali extraction.

1.3 Unconventional Fossil Fuel Sources

1.3.1 Natural Gas

Natural gas (NG) was known in England as early as 1659, but did not replace coal gas as an important source of energy in the world until after World War II. Usefulness of NG has been known for hundreds of years. The Chinese discovered a very long time ago that the energy in natural gas could be harnessed, and used to heat water. In the early days of the natural gas industry, the gas was mainly used to light streetlamps, and the occasional house.

NG is expected to be the fastest growing component of global primary energy consumption. Consumption of NG is projected to increase by an average of 2.8% annually from 2001 to 2025, compared with projected annual growth rates of 1.8% for oil consumption and 1.5% for coal (Demirbas, 2006b).

NG is found to consist mainly of the lightweight alkanes, with varying quantities of carbon dioxide, carbon monoxide, hydrogen, nitrogen and oxygen, in some cases also hydrogen sulfide and possibly ammonia. A typical sample of NG when it is collected at its source contains 80% methane (CH_4), 7% ethane (C_2H_6), 6% propane (C_3H_8), 4% butane and isobutane (C_4H_{10}), and 3% pentane (C_5H_{12}).

The reserves and resources of conventional NG are comparable in size to those of conventional oil, but global gas consumption is still considerably lower than that of oil. NG is found around the world, but the largest reserves are in the former Soviet Union and the Middle East. The proved gas reserves are unevenly distributed around the globe: 41% are in the Middle East and 27% in Russia. Global natural gas reserves by country are shown in Table 1.9.

Around the world, NG use is increasing for a variety of reasons including prices, environmental concerns, fuel diversification and/or energy security issues, market deregulation, and overall economic growth. Figure 1.3 shows production and consumption trends of natural gas in the last decades. In NG consumption, the United States ranks first, the former USSR region ranks second, and Europe ranks third. The largest NG producer is Russia, which is also the largest supplier of NG to

Western Europe. Asia and Oceania import NG to satisfy their demands. The other regions are relatively minor producers and consumers of natural gas.

NG is one of the most widely used forms of energy today. It is commonly used to heat and cool homes and businesses. NG is a very versatile fuel, which can be used for space and water heating, which have traditionally been the predominant uses of natural gas, representing about 75% and 15% of natural gas use, respectively. Furnaces which burn NG to provide space heating and hot water for homes and businesses are the largest users of NG. Ultrahigh-efficiency gas furnaces can convert over 90% of the potential energy in NG into useful heat. The generation of electricity is the other main use of natural gas. NG can be used to generate electricity in many different ways. NG power plants generate more than a couple of hundred megawatts and use the same technology as coal-fired power plants. NG is burned to produce heat, which boils water, creating steam which passes through a turbine to generate electricity.

Natural gas vehicles (NGVs) offer many benefits, from improving public health and the environment to aiding the transition to fuel cell vehicles. The following describes how NGVs are leading the way to a better tomorrow. Compared with vehicles fueled by conventional diesel and gasoline, NGVs can produce significantly lower amounts of harmful emissions such as nitrogen oxides, particulate matter, and toxic and carcinogenic pollutants. NGVs can also reduce emissions of carbon dioxide, the primary greenhouse gas. Fuel cell vehicles powered by hydrogen are the future of transportation. NG vehicle and infrastructure development can

Table 1.9 World natural gas reserves by country. Adapted from Demirbas (2008)

Country	Reserves (trillion cubic meter)	Percentage of world total
Russian Federation	48.1	33.0
Iran	23.0	15.8
Qatar	8.5	5.8
United Arab Emirates	6.0	4.1
Saudi Arabia	5.8	4.0
United States	4.7	3.3
Venezuela	4.0	2.8
Algeria	3.7	2.5
Nigeria	3.5	2.4
Iraq	3.1	2.1
Turkmenistan	2.9	2.0
Malaysia	2.3	1.6
Indonesia	2.0	1.4
Uzbekistan	1.9	1.3
Kazakhstan	1.8	1.3
Rest of World	23.8	16.5

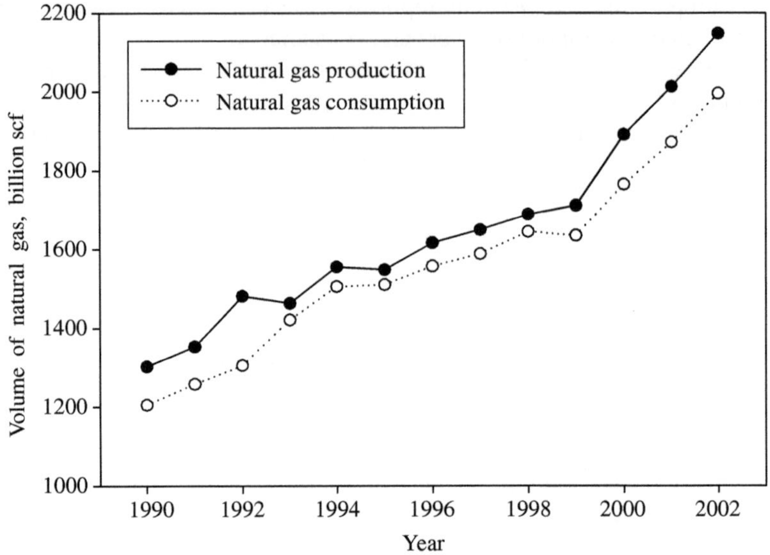

Fig. 1.3 Production and consumption trends of natural gas in the world

facilitate the transition to this technology. Despite the large difference in volumetric energy density between gasoline and compressed natural gas (CNG), the impact of CNG energy density on engine performance is less dramatic. To use CNG safely, technicians and drivers need to know what the differences are and how to work with them. In general, the use of NG results in cleaner and longer-lasting engines. Less carbon builds up on spark plugs, engine oil, and in the combustion chamber. Natural gas has higher ignition temperatures than gasoline, which increases the importance of maintaining proper ignition system operation.

Natural gas-fired combined cycles offering low costs, high efficiency, and low environmental impacts are being chosen wherever natural gas is readily available—in some countries even displacing large new hydropower projects. Co-generation is more cost-effective and can play a much larger role in the energy economy—if based on gas turbines and combined cycles rather than on steam turbines.

1.3.1.1 Gas Hydrates

Natural gas (methane) can be obtained from gas hydrates. Gas hydrates are also called clathrates or methane hydrates. Gas hydrates are potentially one of the most important energy resources for the future. Methane gas hydrates are increasingly considered a potential energy resource. Methane gas hydrates are crystalline solids formed by combination of methane and water at low temperatures and high pressures. Gas hydrates have an ice-like crystalline lattice of water molecules with methane molecules trapped inside. Enormous reserves of hydrates can be found under continental shelves and on land under permafrost. The amount of organic

Table 1.10 Worldwide amounts of organic carbon sources. Adapted from Hacisalihoglu et al. (2008)

Source of organic carbon	Amount (gigaton)
Gas hydrates (onshore and offshore)	10 000–11 000
Recoverable and non-recoverable fossil fuels (oil, coal, gas)	5000
Soil	1400
Dissolved organic matter	980
Land biota	880
Peat	500
Others	70

carbon in gas hydrates is estimated to be twice that in all other fossil fuels combined. However, due to the solid form of the gas hydrates, conventional gas and oil recovery techniques are not suitable. Table 1.10 shows worldwide amounts of organic carbon sources. The recovery of methane generally involves dissociating or melting in situ gas hydrates by heating the reservoir above the temperature of hydrate formation, or decreasing the reservoir pressure below hydrate equilibrium (Lee and Holder, 2001).

Models have been developed to evaluate natural gas production from hydrates by both depressurization and heating methods. There are three methods to obtain methane from gas hydrates: (1) the depressurization method, (2) the thermal stimulation method, and (3) the chemical inhibition method. The thermal stimulation method and the chemical inhibitor injection method are both costly procedures, whereas the depressurization method may prove useful when applied to more than one production.

1.3.2 Oil Shale (Shale Oil)

Oil shale is a sedimentary rock that contains solid hydrocarbon wax kerogen in tightly packed limy mud and clay. The kerogen may be decomposed at elevated temperatures (723 K), resulting in an oil suitable for refinery processing (Dorf, 1977). The oil shale layer was not hot enough to complete the oil generation. For the final step the kerogen must be heated up to 775 K and molecularly combine with additional hydrogen to complete the oil formation. This final process must be performed in the refinery and needs huge amounts of energy which otherwise were provided by the geological environment during oil formation. The kerogen is still in the source rock and could not accumulate in oil fields. Typically, the ratio of kerogen to waste material is very low, making the mining of oil shales unattractive. Hence, due to a combination of environmental and economic concerns, it is very unlikely that oil shale mining will ever be performed on a large scale, though in some places it is has been utilized in small quantities. Interestingly, the shale oil reserves in the world are greater than crude oil or natural gas.

1.3.3 Tar Sand Bitumen

Tar sands are oil traps not deep enough in the Earth to allow for geological conversion into the conventional oil. This oil was not heated enough to complete the process of molecular breakage to reduce the viscosity. The oil has the characteristics of bitumen and is mixed with large amounts of sand due to the proximity to the surface. The tar sand is mined, flooded with water in order to separate the heavier sand, and then processed in special refineries to reduce its high sulfur content (the original oil usually has 3–5% sulfur) and other components. This process requires huge amounts of energy and water. Only oil deposits in deep layers below 75 m are mined in situ (COSO, 2007).

1.4 Renewable Energy Sources

There are five principal renewable energy sources (RES): biomass, flowing water, heat from within the earth, the wind, and the Sun. Biomass shortly describes all plants, trees and organic matter on the Earth. Flowing water is a product of the planet's climate and geography, where modern hydroturbines use water power to generate hydroelectricity. Heat from deep within the Earth is called geothermal energy, and where the temperatures are hot enough, geothermal energy can be used to generate electricity. Natural conditions of climate and geography produce the winds that blow across the landscape, and like hydropower, modern day wind turbines produce electricity. Another renewable source is solar energy, which is simply energy that comes directly from the Sun. The Sun is a constant natural source of heat and light, and its radiation can be converted into electricity. Figure 1.4 shows the percentage of each renewable energy source used in the world (Demirbas, 2006a). The distribution of renewable energy consumption as percentage of total renewable energy in the world is: biomass 46%, hydroelectric 45%, geothermal 6%, wind 2%, and solar 1%.

All renewable energy sources can be used to produce electricity. Solar energy and geothermal energy can supply both electricity and heat. Biomass can supply all three forms of useful energy. RES are derived from those natural, mechanical, thermal and growth processes that repeat themselves within our lifetime and may be relied upon to produce predictable quantities of energy when required. Table 1.11 shows the renewable energy sources and their forms of usage (Demirbas, 2006a).

Renewable energy scenarios depend on environmental protection, which is an essential characteristic of sustainable developments. Worldwide biomass ranks fourth as an energy resource, providing approximately 14% of the world's energy needs. Biomass, mainly now represent only 3% of primary energy consumption

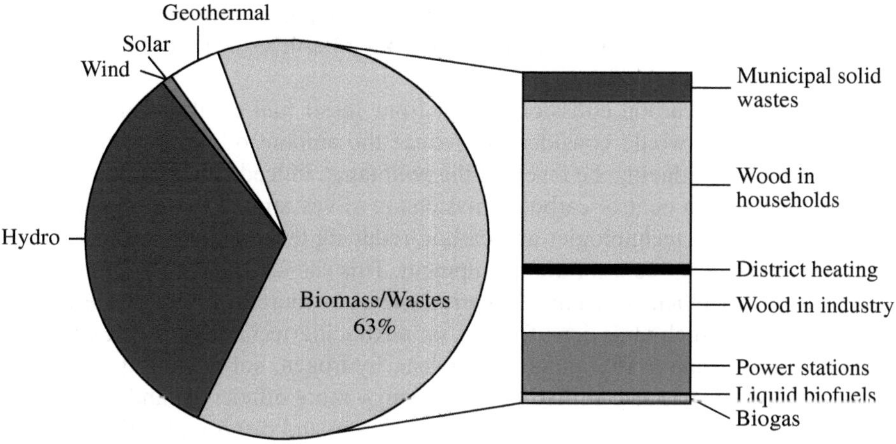

Fig. 1.4 Distribution of each renewable energy source in 1995

Table 1.11 The main renewable energy sources and their usage

Energy source	Energy conversion and usage options
Hydropower	Power generation
Modern biomass	Heat and power generation, pyrolysis, gasification, digestion
Geothermal	Urban heating, power generation, hydrothermal, hot dry rock
Solar	Solar home system, solar dryers, solar cookers
Direct solar	Photovoltaics, thermal power generation, water heaters
Wind	Power generation, wind generators, windmills, water pumps
Wave	Numerous designs
Tidal	Barrage, tidal stream

in industrialized countries. However, much of the rural population in developing countries, which represents about 50% of the world's population, is reliant on biomass, mainly in the form of wood, for fuel (Ramage and Scurlock, 1996).

Since the oil crisis of 1973, considerable progress has been made in the search for alternative energy sources. In the early 1970s, with the first oil price shock, interest in renewable energy resources re-emerged.

Renewable fuels (re-fuels) are the major alternatives to conventional fossil fuels. Renewable energy is a clean or inexhaustible energy like hydrogen energy or nuclear energy. The most important benefit of renewable energy systems is the decrease of environmental pollution. The uses of renewable energy sources (RES) instead of conventional energy sources have controlled social, economical and environmental problems (Akella et al., 2009). RES contributed 2% of the world's energy consumption in 1998, including 7 exajoules (EJ) from modern biomass and

2 EJ for all other renewables (UNDP, 2004). RES are readily available in nature. Increasing atmospheric concentrations of greenhouse gasses increase the amount of heat trapped (or decrease the heat radiated from the Earth's surface), thereby raising the surface temperature of the Earth.

About 98% of carbon emissions result from fossil fuel combustion. Reducing use of fossil fuels would considerably reduce the amount of carbon dioxide produced, as well as reducing the levels of the pollutants. Indeed, much of the variation in cost estimates to control carbon emissions revolves around the availability and cost of carbon-free technologies and carbon-reducing technologies, such as energy efficiency and energy conservation equipment. This can be achieved by either using less energy altogether, or using alternative energy resources. Much of the current effort to control such emissions focuses on advancing technologies that emit less carbon or no carbon at all, including nuclear, hydrogen, solar, wind, geothermal, other RES. Focus has also shifted to using energy more efficiently, and to developing innovative technologies and strategies to capture and dispose of carbon dioxide emitted during fossil fuel combustion.

Renewable energy is a promising alternative solution because it is clean and environmentally safe. They also produce lower or negligible levels of greenhouse gases and other pollutants when compared with the fossil energy sources they replace. Table 1.12 shows the global renewable energy scenario until 2040. Approximately, half of the global energy supply will be from renewables by 2040, according to the European Renewable Energy Council (EREC, 2006). The most significant developments in renewable energy production are observed in photovoltaics (from 0.2 to 784 Mtoe) and wind energy (from 4.7 to 688 Mtoe) between 2001 and 2040.

Table 1.12 Future global renewable energy estimates

	2001	2010	2020	2030	2040
Total consumption (million ton oil equivalent)	10 038	10 549	11 425	12 352	13 310
Biomass	1080	1313	1791	2483	3271
Large hydropower	22.7	266	309	341	358
Geothermal	43.2	86	186	333	493
Small hydropower	9.5	19	49	106	189
Wind	4.7	44	266	542	688
Solar thermal	4.1	15	66	244	480
Photovoltaic	0.2	2	24	221	784
Solar thermal electricity	0.1	0.4	3	16	68
Marine (tidal/wave/ocean)	0.05	0.1	0.4	3	20
Total renewable energy sources	1365.5	1745.5	2694.4	4289	6351
Renewable energy sources contribution (%)	13.6	16.6	23.6	34.7	47.7

RES are available throughout the world. The RES are peaceful energy sources for every country—poor or rich, developing or developed. The RES offer a number of technical and environmental benefits over conventional fossil fuels, which make them attractive as alternatives for energy needs. As before, the energy system based on fossil fuels is unsustainable due to equity issues as well as environmental, economic, and geopolitical concerns that will have implications far into the future. Hence, sustainable RES such as biomass, hydropower, wind, solar, geothermal, and marine energy sources will play an important role in the world's future energy supply.

1.4.1 Biomass

Biomass resources include wood and wood wastes, agricultural crops and their waste byproducts, municipal solid waste, animal wastes, waste from food processing and aquatic plants and algae. Biomass has been recognized as a potential source for the renewable energy to substitute the declining fossil fuel resources. The majority of biomass energy is produced from wood and wood wastes (64%), followed by municipal solid waste (24%), agricultural waste (5%), and landfill gases (5%). Most biomass consists of hemicelluloses, cellulose, lignin and minor amounts of other organics.

Biomass is used to meet a variety of energy needs, including generating electricity, heating homes, fueling vehicles and providing process heat for industrial facilities. The conversion technologies for utilizing biomass can be separated into four basic categories: direct combustion processes, thermochemical processes, biochemical processes and agrochemical processes. Thermochemical conversion processes can be subdivided into gasification, pyrolysis, direct liquefaction and supercritical fluid liquefaction.

There are three ways to use biomass. It can be burned to produce heat and electricity, changed to gas-like fuels such as methane, hydrogen and carbon monoxide, or converted to a liquid fuel. Liquid fuels, also called biofuels, are predominantly two forms of alcohol: ethanol and methanol. Because biomass can be converted directly into a liquid fuel, it could someday supply much of our transportation fuel needs for cars, trucks, buses, airplanes, and trains. This is very important because nearly one-third of our nation's energy is now used for transportation.

Biomass today represents only 3% of the energy consumed in industrialized countries. However, much of the rural population in developing countries, which represents about 50% of the world's population, relies on biomass, mainly in the form of wood, for fuel. Biomass accounts for 35% of primary energy consumption in developing countries, raising the world total to 14% of primary energy consumption (Demirbas, 2005a). The importance of biomass varies significantly in different world regions, as shown in Table 1.13. In Europe, North America and the Middle East, the share of biomass averages 2 to 3% of total final energy consumption, whereas in Africa, Asia and Latin America, which together account for three-quarters of the world's population, biomass provides a substantial share of the energy

Table 1.13 The importance of biomass in different world regions

Region	Share of biomass in final energy consumption
Africa (average)	62.0
Burundi	93.8
Ethiopia	85.6
Kenya	69.6
Somalia	86.5
Sudan	83.7
Uganda	94.6
South Asia (average)	56.3
East Asia (average)	25.1
China	23.5
Latin America (average)	18.2
Europe (average)	3.5
North America (average)	2.7
Middle East (average)	0.3

needs: a third on average, but as much as 80–90% in some of the poorest countries of Africa and Asia (e.g., Angola, Ethiopia, Mozambique, Tanzania, Democratic Republic of Congo, Nepal and Myanmar). Indeed, for large portions of the rural populations of developing countries, and for the poorest sections of urban populations, biomass is often the only available and affordable source of energy for basic needs such as cooking and heating (Demirbas, 2005a).

The term "modern biomass" is generally used to describe the traditional biomass use through the efficient and clean combustion technologies and sustained supply of biomass resources, environmentally sound and competitive fuels, heat and electricity using modern conversion technologies. Ethanol and fatty acid (m)ethylester (biodiesel), as well as diesel produced from biomass by Fischer–Tropsch synthesis, are modern biomass-based transportation fuels. Modern biomass produced in a sustainable way excludes traditional uses of biomass as fuelwood and includes electricity generation and heat production, as well as transportation fuels, from agricultural and forest residues and solid waste. On the other hand, "traditional biomass" is produced in an unsustainable way and it is used as a non-commercial source—usually with very low efficiencies for cooking in many countries (Goldemberg and Coelho, 2004).

Biomass can be thermochemically converted into liquid fuel, gases, such as methane, carbon monoxide, or hydrogen by pyrolysis. Bioethanol can be obtained from cellulosic biomass by fermenting and distilling sugar solutions. Vegetable oils such as soybean and canola oils can be chemically converted into liquid fuel known as biodiesel. These fuels can be used as diesel fuel and gasoline in conventional engines with little modification to the system. Certain organic compounds, specifically municipal biosolids (sewage) and animal wastes (manures) can be biochemically converted into methane by anaerobic digestion. Energy crops, especially liquid biofuel (vegetable oils and biodiesels) crops have the potential to be substituted for a fraction of petroleum distillates and petroleum-based petrochemicals in the near future (Hamelinck and Faaij, 2002).

In the future, biomass has the potential to provide a cost-effective and sustainable supply of energy, while at the same time aiding countries in meeting their greenhouse gas reduction targets. In the short to medium term, biomass is expected to dominate energy supply. For the generation of electricity and heat, while using advanced combustion technology, organic wastes can be used as modern biomass. Also a number of crops and crop residues may fit modern bioenergy chains (Pimentel et al., 1981; Haberl and Geissler, 2000; Hoogwijk et al., 2003).

In industrialized countries, the main biomass processes expected to be utilized in the future are direct combustion of residues and wastes for electricity generation, bioethanol and biodiesel as liquid fuels, and combined heat and power production from energy crops. In the short to medium term, biomass waste and residues are expected to dominate the biomass supply, to be substituted by energy crops in the longer term. The future of biomass electricity generation lies in biomass integrated gasification/gas turbine technology, which offers high energy conversion efficiencies. The electricity is produced by direct combustion of biomass, advanced gasification and pyrolysis technologies, which are almost ready for commercial-scale use. Biomass is burned to produce steam and the steam turns a turbine and drives a generator, producing electricity. Because of potential ash build-up (which fouls boilers, reduces efficiency and increases costs), only certain types of biomass materials are used for direct combustion. Gasifiers are used to convert biomass into a combustible gas (biogas). The biogas is then used to drive a high-efficiency, combined-cycle gas turbine. Heat is used to thermochemically convert biomass into a pyrolysis oil. The oil, which is easier to store and transport than solid biomass material, is then burned like petroleum to generate electricity. Pyrolysis can also convert biomass into phenolic oil, a chemical used to make wood adhesives, molded plastics and foam insulation. Wood adhesives are used to glue together polywood and other composite wood products. Biomass can also be converted into transportation fuels such as ethanol, methanol, biodiesel and additives for reformulated gasoline. Biofuels are used in pure form or blended with gasoline (Demirbas, 2004).

1.4.1.1 Biohydrogen from Biomass

Since hydrogen is not a primary fuel, it must be manufactured from water with either fossil or non-fossil energy sources. Widespread use of hydrogen as an energy source could improve global climate change, energy efficiency, and air quality. The thermochemical conversion processes, such as pyrolysis, gasification and steam gasification are available for converting the biomass to a more useful energy. The yield from steam gasification increases with increasing water-to-sample ratio. The yields of hydrogen from the pyrolysis and the steam gasification increase with increasing of temperature. The list of some biomass material used for hydrogen production is given in Table 1.14. Hydrogen-powered fuel cells are important in enabling technology for the hydrogen future and more efficient alternatives to the combustion of gasoline and other fossil fuels. Hydrogen has the potential to solve two major energy problems: reducing dependence on petroleum and reducing pollution and greenhouse gas emissions.

Hydrogen is currently more expensive than conventional energy sources. There are different technologies presently being practiced to produce hydrogen economically from biomass. Hydrogen can be produced by pyrolysis from biomass (Arni, 2004). It can be burned to produce heat or passed through a fuel cell to produce electricity. Biohydrogen technology will play a major role in future because it can utilize renewable sources of energy (Nath and Das, 2003).

Biological generation of hydrogen (biohydrogen) technologies provide a wide range of approaches to generate hydrogen, including direct biophotolysis, indirect biophotolysis, photofermentations, and dark fermentation (Levin et al., 2004). Biological hydrogen production processes are found to be more environmental friendly and less energy intensive as compared to thermochemical and electrochemical processes (Das and Veziroglu, 2001). Researchers have been investigating hydrogen production with anaerobic bacteria since the 1980s (Nandi and Sengupta, 1998; Chang et al., 2002).

The three types of microorganisms that generate hydrogen are: cyanobacteria, anaerobic bacteria, and fermentative bacteria. The cyanobacteria directly decompose water into hydrogen and oxygen in the presence of light energy by photosynthesis. Photosynthetic bacteria use organic substrates like organic acids. Anaerobic bacteria use organic substances as the sole source of electrons and energy, converting them into hydrogen. Biohydrogen can be generated using bacteria such as *Clostridia*, by adjusting temperature, pH control, reactor hydraulic retention time (HRT) and other factors of the treatment system.

Biological hydrogen can be generated from plants by biophotolysis of water using microalgae (green algae and cyanobacteria), fermentation of organic com-

Table 1.14 List of some biomass material used for hydrogen production

Biomass species	Main conversion process
Bionut shell	Steam gasification
Olive husk	Pyrolysis
Tea waste	Pyrolysis
Crop straw	Pyrolysis
Black liquor	Steam gasification
Municipal solid waste	Supercritical water extraction
Crop grain residue	Supercritical fluid extraction
Pulp and paper waste	Microbial fermentation
Petroleum basis plastic waste	Supercritical fluid extraction
Manure slurry	Microbial fermentation

pounds, and photodecomposition of organic compounds by photosynthetic bacteria. To produce hydrogen by fermentation of biomass, a continuous process using a non-sterile substrate with a readily available mixed microflora is desirable (Hussy et al., 2005). A successful biological conversion of biomass to hydrogen depends strongly on the processing of raw materials to produce feedstock, which can be fermented by the microorganisms (de Vrije et al., 2002).

Hydrogen production from the bacterial fermentation of sugars has been examined in a variety of reactor systems. Hexose concentration has a greater effect on H_2 yields than the HRT. Flocculation also was an important factor in the performance of the reactor (Van Ginkel and Logan, 2005).

Hydrogen gas is a product of the mixed acid fermentation of *Escherichia coli*, the butylene glycol fermentation of *Aerobacter*, and the butyric acid fermentations of *Clostridium* spp. (Aiba et al., 1973). Anaerobic fermentation was conducted to improve hydrogen fermentation of food waste in a leaching bed reactor by heat-shocked anaerobic sludge, and also to investigate the effect of dilution rate on the production of hydrogen and metabolites in hydrogen fermentation (Han and Shin, 2004).

1.4.2 Hydropower

The water in rivers and streams can be captured and turned into hydropower (HP), also called hydroelectric power. HP currently provides about 17% of the world's electricity supply, virtually all of Norway's electricity and more than 40% of the electricity used in developing countries. However, there is great potential in hydropower worldwide. Norway produces more than 99% of its electricity with hydropower. New Zealand uses hydropower for 75% of its electricity. HP provides more than 97% of all electricity generated by renewable sources. Other sources, biomass, geothermal, solar and wind account for less than 3% of renewable electricity production. When the electricity share of total energy consumption is considered, the increase becomes even more dramatic.

Hydroelectricity, at present the most important of the clean, economically feasible, renewable energy options, can benefit from a water resources development project; however, that is seldom the only benefit. It is clear, therefore, that hydropower has an important role to play in the future, both in terms of energy supply and water resources development.

There is no universal consensus on the definition of small hydropower. Some countries of the European Union such as Portugal, Spain, Ireland, Greece, and Belgium accept 10 MW as the upper limit for installed capacity. In Italy the limit is 3 MW, in France 8 MW, in the UK 5 MW, in Canada 20–25 MW, in the United States 30 MW; however, a value of up to 10 MW total capacity is becoming the generally accepted amount considered as small hydropower in the rest of the world. If total installed capacity of any hydropower system is bigger than 10 MW, it is generally accepted as a large hydropower system (Demirbas, 2006a). Small hydropower can be further subdivided into minihydro, usually defined as less than 500 kW and microhydro, less than 100 kW. Small hydropower is one of the most valuable energy sources used for electricity in rural comminutes. Small hydroelectricity growth is to decrease the gap of decentralized production for private sector and municipal activity production. Small-scale hydropower systems supply energy from the flowing or running water and convert it to electrical energy. Small hydropower systems depend on the water flow where the resource exists. If a well-designed small hydropower system is established anywhere, it can fit in with its surroundings and will have minimal negative impacts on the environment. Small hydropower systems allow for self-sufficiency by using natural water resources, which are sometimes scarce. These systems supply low-cost energy production applied in many developing countries in the world.

There are two small-scale HP systems: microhydropower systems (MHP) with capacities below 100 kW and SHP systems with capacity between 101 kW and 1 MW. The MHP systems, which use cross-flow turbines and Pelton wheels can provide both direct mechanical energy and electrical energy. However, due to design constraints, turbines up to a capacity of only 30 kW are suitable for extracting mechanical energy. Of the total installed capacity of about 12 MW of MHP systems, half is used solely for crop processing. The most popular of the MHP systems is the peltric set, which is an integrated Pelton turbine and electricity generation unit with an average capacity of 1 kW. MHP systems are sometimes described as those having capacities below 100 kW, minihydropower plants are those ranging from 100 to 1000 kW, and the SHP plants are those that produce from 1 to 30 MW.

Dams are individually unique structures and dam construction represents the largest structures of basic infrastructure in all nations. The construction of a dam and power plant, along with the impounding of a reservoir, creates certain social and physical changes. The total installed capacity of HP is 640 000 MW (26% of the theoretical potential), generating an estimated 2380 TWh/year in the world, producing nearly 20% of the total global supply of electricity.

Hydropower provides unique benefits, rarely found in other sources of energy. These benefits can be attributed to the electricity itself, or to side-benefits, often associated with reservoir development. Principles of sustainable development of particular relevance to energy provision include: improving the well-being of entire populations, ensuring that development is people-centered, participatory and equitable; integrating environmental concerns into decision-making processes; and taking into account the full range of costs and benefits of development.

Investment costs for SHP plants vary according to site-specific and local characteristics. The most important system and cost elements are: (1) civil engineering, (2) equipment, and (3) turbine. The electrical generator represents less than 5% of the total cost of a power plant and the efficiency of generators for new plants is already close to 100%. However, standardization of generator equipment for small hydropower could further reduce installation and maintenance costs.

Despite the recent debates, few would disclaim that the net environmental benefits of hydropower are far superior to fossil-based generation. Hydroelectricity is produced for an average of 0.85 US¢ per kWh. In comparison with hydropower, thermal plants take less time to design, obtain approval, build and recover investment.

The remaining economically exploitable potential is about 5400 TWh/yr. An investment of at least US$ 1500 billion would be necessary to realize such a program. For a mean level of hydropower plants capacity in the range of 50 to 100 MW, some 20 000 plants would need to be built. In order to implement a plant construction program of this magnitude, a great deal of work (technical, financial and political) would need to be accomplished by all the investments involved, particularly in Asia, South America and Africa (Demirbas, 2006a).

1.4.3 Geothermal

As an energy source, geothermal energy has come of age. Geothermal energy for electricity generation has been produced commercially since 1913, and for four decades on the scale of hundreds of megawatts both for electricity generation and direct use. Utilization has increased rapidly during the last three decades, where it can be put to work in various forms such as electricity generation, direct use, space heating, heat pumps, greenhouse heating, and industrial. In 2000, geothermal resources have been identified in over 80 countries and there are quantified records of geothermal utilization in 58 countries (Fridleifsson, 2001). The world's top countries using geothermal energy directly are given in Table 1.15. In Tuscany,

Table 1.15 The world's top countries using geothermal energy in direct uses. Adapted from Demirbas (2008)

Country	Installed MWt	Production (GWh/yr)
China	2282	10 531
Japan	1167	7482
USA	3766	5640
Iceland	1469	5603
Turkey	820	4377
New Zealand	308	1967
Georgia	250	1752
Russia	308	1707
France	326	1360
Sweden	377	1147
Hungary	473	1135
Mexico	164	1089
Italy	326	1048
Romania	152	797
Switzerland	547	663

Italy, a geothermal plant has operated since the early 1900s. There are also geothermal power stations in the USA, New Zealand and Iceland. In Southampton (UK) there is a district heating scheme based on geothermal energy, where hot water is pumped up from about 1800 m below ground. This and other methods have added to the rapid increase of utilization of geothermal energy in the past three decades.

Geothermal energy is clean, cheap and renewable, and can be utilized in various forms such as space heating and domestic hot water supply, CO_2 and dry-ice production process, heat pumps, greenhouse heating, swimming and balneology (therapeutic baths), industrial processes and electricity generation. In terms of using geothermal energy directly, the main types of use are bathing, swimming and balneology (42%), space heating (35%), greenhouses (9%), fish farming (6%), and industry (6%) (Fridleifsson, 2001). Electricity is produced with geothermal steam in 21 countries spread over all continents, where electricity generation is improving faster in geothermal energy-rich countries. Low-temperature geothermal energy is exploited in many countries to generate heat, with an estimated thermal capacity of about 10 000 MW.

From above, direct application of geothermal energy can involve a wide variety of end-uses, such as heating and cooling homes, industrial processes, greenhouses, fish farming, and health spas. It uses mostly existing technology and straightforward engineering. The technology, reliability, economics, and environmental acceptability of direct use of geothermal energy have been demonstrated throughout the world.

The total global installed geothermal electric capacity was 7304 MWe in 1996. In much of the world, electricity from fossil fuel-burning electricity plants can be constructed at half the cost of new geothermal electricity. A comparison of the renewable energy sources shows the current electrical energy cost to be 2–10 US¢ per kWh for geothermal and hydropower, 5–13 US¢ per kWh for wind, 5–15 US¢ per kWh for biomass, 25–125 US¢ per kWh for solar photovoltaic and 12–18 US¢ per kWh for solar thermal electricity. Of the total electricity production from renewables of 2826 TWh in 1998, 92% came from hydropower, 5.5% from biomass, 1.6% from geothermal and 0.6% from wind. Solar electricity contributed 0.05% and tidal 0.02%. Geothermal energy is a power source that produces electricity with minimal environmental impact, and with its proven technology and abundant resources, can make a significant contribution towards reducing the emission of greenhouse gases (Demirbas, 2006a).

1.4.4 Wind

Energy from the wind has been used for centuries to power windmills to mill wheat or pump water. More recently large wind turbines have been designed to generate electricity. This source of energy is non-polluting and freely available in many areas. Wind turbines are becoming more efficient, and therefore the cost of the electricity they generate is falling.

Wind power has long been used for grain-milling and water-pumping applications. Significant technical progress since the 1980s, driven by advances in aerodynamics, materials, design, controls, and computing power, has led to economically competitive electrical energy production from wind turbines. Technology development, favorable economic incentives (due to its early development status and environmental benefits), and increasing costs of power from traditional fossil sources have led to significant worldwide sales growth since the early 1980s (Kutz, 2007).

There are wind farms around the world. Because the UK is on the edge of the Atlantic Ocean it has one of the best wind resources in Europe. Offshore wind farms in coastal waters are being developed because winds are often stronger blowing across the sea. Turbines can produce between 500 kW and 1 MW of electricity. Production of wind-generated electricity has risen from practically zero in the early 1980s to more than 7.5 TWh per year in 1995. This increase led to a peak in the global cumulative generating capacity of 6500 MW in late 1997 (Garg and Datta, 1998).

Wind energy is a significant resource; it is safe, clean, and abundant. Wind energy is an indigenous supply, permanently available in virtually every nation in the world. Figure 1.5 shows the growth in the global wind turbine installed capacity. Globally, wind power generation more than quadrupled between 1999 and 2005. Wind energy is abundant, renewable, widely distributed, clean, and mitigates the greenhouse effect if it is used to replace fossil fuel-derived electricity. Wind energy has limitations based on geography and meteorology, plus there may be political or environmental problems (e.g., dead birds) with erecting turbines (Garg and Datta, 1998). On the other hand, wind can cause air pollution by degradation and distribution of pieces of pollutants such as waste paper, straw, etc. Figure 1.6 shows the growth scenarios for global installed wind power (IEA, 2006).

An advantage of wind turbines over some forms of renewable energy is that they can produce electricity whenever the wind blows (at night and also during the

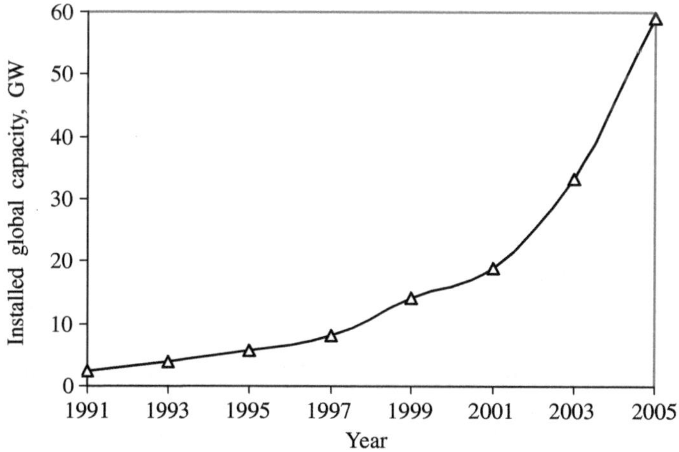

Fig. 1.5 Growth in world wind turbine installed capacity

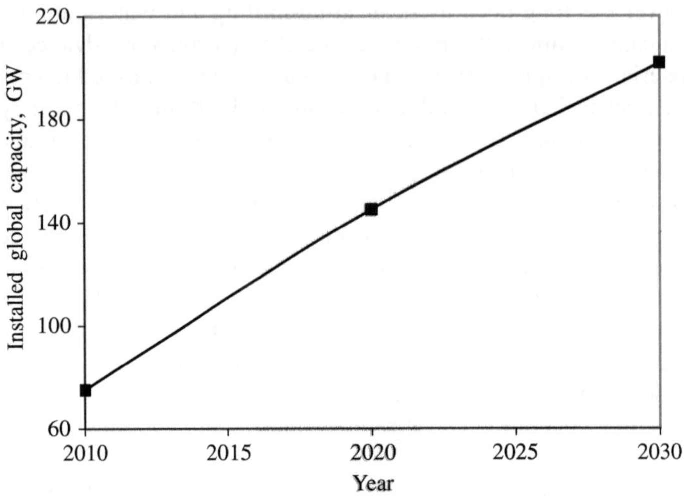

Fig. 1.6 Growth scenarios for global installed wind power

day). In theory, wind systems can produce electricity 24 hours every day, unlike PV systems that can not make power at night. However, even in the windiest places, the wind does not blow all the time. So while wind farms do not need batteries for backup storage of electricity, small wind systems do need backup batteries. Wind power in coastal and other windy regions is promising as well. By any measure the power in the wind is no longer an alternative source of energy.

1.4.5 Solar

Solar energy is defined as the radiant energy transmitted by the Sun and intercepted by Earth. It is transmitted through space to Earth by electromagnetic radiation with wavelengths ranging between 0.20 and 15 μm. The availability of solar flux for terrestrial applications varies with season, time of day, location, and collecting surface orientation. In this section, we shall treat these matters analytically (Kutz, 2007).

Energy comes from processes called solar heating (SH), solar home heating (SHH), solar dryer (SD), solar cooker (SC), solar water heating (SWH), solar photovoltaic (SPV) which converts sunlight directly into electricity, and solar thermal electric power (STEP), when the Sun's energy is concentrated to heat water and produce steam, which is used to generate electricity. The major component of any solar energy system is the solar collector. Solar energy collectors are special kinds of heat exchangers that transform solar radiation energy to internal energy of the transport medium.

Solar dryers are used for drying fruits and spices. The three most popular types of SD are box type, cabinet type, and tunnel type. Box type uses direct heat for dehydration. In cabinet-type dryers, air heated by the collector dehydrates the food product, whereas in tunnel-type dryers forced air circulation is used to distribute heat for dehydration. Cabinet and tunnel type dryers yield a high quality of dried products but they are very bulky and costly compared to the box-type dryers. Of about 800 dryers disseminated so far, 760 are of the box type (Pokharel, 2003).

Solar energy systems are solar home systems, solar photovoltaic (SPV) systems, solar water heating (SWH) systems, solar dryers, and solar cookers. These systems are installed and managed by a household or a small community. A solar home system is a PV system with a maximum capacity of 40 W.

One of the most abundant energy resources on the surface of the Earth is sunlight. Today, solar energy has a tiny contribution to the total global primary energy supply of less than 1%. Photovoltaic (PV) systems, other than solar home heating systems, are used for communication, water pumping for drinking and irrigation, and electricity generation. The total installed capacity of such systems is estimated at about 1000 kW. A solar home heating system is a solar PV system with a maximum capacity of 40 W. These systems are installed and managed by a household or a small community (Garg and Datta, 1998).

Like wind power markets, PV markets have seen rapid growth and costs have fallen dramatically. The total installed capacity of such systems is estimated at about 1000 kW. The PV installed capacities are growing at a rate of 30% per year (Demirbas, 2005a). Solar PV systems have been found to be a promising energy in future. One of the most significant developments in renewable energy production is observed in PVs (EWEA, 2005; Reijnders, 2006). The PV will then be the largest renewable electricity source with a production of 25.1% of global power generation in 2040 (EWEA, 2005).

A solar thermal electricity power system is a device that uses solar radiation for the generation of electricity through the solar thermal conversion (Xiao et al., 2004). Figure 1.7 shows the plot for electricity costs from solar thermal power plants. Solar thermal electricity may be defined as the result of a process by which directly collected solar energy is converted to electricity through the use of some sort of heat to electricity conversion device (Mills, 2004). The last three decades have witnessed a trend in solar thermal electricity generation of increasing the concentration of sunlight (Kribus, 2002). There are three main systems of solar thermal electricity: solar towers, dishes and the parabolic troughs (Zhang et al., 1998). Solar thermal power stations based on parabolic and heliostat trough concentrating collectors can soon become a competitive option on the world's electricity market (Trieb, 2000). Table 1.16 shows the economics and emissions of conventional technologies compared with solar power generation.

It has been estimated that a proposed solar system has a net fuel-to-electricity efficiency higher than 60% even when both the energy to produce high-pressure oxygen and to liquefy the captured CO_2 are taken into account (Kosugi and Pyong, 2003). Based on future trends, it is foreseen that by the year 2025, the PV electricity would be more economical than fossil fuel electricity (Muneer et al., 2005). A mathematical model for simulating an innovative design of a solar-heated anaerobic digester has been developed by Axaopoulos et al. (2001).

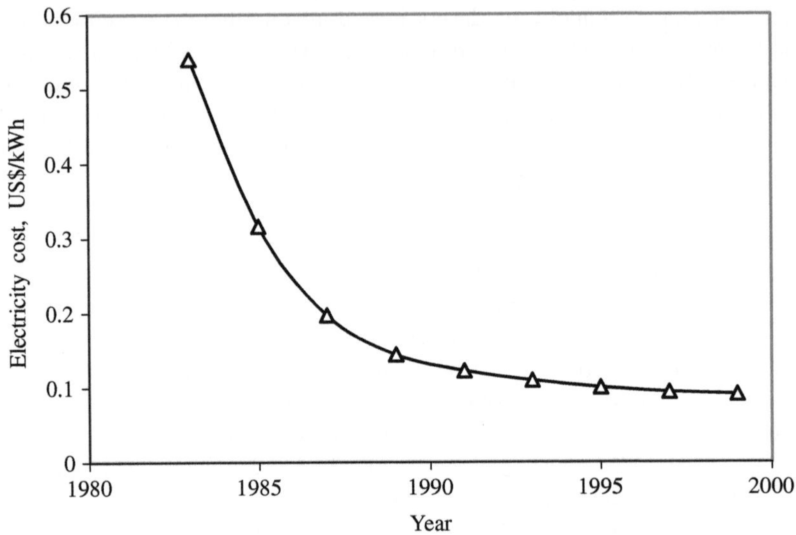

Fig. 1.7 Plot for electricity costs from solar thermal power plants

Table 1.16 Economics and emissions of conventional technologies compared with solar power generation. Adapted from Demirbas (2006a)

	Carbon emissions	Generation costs
Electricity generation technology	g C/kWh	US ¢ per kWh
Solar thermal and solar PV systems	0	9–40
Pulverized coal—natural gas turbine	100–230	5–7

Table 1.17 Properties of common solar collectors

Type	Absorber	Motion	Temperature range, K
Flat plate collector	Flat	Stationary	300–350
Compound parabolic collector	Tubular	Stationary	330–510
Cylindrical trough collector	Tubular	Single-axis tracking	330–575
Parabolic trough collector	Tubular	Single-axis tracking	330–575
Parabolic dish collector	Point	Two-axis tracking	275–775
Heliostat trough collector	Point	Two-axis tracking	425–1260

A historical introduction into the uses of solar energy was attempted followed by a description of the various types of collectors including flat-plate, compound parabolic, evacuated tube, parabolic trough, Fresnel lens, parabolic dish and heliostat field collectors (Kalogirou, 2004a, 2004b). Solar thermal electricity systems utilize solar radiation to generate electricity by the photothermal conversion method. There are basically two types of solar collectors: non-concentrating (stationary type) and concentrating. The properties of common solar collectors are given in Table 1.17. Temperatures exceeding about 2300 K should be feasible, given appropriate concentration of the incident sunlight. The theoretical limit on the concentration of sunlight is about 40 000 (Kribus, 2002). A solar thermal conversion system with heliostat trough collector can achieve temperatures in excess of 2000 K.

Solar-supported power plants (SSPPs) use technology that is similar to that used in coal-fired power plants. For example, SSPPs use similar steam turbine generators and fuel delivery systems. Electricity costs are in the 9 to 40 US¢ per kWh range (Table 1.16). The feasibility of combining gas and steam expansion in a power cycle has been extensively explored (Sorensen, 1983). Concentrated solar energy is used to produce steam, the steam turns a turbine and drives a generator, producing electricity. Solar radiation can be used as a primary energy source or as a secondary energy source to power gas turbines.

1.4.6 Other Renewable Energy Sources

Wave power, tidal power, municipal solid waste, gas from animal wastes (biogas), landfill, peat energy and ocean thermal energy conversion (OTEC) are the other renewable energy sources (RES). Water energy sources are hydropower, tidal and wave technologies.

Marine energy sources include current, tidal, ocean thermal energy conversion (OTEC) and wave technologies. The world wave resource is between 200 and 5000 GW mostly found in offshore locations (Garg and Datta, 1998). Wave energy converters fixed to the shoreline are likely to be the first to be fully developed and deployed, but waves are typically two to three times more powerful in deep offshore waters than at the shoreline. Wave energy can be harnessed in coastal areas, close to the shore. The first patent for a wave energy device was filed in Paris in 1799, and by 1973 there were 340 British patents for wave energy devices. By comparison to wind and PV, wave energy and tidal stream are very much in their infancy. Currently, around 1 MW of wave energy devices are installed worldwide, mainly from demonstration projects.

Water is a renewable energy source, which can be used in electricity generation by using its lifting force (buoyant force). It is an important electricity generating apparatus, using gravity and buoyancy, which can curtail costs of power generation and prevent environmental pollution and prevent destruction of an ecosystem. The wave conversion plant using buoyancy chambers is another solution for electricity generation using the water lifting force. There are a good many reasons why the water lifting force will be used in the future at many scales of electricity generation. The hydraulic ram is another attractive solution for electricity generation where a large gravity flow exists. Hydraulic ram (hydram) pumps are water lifting or water pumping devices that are powered by filling water. The hydram pumps have been used for over two centuries in many parts of the world. It is a useful device that can pump water uphill from a flowing source of water above the source with no power requirement except the force of gravity.

The ocean thermal energy conversion (OTEC) is an energy technology that converts solar radiation to electric power. OTEC systems use the ocean's natural thermal gradient to drive a power-producing cycle. As long as the temperature between the warm surface water and the cold deep water differs by about 20 K, an OTEC system can produce a significant amount of power. The oceans are thus a vast renewable resource, with the potential to help us produce billions of watts of electric power.

Anaerobic digestion in landfills occurs in a series of stages, each of which is characterized by the increase or decrease of specific bacterial populations and the formation and utilization of certain metabolic products. Landfill gas is typically composed of about 50% methane by volume. Producing energy from landfill gas improves local air quality, eliminates a potential explosion hazard and reduces greenhouse gas emissions to the atmosphere. Hydrogen, produced by passing an electrical current through water, can be used to store solar energy and regenerate it when needed for nighttime energy requirements. The solid waste management practices are collection, recovery and disposal, together with the results of cost analysis.

Biogas can be obtained from digesting the organic material of municipal solid wastes (MSW). The composition of MSW varies by the source of waste; however, in all cases the major constituents of MSW are organic in nature and the organics accounts for more than 50% of MSW. The economic exploitation of methane is worthwhile after one year from the start of the landfill operation.

Biogas can be obtained from several sources. It is obtained from decomposing organic material. Biogas is composed of methane (CH_4), carbon dioxide (CO_2), air, ammonia, carbon monoxide, hydrogen, sulfur gases, nitrogen and oxygen. Among its components, methane is the most important one, particularly for the combustion process in vehicle engines. A typical analysis of raw landfill gas is given in Table 1.18. CH_4 and CO_2 make up around 90% of the gas volume produced. Typically, the main constituents of landfill gas are methane and carbon dioxide, both of which are major contributors to global warming. But, due to the widely varying nature of the contents of landfill sites the constituents of landfill gases vary widely.

Table 1.18 Typical analysis of raw landfill gas

Component	Chemical formula	Content
Methane	CH_4	40–60 (% by vol.)
Carbon dioxide	CO_2	20–40 (% by vol.)
Nitrogen	N_2	2–20 (% by vol.)
Oxygen	O_2	< 1 (% by vol.)
Heavier hydrocarbons	C_nH_{2n+2}	< 1 (% by vol.)
Hydrogen sulfide	H_2S	40–100 ppm
Complex organics	–	1000–2000 ppm

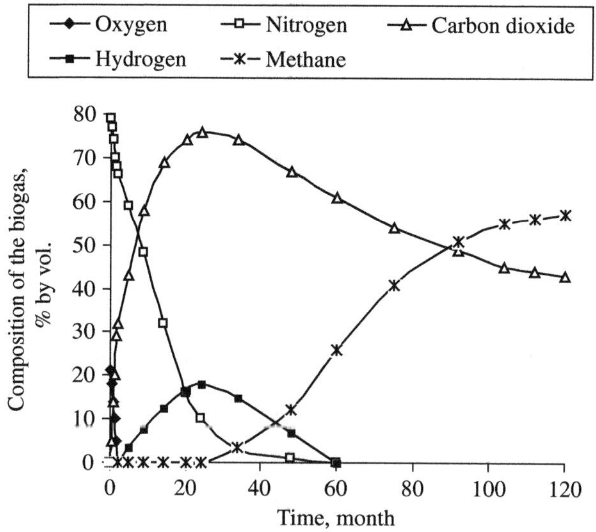

Fig. 1.8 Production of biogas components with time in landfill

The generation of municipal solid wastes (MSW) has increased in parallel to rapid industrialization. Approximately 16% of all discarded MSW is incinerated; the remainder is disposed of in landfills. Disposal of MSW in sanitary landfills is usually associated with soil, surface water and groundwater contamination when the landfill is not properly constructed. The flow rate and composition of leachate vary from site to site, seasonally at each site and depending on the age of the landfill. The processing of MSW (i.e., landfill, incineration, aerobic composting) secures many advantages and limitations. The greenhouse gas emissions can be reduced by the uncontrolled releasing of methane from improperly disposed organic waste in a large landfill.

Decomposition in landfills occurs in a series of stages, each of which is characterized by the increase or decrease of specific bacterial populations and the formation and utilization of certain metabolic products. The first stage of decomposition, which usually lasts less than a week, is characterized by the removal of oxygen from the waste by aerobic bacteria. In the second stage, which has been termed the anaerobic acid stage, a diverse population of hydrolytic and fermentative bacteria hydrolyzes polymers, such as cellulose, hemicellulose, proteins, and lipids, into soluble sugars, amino acids, long-chain carboxylic acids, and glycerol. The main components of landfill gas are byproducts of the decomposition of organic material, usually in the form of domestic waste, by the action of naturally occurring bacteria under anaerobic conditions. Figure 1.8 shows the production of biogas components with time in a landfill.

Methods developed for treatment of landfill leachates can be classified as physical, chemical and biological, which are usually used in combinations in order to improve the treatment efficiency. Biological treatment methods used for the leachate treatment can be classified as aerobic, anaerobic and anoxic processes which are widely used for the removal of biodegradable compounds (Kargi and Pamukoglu, 2004a). Biological treatment of landfill leachate usually results in low nutrient removals because of high chemical oxygen demand (COD), high ammonium-N content and the presence of toxic compounds such as heavy metals (Uygur and Kargi, 2004). Landfill leachate obtained from the solid waste landfill area contained high COD and ammonium ions which resulted in low COD and ammonium removals by direct biological treatment. Several anaerobic and aerobic treatment systems have been studied in landfill leachate. Leachates contain non-biodegradable substrates, which are not removed by biological treatment alone and an increase of leachate input may cause a reduction in substrate removal (Cecen et al., 2003). Raw landfill leachate has been subjected to pre-treatment by coagulation–flocculation and air stripping of ammonia before biological treatment in order to improve biological treatability of the leachate (Kargi and Pamukoglu, 2004b). In other forms of treatment, natural zeolite and bentonite can be utilized as a novel landfill liner material (Kayabali, 1997).

1.5 Nuclear Fuel Sources

Nuclear power plants are based on uranium mined in surface mines, or by in situ leaching.

Nuclear energy has been used to produce electricity for more than half a century. Worldwide, nuclear energy accounts for 6% of energy and 16% of electricity, and 23% in OECD countries (UNDP, 2004). OECD countries produce almost 55% of the world's uranium. The global nuclear energy consumption increased rapidly from 0.1% in 1970 to 7.4% in 1998. This increase was especially high in the 1980s.

Fissile materials are defined as materials that are fissionable by neutrons with zero kinetic energy. In nuclear engineering, a fissile material is one that is capable of sustaining a chain reaction of nuclear fission. Nuclear power reactors are mainly fueled with uranium, the heaviest element that occurs in nature in more than trace quantities. The principal nuclear energy sources are uranium-235, plutonium-239, uranium-233 and thorium.

Thorium was discovered by Berzelius in 1828. Thorium occurs in thorite and in thorianite. Large deposits of thorium minerals have been reported in New England and elsewhere, but these have yet to be exploited. Thorium is now thought to be about three times as abundant as uranium and about as abundant as lead or molybdenum. The metal is a source of nuclear power. There is probably more energy available for use from thorium in the minerals of the Earth's crust than from both uranium and fossil fuels. Thorium might serve as a matrix for fissioning or transmuting minor actinides should this be required. The necessity and case for this objective, or for preferring thorium to other matrices, both remain to be established.

The essence of a conventional nuclear reactor is the controlled fission chain reaction of U-235 and Pu-239. This produces heat, which is used to make steam which drives a turbine. The chain reaction depends on having a surplus of neutrons to keep it going.

Work has been done in developing thorium cycle converter-reactor systems. Several prototypes, including the high-temperature gas-cooled reactor (HTGR) and molten salt converter reactor experiment (MSRE), have operated. While the HTGR reactors are efficient, they are not expected to become important commercially for many years because of certain operating difficulties. Thorium is recovered commercially from the mineral monazite, which contains from 3 to 9% ThO_2 along with rare-earth minerals. Much of the internal heat the Earth produces has been attributed to thorium and uranium.

In the long-term, thorium may well provide a means of resource extension, in parallel with fast-reactor developments rather than as a substitute for them. Meanwhile the most valuable contribution to industry that it can make appears likely to be in extending fuel irradiation where recycling of recovered fissile and fertile values is considered impracticable or undesirable.

Thorium has long been recognized as a possible nuclear fuel, since although the natural element consists almost entirely of the non-fissile Th-232, it is fertile in being capable of transmutation by a neutron flux into the fissile uranium-233. In favorable circumstances, given a start with some other neutron source such as U-235, plutonium or an accelerator-driven generator, it can sustain a thermal breeding cycle in which as much fissile material is generated as consumed. This cycle has been adequately demonstrated in principle. Thorium, particularly with breeding, could therefore add to the available nuclear fuel currently dependent on the established uranium cycle. For neutronic reasons, thorium is also somewhat more suited than uranium to HTGRs, which are capable of better thermodynamic efficiencies than the currently dominant water-cooled types, or of serving as sources of process heat. However, uranium is still plentiful, HTGRs have never been commercially deployed, and the thorium cycle has not been widely adopted (Thorn et al., 1983).

The technology to utilize thorium in nuclear reactors was thought to be similar to that of uranium. Thorium resources are larger than those of uranium, and neutron yields of U-233 in the thermal and epithermal regions are higher than for Pu-239 in the uranium/plutonium fuel cycle. The introduction of the thorium-based nuclear fuel cycle would therefore vastly enlarge the fissile resources by breeding U-233. Other reasons identified in past studies are the potential for fuel cycle cost reduction, the reduction in U-235 enrichment requirements, safer reactor operation because of lower core excess reactivity requirements, and safer and more reliable operation of ThO2 fuel as compared to UO2 fuel at high burn-up due to the former's higher irradiation and corrosion resistance (IAEA, 2002).

For many years there has been interest in utilizing thorium (Th-232) as a nuclear fuel since it is three times as abundant in the Earth's crust as uranium. Also, all of the mined thorium is potentially useable in a reactor, compared with the 0.7% of natural uranium, so some 40 times the amount of energy per unit mass might be available. A thorium reactor would work by having Th-232 capture a neutron to become Th-233 which decays to uranium-233, which fissions. The problem is that insufficient neutrons are generated to keep the reaction going.

Thorium is also an alpha emitter. Th-232 goes through six alpha and four beta decay steps before becoming the stable isotope Pb-208. Th-232 is sufficiently radioactive to completely expose a photographic plate in a few hours. Thorium disintegrates with the production of "thoron" (Rn-220), which is an alpha emitter and presents a radiation hazard. Good ventilation of areas where thorium is stored or handled is therefore essential.

Several methods are available for producing thorium metal; it can be obtained by reducing thorium oxide with calcium, by electrolysis of anhydrous thorium chloride in a fused mixture of sodium and potassium chlorides, by calcium reduction of thorium tetrachloride mixed with anhydrous zinc chloride, and by reduction of thorium tetrachloride with an alkali metal.

The following is a comparison of thorium with uranium as nuclear fuels:

- Weapons-grade fissionable material (U-233) is harder to retrieve safely and clandestinely from the thorium reactor than plutonium is from the uranium breeder reactor.
- Thorium produces 10 to 10 000 times less long-lived radioactive waste than uranium or plutonium reactors.
- Thorium comes out of the ground as a 100% pure, usable isotope, which does not require enrichment, whereas natural uranium contains only 0.7% fissionable U-235.
- Because thorium does not sustain chain reaction, fission stops by default if we stop priming it, and a runaway chain reaction accident is improbable.

A neutron is captured by ${}_{90}Th^{232}$, which makes it ${}_{90}Th^{233}$:

$$ {}_{90}Th^{232} + {}_{0}n^{1} \rightarrow {}_{90}Th^{233} \tag{1.9}$$

Th-233 spontaneously emits a beta particle, leaving behind one additional proton, and one fewer neutron. This is called beta decay.

$$ {}_{90}Th^{233} \rightarrow {}_{91}Pa^{233} + \beta \tag{1.10}$$

The element with 91 protons is protactinium (Pa). The isotope ${}_{91}PA^{233}$ also undergoes beta decay,

$$ {}_{91}Pa^{233} \rightarrow {}_{92}U^{233} + \beta \tag{1.11}$$

The U-233 isotope that is produced in step (3) is fissionable, but has fewer neutrons than its heavier cousin, U-235, and its fission releases only 2 neutrons, not 3.

$$ {}_{92}U^{233} + {}_{0}n^{1} \rightarrow \text{fission fragments} + 2{}_{0}n^{1} \tag{1.12}$$

The priming neutrons are emitted in a process called spallation (interaction of particles with a target), which is the induced splitting of an otherwise non-fissionable large nucleus. The capability of high-current, high-energy accelerators to produce neutrons by spallation from heavy elements has been used in the structural research of such materials. In this process a beam of high-energy protons (usually > 500 MeV) is directed at a high atomic number target such as tungsten, tantalum, depleted uranium, thorium, zirconium, lead, lead-bismuth, mercury and up to one neutron can be produced per 25 MeV of the incident proton beam.

The idea is to use a particle accelerator producing neutrons by spallation to feed a fuel/moderator assembly where the neutrons multiply by fission chain reactions.

If the energy liberated becomes substantially greater than that needed to drive the accelerator, the process has a net gain and becomes self supporting. Hence this is called an energy amplifier (EA). In the energy amplifier (EA), a proton beam impinges on lead, the high-energy protons splitting lead nuclei, leading to the release of neutrons. If the spallation target is surrounded by a blanket assembly of nuclear fuel, such as fissile isotopes of uranium or plutonium, there is a possibility of sustaining a fission reaction. This is described as an accelerator-driven system (ADS). In this, up to 10% of the neutrons could come from the spallation, though it would normally be less, even where actinide incineration is the main objective (Demirbas, 2005b).

Summary

Energy sources are classified into three groups: fossil, fissile and renewable. The term fossil refers to an earlier geologic age. Fossil fuels were formed a great many years ago and are not renewable. The fossil energy sources are petroleum, coal, bitumens, natural gas, oil shales, and tar sands. The main fissile energy sources are uranium and thorium. Conventional fuel sources are petroleum and coal (including lignite). Our modern way of life is intimately dependent upon fossil fuels or mineral fuels. Conventional energy sources based on petroleum and coal have proven to be highly effective drivers of economic progress, but at the same time damaging to the environment and to human health. Renewable energy sources such as wind, solar, biomass and hydropower energy are important today, and they will be important in the future.

Hydrogen is seen as the energy carrier of the future. Hydrogen is storable and transferable, has high heat value per mass unit and its sources are globally distributed. Hydrogen is strategically important as it has low emission, is environmentally benign, and represents a cleaner and more sustainable energy system. Hydrogen seems to be the future energy carrier by the virtue of being renewable.

Hydrogen is not a primary fuel. It must be manufactured from water with either fossil or non-fossil energy sources. Widespread use of hydrogen as an energy source could improve global climate change, energy efficiency, and air quality. The thermochemical conversion processes, such as pyrolysis, gasification and steam gasification are available for converting the biomass to a more useful energy.

References

Aiba, S., Humphrey, A.E., Milis, N.F. 1973. Biochemical engineering, 2nd edn. Academic, New York.

Akella, A.K., Saini, R.P., Sharma, M.P. 2009. Social, economical and environmental impacts of renewable systems. Renew Energy 34:390–396.

Annual Energy Review (AER) 1999. 2000. Energy Information Administration, US Department of Energy, Washington, DC.

Arni, S. 2004. Hydrogen-rich gas production from biomass via thermochemical pathways. Energy Edu Sci Technol 13:47–54.

Axaopoulos, P., Panagakis, P., Tsavdaris, A., Georgakakis, D. 2001. Simulation and experimental performance of a solar heated anaerobic digester. Solar Energy 70:155–164.

Cecen, F., Erdincler, A., Kilic, E. 2003. Effect of powdered activated carbon addition on sludge dewaterability and substrate removal in landfill leachate treatment. Advances Environ Res 7:707–713.

Chang, J.-S., Lee, K.-S., Lin, P.-J. 2002. Biohydrogen production with fixed-bed bioreactors. Int J Hydrogen Energy 27:1167–1174.

COSO (Crude Oil Supply Outlook). 2007. Report to the Energy Watch Group EWG-Series No. 3/2007. http://www.energywatchgroup.org/fileadmin/global/pdf/EWG-Oilreport-10-2007.pdf. Accessed 2009.

Das, D., Veziroglu, T.N. 2001. Hydrogen production by biological processes: a survey of literature. Int J Hydrogen Energy 26:13–28.

Demirbas, A. 1999. Evaluation of biomass materials as energy sources: upgrading of tea waste by briquetting process. Energy Sources 21:215–220.

Demirbas, A. 2004. The importance of biomass. Energy Sources 26:361–366.

Demirbas, A. 2005a. Potential applications of renewable energy sources, biomass combustion problems in boiler power systems and combustion related environmental issues. Progress Energy Combus Sci 31:171–192.

Demirbas, A. 2005b. Options and trends of thorium fuel utilization in Turkey. Energy Sources 27:597–603.

Demirbas, A. 2006a. Energy priorities and new energy strategies. Energy Edu Sci Technol 16:53–109.

Demirbas, A. 2006b. The importance of natural gas in the world. Energy Sources Part B:413–420.

Demirbas, A. 2007. Utilization of coals as sources of chemicals. Energy Sources 29:677–684.

Demirbas, A. 2008. Biodiesel: a realistic fuel alternative for diesel engines. Springer, London.

Demirbas, A., Güllü, D. 1998. Acetic acid, methanol and acetone from lignocellulosics by pyrolysis. Energy Edu Sci Technol 2:102–110.

de Vrije, T., de Haas, G.G., Tan, G.B., Keijsers, E.R.P., Claassen, P.A.M. 2002. Pretreatment of *Miscanthus* for hydrogen production by *Thermotoga elfii*. Int J Hydrogen Energy 27:1381–1390.

Dorf, R.C. 1977. Energy resources and policy. Addison-Wesley, Los Angeles, CA.

Dry, M.E. 1999. Fischer–Tropsch reactions and the environment. Appl Catal A General 189:185–190.

EREC (European Renewable Energy Council). 2006. Renewable energy scenario by 2040, EREC Statistics, Brussels.

EWEA (European Wind Energy Association). 2005. Report: large scale integration of wind energy in the European power supply: analysis, issues and recommendations. EWEA, Brussels, Belgium.

Fridleifsson, I.B. 2001. Geothermal energy for the benefit of the people. Renew Sustain Energy Rev 5:299–312.

Garg, H.P., Datta, G. 1998. Global status on renewable energy, *in* Solar Energy Heating and Cooling Methods in Building, International Workshop, Iran University of Science and Technology, 19–20 May 1998.

Goldemberg, J., Coelho, S.T. 2004. Renewable energy: traditional biomass vs. modern biomass. Energy Policy 32:711–714.

Gorman, J. 2002. Hydrogen: The next generation. Science News 162:235–236.

Haberl, H., Geissler, S. 2000. Cascade utilization of biomass: strategies for a more efficient use of a scarce resource. Ecol Eng 16:S111–S121.

Hacisalihoglu, B., Demirbas, A.H., Hacisalihoglu, S. 2008. Hydrogen from gas hydrate and hydrogen sulfide in the Black Sea. Energy Edu Sci Technol 21:109–115.

Hamelinck, C.N., Faaij, A.P.C. 2002. Future prospects for production of methanol and hydrogen from biomass. J Power Sources 111:1–22.

Han, S.-K., Shin, H.-S. 2004. Biohydrogen production by anaerobic fermentation of food waste. Int J Hydrogen Energy 29:569–577.

Hoogwijk, M., Faaij, A., van den Broek, R., Berndes, G., Gielen, D., Turkenburg, W. 2003. Exploration of the ranges of the global potential of biomass for energy. Biomass Bioenergy 25:119–133.

Hussy, I., Hawkes, F.R., Dinsdale, R., Hawkes, D.L. 2005. Continuous fermentative hydrogen production from sucrose and sugarbeet. Int J Hydrogen Energy 30:471–483.

IAEA (International Atomic Energy Agency). 2002. Thorium fuel utilization: options and trends. Proceedings of three IAEA meetings held in Vienna in 1997, 1998 and 1999.

IEA (International Energy Annual). 2000. Energy Information Administration, US Department of Energy, Washington, DC.

IEA (International Energy Agency). 2006. Key world energy statistics. IEA, Paris. http://www.iea.org/textbase/nppdf/free/2006/key2006.pdf. Accessed 2009.

Kalogirou, A.S. 2004a. Environmental benefits of domestic solar energy systems. Energy Convers Manage 45:3075–3092.

Kalogirou, A.S. 2004b. Solar thermal collectors and applications. Progress Energy Combus Sci 30:231–295.

Kargi, F., Pamukoglu, M.Y. 2004a. Adsorbent supplemented biological treatment of pre-treated landfill leachate by fed-batch operation. Biores Technol 94:285–291.

Kargi, F., Pamukoglu, M.Y. 2004b. Repeated fed-batch biological treatment of pre-treated landfill leachate by powdered activated carbon addition. Enzyme Microbial Technol 34:422–428.

Kayabali, K. 1997. Engineering aspects of a novel landfill liner material: bentonite-amended natural zeolite. Eng Geology 46:105–114.

Kosugi, T., Pyong, S.P. 2003. Economic evaluation of solar thermal hybrid H_2O turbine. Energy 28:185–198.

Kribus, A. 2002. A high-efficiency triple cycle for solar power generation. Solar Energy 72:1–11.

Kutz, M. (Ed.) 2007. Environmentally conscious alternative energy production. Wiley, Hoboken, NJ.

Lee, S.-Y., Holder, G.D. 2001. Methane hydrates potential as a future energy source. Fuel Proc Technol 71:181–186.

Levin, D.B., Pitt, L., Love, M. 2004. Biohydrogen production: prospects and practical application. Int J Hydrogen Energy 29:173–185.

Mills, D. 2004. Advances in solar thermal electricity technology. Solar Energy 76:19–31.

Muneer, T., Asif, M., Munawwar, S. 2005. Sustainable production of solar electricity with particular reference to the Indian economy. Renew Sustain Energy Rev 9:444–473.

Nandi, R., Sengupta, S. 1998. Microbial production of hydrogen: an overview. Critical Rev Microbiol 24:61–84.

Nath, K., Das, D. 2003. Hydrogen from biomass. Current Sci 85:265–271.

Pimentel, D., Moran, M.A., Fast, S., Weber, G., Bukantis, R., Balliett, L., Boveng, P., Cleveland, C., Hindman, S., Young, M., 1981. Biomass energy from crop and forest residues. Science 212:1110–1115.

Pokharel, S. 2003. Promotional issues on alternative energy technologies in Nepal. Energy Policy 31:307–318.

Ramage, J., Scurlock, J. 1996. Biomass, *in* Renewable energy-power for a sustainable future, G. Boyle (Ed.). Oxford University Press, Oxford.

Reijnders, L. 2006. Conditions for the sustainability of biomass-based fuel use. Energy Policy 34:863–876.

Schobert, H.H., Song, C. 2002. Chemicals and materials from coal in the 21st century. Fuel 81:15–32.

Schulz, H. 1999. Short history and present trends of FT synthesis. Appl Catal A General 186:1–16.

Sorensen, H.A. 1983. Energy conversion systems. Wiley, New York.

Thorn, J.D., John, C.T., Burstall, R.F. 1983. Nuclear power technology, W. Marshall (Ed.), Vol. 2, Chap. 17, pp. 377–381. Oxford University Press, Oxford.

Trieb, F. 2000. Competitive solar thermal power stations until 2010: the challenge of market introduction. Renew Energy 19:163–171.

UNDP (United Nations Development Programme). 2004. World Energy Assessment 2004: Energy and the Challenge of Sustainability. UNDP, New York.

Uygur, A., Kargi, F. 2004. Biological nutrient removal from pre-treated landfill leachate in a sequencing batch reactor. J Environ Manage 71: 9–14.

WEC (World Energy Council). 2004. Survey of energy resources. WEC, London.

Xiao, C., Luo, H., Tang, R., Zhong, H. 2004. Solar thermal utilization in China. Renew Energy 29:1549–1556.

Van Ginkel, S.W., Logan, B. 2005. Increased biological hydrogen production with reduced organic loading. Water Res 39:3819–3826.

Veziroglu, T. 1975. Hydrogen energy, Part B. Plenum, New York.

Vijayaraghavan, K., Soom, M.A.M. 2006. Trends in biohydrogen generation: a review. Environ Sci 3:255–271.

Zhang, Q.C., Zhao, K., Zhang, B.C., Wang, L.F., Shen, Z.L., Zhou, Z.J., Lu, D.Q., Xie, D.L., Li, B.F. 1998. New cermet solar coatings for solar thermal electricity applications. Solar Energy 64:109–114.

Chapter 2
Fuels from Biomass

2.1 Introduction

The term biomass (Greek *bio* meaning *life* + *maza* meaning *mass*) refers to non-fossilized and biodegradable organic material originating from plants, animals and microorganisms. The biomass includes products, byproducts, residues and waste from agriculture, forestry and related industries as well as the non-fossilized and biodegradable organic fractions of industrial and municipal solid wastes. Biomass also includes gases and liquids recovered from the decomposition of non-fossilized and biodegradable organic material.

According to another definition, the term biomass refers to wood, short-rotation woody crops, agricultural wastes, short-rotation herbaceous species, wood wastes, bagasse, industrial residues, waste paper, municipal solid waste, sawdust, biosolids, grass, waste from food processing, aquatic plants and algae animal wastes, and a host of other materials (Demirbas, 2008a). Biomass is the plant and animal material, especially agricultural waste products, used as a source of fuel. It is a rather simple term for all organic materials that comes from plants, trees, crops and algae.

Biomass is commonly recognized as an important renewable energy, where solar energy is stored as chemical energy via photosynthesis during the growth of plants and trees, which can be released via direct or indirect combustion. Figure 2.1 shows the main steps of photosynthetic biomass growth.

A useful way to analyze the biosphere system is to trace the work potential, or exergy, of the various energy flows from the sources, such as the Sun, to its eventual destruction in natural decay or human utilization. Photosynthesis is a natural mechanism for capturing and concentrating carbon dioxide directly from the air. Chemical energy and organic carbon are obtained by organisms either directly or indirectly via the photosynthetic conversion of solar energy. These organisms have evolved metabolic machineries for the photochemical reduction of carbon dioxide to organic matter. The majority of the bioengineering strategies for biochemically derived fuels involve options for the disposition of organic matter produced via photosynthate. The bulk of the presently exploited photosynthate is directed toward the production of wood, food, and feed. During processing and consumption, waste organic materials are generated, which can be used for energy production via combustion, pyrolysis or biochemical conversions to ethanol, hydrogen, methane, and isopropanol.

All biomass is produced by green plants converting sunlight into plant material through photosynthesis. Photosynthesis is the fundamental process in biomass energy whereby energy in sunlight is captured and stored in the chemical bonds of the tissues of living plants. Photosynthesis is a carbon fixation reaction by the electrochemical reduction of carbon dioxide. The fixation or reduction of carbon dioxide is a light-independent process. Although some of the steps in photosynthesis are still not completely understood, the overall photosynthetic equation has been known since the 1800s. The productivity of biomass depends on the presence of the required amounts of sunlight, water, and various nutrients, as well as proper environmental conditions including temperature and humidity.

The photosynthetic apparatus and the mechanisms by which it operates have been intensively investigated over the past 40 to 50 years. The photosynthesis is that of three series of interconnected oxidation-reduction reactions: The first involves the evolution of oxygen from water. The second is the transfer of H atoms to a primary hydrogen acceptor. The third is the reduction of CO_2 to carbohydrates by the primary hydrogen acceptor. The light energy required for photosynthesis is used to drive the H atoms against the potential gradient.

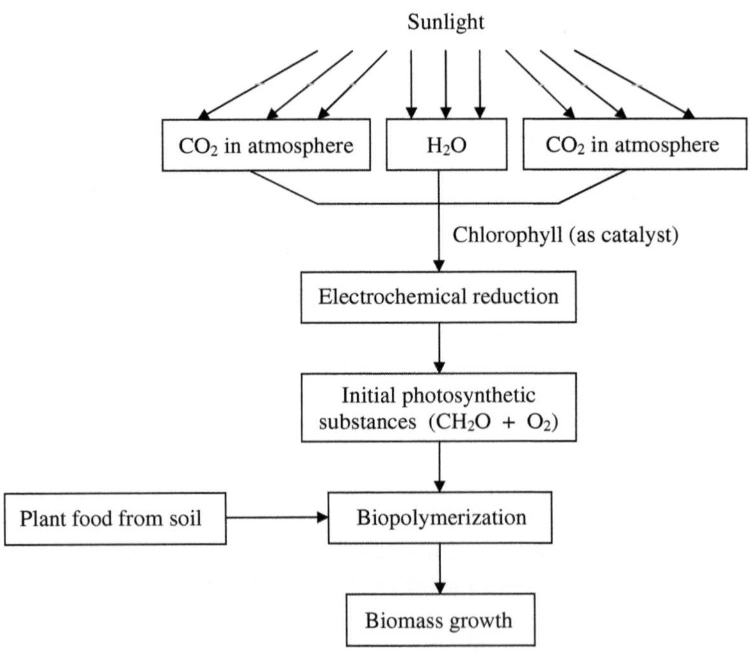

Fig. 2.1 The main steps of photosynthetic biomass growth

The photochemical stage of photosynthesis consists of two separate steps, I and II. The products of light reaction II are an intermediate oxidant and a strong oxidant which is capable of oxidizing water to oxygen. An intermediate oxidant and a strong reductant that can reduce carbon dioxide are produced in light reaction I. The two light reactions involve two pigment systems, photosystems I and II, interconnected by enzymatic reactions coupled with photophosphorylation yielding adenosine triphosphate (ATP).

2.2 Biomass Feedstocks

The biosphere processes large amounts of energy of varying types and quality. Biomass feedstocks are marked by their tremendous diversity, which makes them rather difficult to characterize as a whole. Feedstocks that can be utilized with conversion processes are primarily the organic materials now being landfilled. These include forest products wastes, agricultural residues, organic fractions of municipal solid wastes, paper, cardboard, plastic, food waste, green waste, and other waste. Non-biodegradable organic feedstocks, such as most plastics, are not convertible by biochemical processes. Bio-based materials require pre-treatment by chemical, physical, or biological means to open up the structure of biomass. The major categories of biomass feedstock are given in Table 2.1.

Biomass is the world's fourth largest energy source worldwide, following coal, oil and natural gas. Biomass appears to be an attractive feedstock for three main

Table 2.1 Major categories of biomass feedstocks

Forest products	Wood, logging residues, trees, shrubs and wood residues, sawdust, bark, etc.
Biorenewable wastes	Agricultural wastes, crop residues, mill wood wastes, urban wood wastes, urban organic wastes
Energy crops	Short-rotation woody crops, herbaceous woody crops, grasses, starch crops, sugar crops, forage crops, oilseed crops, switchgrass, miscanthus
Aquatic plants	Algae, water weed, water hyacinth, reed and rushes
Food crops	Grains, oil crops
Sugar crops	Sugar cane, sugar beets, molasses, sorghum
Landfill	Hazardous waste, non-hazardous waste, Inert waste, liquid waste
Organic wastes	Municipal solid waste, industrial organic wastes, municipal sewage and sludges
Algae	Prokaryotic algae, eukaryotic algae, kelps
Mosses	Bryophyta, polytrichales
Lichens	Crustose lichens, foliose lichens, fruticose lichen

reasons. First, it is a renewable resource that could be sustainably developed in the future. Second, it appears to have formidably positive environmental properties by reducing GHG emissions, possibly reducing NO_x and SO_x depending on the fossil fuels displaced. However, there are also negative impacts, such as the release of polycyclic aromatic hydrocarbons, dioxins, furans, volatile organic compounds, and heavy metals especially when combusted in traditional stoves. Third, it appears to have significant economic potential provided that fossil fuel prices increase in the future.

Biomass is a sustainable feedstock for chemicals and energy products that could potentially enhance the energy independence of the world. As an energy source that is highly productive, renewable, carbon-neutral, and easy to store and transport, biomass has drawn worldwide attention recently. Biomass offers important advantages as a combustion feedstock due to the high volatility of the fuel and the high reactivity of both the fuel and the resulting char. However, it should be noted that in comparison with solid fossil fuels, biomass contains much less carbon and more oxygen and has a low heating value. With higher oxygen content than fossil fuels, biomass feedstocks have fundamentally lower energy content. The maximum useful energy that can be theoretically extracted from biomass feedstock is its chemical exergy. Table 2.2 shows the elemental analyses and higher heating value (HHV) of biomass samples.

The waste products of a home include paper, containers, tin cans, aluminum cans, and food scraps, as well as sewage. The waste products of industry and commerce include paper, wood, and metal scraps, as well as agricultural waste products. Biodegradable wastes, such as paper fines and industrial biosludge, into mixed alcohol fuels (e.g., isopropanol, isobutanol, isopentanol). The wastes are first treated with lime to enhance reactivity. Then, they are converted to volatile fatty acids (VFAs) such as acetic acid, propionic acid, and butyric acid, using a mixed culture of microorganisms derived from cattle rumen or anaerobic waste treatment facilities.

Typical solid wastes include wood material, pulp and paper industry residues, agricultural residues, organic municipal material, sewage, manure, and food processing byproducts. Biomass is considered one of the main renewable energy resources of the future due to its large potential, economic viability and various social and environmental benefits. It is estimated that by 2050 biomass could provide nearly 38% of the world's direct fuel use and 17% of the world's electricity. If biomass is produced more efficiently and used with modern conversion technologies, it can supply a considerable range and diversity of fuels at small and large scales. Municipal solid waste (MSW) is defined as waste durable goods, non-durable goods, containers and packaging, food scraps, yard trimmings, and miscellaneous inorganic wastes from residential, commercial, and industrial sources (Demirbas, 2004).

Forests are a principal global economic, as well as ecological, resource. Forests have played a big role in the development of human societies. The prime marketable product of most forests is wood for use as timber, fuelwood, pulp and paper, providing some 3.4 billion cubic meters of timber-equivalent per year globally. Asia and Africa use 75% of global wood fuels. The lumber, plywood, pulp, and paper industries burn their own wood residues in large furnaces and boilers to supply 60% of the energy needed to run factories (Demirbas, 2003). Figure 2.2 shows the use

Table 2.2 Elemental analyses (wt.%) and calorific values (HHV, MJ/kg) of biomass samples

Biomass sample	C	H	O	N	Ash	Higher heating value (MJ/kg)
Olive husk	50.0	6.2	42.2	1.6	3.6	19.0
Hazelnut shell	52.9	5.6	42.7	1.4	1.4	19.3
Hazelnut seedcoat	51.0	5.4	42.3	1.3	1.8	19.3
Softwood	52.1	6.1	41.0	0.2	1.7	20.0
Hardwood	48.6	6.2	41.1	0.4	2.7	18.8
Wheat straw	45.5	5.1	34.1	1.8	13.5	17.0
Wood bark	53.1	6.1	40.6	0.2	1.6	20.5
Waste material	48.3	5.7	45.3	0.7	4.5	17.1
Water hyacinth	39.8	5.0	34.3	1.9	19.0	14.6
Corncob	49.0	5.4	44.6	0.4	1.0	18.4
Corn stover	45.1	6.0	43.1	0.9	4.9	17.4
Brown kelp	28.4	4.1	24.3	4.8	38.4	10.8
Tea waste	48.6	5.5	39.5	0.5	1.4	17.1
Bagasse	45.3	5.1	40.2	0.1	9.3	16.9
Spruce wood	51.9	6.1	40.9	0.3	1.5	20.1
Beech wood	49.5	6.2	41.2	0.4	1.4	19.2
Poplar wood	49.0	6.1	42.8	0.1	1.0	18.8
Ailanthus wood	49.5	6.2	41.0	0.3	1.7	19.0

of world wood products including lumber, plywood, paper and paperboard between the years 1970 and 2005. The availability of fuelwood from forests is continually declining at an ever-increasing rate due to indiscriminate deforestation and slow regeneration, as well as afforestation (Jain and Singh, 1999). The fuelwoods generally used by local people were identified and analyzed quantitatively to select a few species with the best fuelwood characteristics so that plantation on non-agricultural lands could be undertaken (Jain, 1992).

Wood biomass involves trees with commercial structure and forest residues not being used in the traditional forest products industries. Available forest residues may appear to be an attractive fuel source. Collection and handling and transport costs are critical factors in the use of forest residues. Although the heat produced from wood wastes is less than that from oil or gas, its cost compared to fossil fuels makes it an attractive source of readily available heat or heat and power. The most effective utilization of wood wastes, particularly in the saw milling and plywood industry, plays an important role in energy efficient production. Table 2.3 shows the sources of available forest and wood manufacturing residues. Wood preparation involves the conversion of roundwood logs into a form suitable for pulping and includes processes for debarking, chipping, screening, handling, and storage.

Forest residues typically refer to those parts of trees such as treetops, branches, small-diameter wood, stumps and dead wood as well as undergrowth and low-value species. The conversion of wood to biofuels and biochemicals has long been a goal of the forest products industry. Forest residues alone count for some 50% of the total forest biomass and are currently left in the forest to rot (Demirbas, 2001).

Agricultural residues, grasses, algae, kelps, lichens and mosses are also important biomass feedstocks in the world. Algae can grow practically in every place where there is enough sunshine. Some algae can grow in saline water. The most significant difference of algal oil is in the yield and hence its biodiesel yield. According to some estimates, the yield (per acre) of oil from algae is over 200 times the yield from the best-performing plant/vegetable oils (Sheehan et al., 1998). Microal-

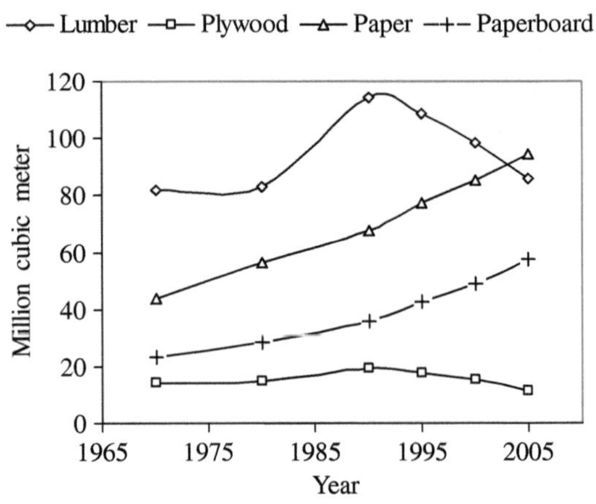

Fig. 2.2 Use of global wood products: lumber, plywood, paper, and paperboard (1970–2005)

Table 2.3 Sources of available forest and wood manufacturing residues

Source of residue	Type of residue
Forest operations	Branches, bark chips, leaves/needles, stumps, roots, and sawdust
Lumber production	Bark, sawdust, clippings, split wood
Saw milling	Bark, trimmings, split wood, sawdust, planer shavings, sander dust
Plywood production	Bark, sawdust, veneer clippings and wastes, panel trim, sander dust
Paper production	Slab chips, pulping reject, sawdust, clippings
Paperboard production	Bark, sawdust, screening fines, panel trim, sander dust

gae are the fastest growing photosynthesizing organisms. They can complete an entire growing cycle every few days. Approximately 46 tons of oil/hectare/year can be produced from diatom algae. Different algae species produce different amounts of oil. Some algae produce up to 50% oil by weight.

2.3 The Chemistry of Biomass

The components of biomass include cellulose, hemicelluloses, lignin, extractives, lipids, proteins, simple sugars, starches, water, hydrocarbons, ash, and other compounds. Two larger carbohydrate categories that have significant value are cellulose and hemicelluloses (holocellulose). The lignin fraction consists of non-sugar type macromolecules. Three structural components are cellulose, hemicelluloses and lignin, which have rough formulae as $CH_{1.67}O_{0.83}$, $CH_{1.64}O_{0.78}$, and $C_{10}H_{11}O_{3.5}$, respectively.

The basic structure of all wood and woody biomass consists of cellulose, hemicelluloses, lignin and extractives. Their relative composition is shown in Table 2.4. Softwoods and hardwoods differ greatly in wood structure and composition. Hardwoods contain a greater fraction of vessels and parenchyma cells. Hardwoods have a higher proportion of cellulose, hemicelluloses and extractives than softwoods, but softwoods have a higher proportion of lignin. Hardwoods are denser than softwoods.

Cellulose is a remarkable pure organic polymer, consisting solely of units of anhydroglucose held together in a giant straight-chain molecule. These anhydroglucose units are bound together by β-(1,4)-glycosidic linkages. Due to this linkage, cellobiose is established as the repeat unit for cellulose chains. Cellulose must be hydrolyzed to glucose before fermentation to ethanol.

By forming intramolecular and intermolecular hydrogen bonds between OH groups within the same cellulose chain and the surrounding cellulose chains, the chains tend to be arranged in parallel and form a crystalline supermolecular structure. Then, bundles of linear cellulose chains (in the longitudinal direction) form a microfibril which is oriented in the cell wall structure. Cellulose is insoluble in most solvents and has a low accessibility to acid and enzymatic hydrolysis (Demirbas, 2008b).

Unlike cellulose, hemicelluloses consist of different monosaccharide units. In addition, the polymer chains of hemicelluloses have short branches and are amorphous. Because of the amorphous morphology, hemicelluloses are partially soluble

Table 2.4 Structural composition of wood (wt.% of dry and ash-free sample)

Wood species	Cellulose	Hemicelluloses	Lignin	Extractives
Hardwood	43–48	27–35	16–24	2–8
Softwood	40–44	24–29	26–33	1–5

or swellable in water. Hemicelluloses (arabinoglycuronoxylan and galactoglucomannans) are related to plant gums in composition, and occur in much shorter molecular chains than cellulose. Hemicelluloses are derived mainly from chains of pentose sugars, and act as the cement material holding together the cellulose micelles and fiber (Demirbas, 2008b).

Among the most important sugar of the hemicelluloses component is xylose. In hardwood xylan, the backbone chain consists of xylose units which are linked by β-(1,4)-glycosidic bonds and branched by α-(1,2)-glycosidic bonds with 4-O-methylglucuronic acid groups. In addition, O-acetyl groups sometime replace the OH groups in position C_2 and C_3. For softwood xylan, the acetyl groups are fewer in the backbone chain. However, softwood xylan has additional branches consisting of arabinofuranose units linked by α-(1,3)-glycosidic bonds to the backbone. Hemicelluloses are largely soluble in alkali and, as such, are more easily hydrolyzed (Demirbas, 2008b).

Lignins are polymers of aromatic compounds. Their functions are to provide structural strength, provide sealing of the water-conducting system that links roots with leaves, and protect plants against degradation. Lignin is a macromolecule, which consists of alkylphenols and has a complex three-dimensional structure. Lignin is covalently linked with xylans in the case of hardwoods and with galactoglucomannans in softwoods. The basic chemical phenylpropane units of lignin (primarily syringyl, guaiacyl and p-hydroxy phenol) as shown in Fig. 2.3 are bonded together by a set of linkages to form a very complex matrix. This matrix comprises a variety of functional groups, such as hydroxyl, methoxyl and carbonyl, which impart a high polarity to the lignin macromolecule (Demirbas, 2008b).

Pyrolysis of lignin, e.g., during combustion, yields a wide range of products, of which the most characteristic ones are methoxy phenols. Of those, the most important are guaiacol and syringol and their derivatives; their presence can be used to trace a smoke source to a wood fire. Lignin biosynthesis begins with the synthesis of monolignols. The starting material is the amino acid phenylalanine. The first reactions in the biosynthesis are shared with the phenylpropanoid pathway, and

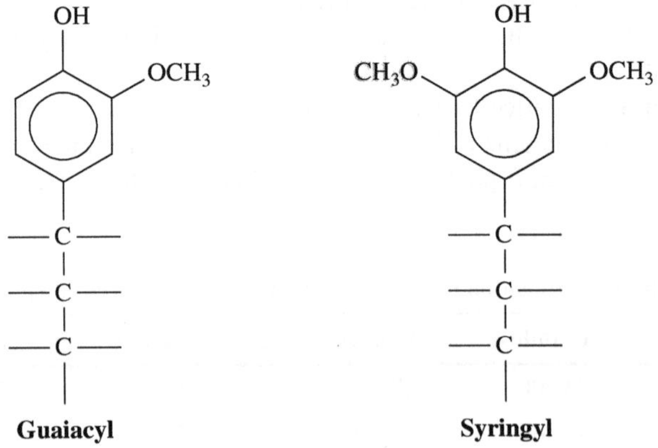

Fig. 2.3 Schematic illustration of building units of lignin

monolignols are considered to be a part of this group of compounds. There are three main types of monolignols: coniferyl alcohol, sinapyl alcohol and paracoumaryl alcohol. Different plants use different monolignols. For example, gymnosperms as Norway spruce have a lignin that consists almost entirely of coniferyl alcohol. Dicotyledonic lignin is a mixture of coniferyl alcohol and sinapyl alcohol (normally more of the latter), and monocotylednic lignin is a mixture of all three monolignols. Some monocotyledons have mostly coniferyl alcohol (as many grasses), while others have mainly sinapyl alcohols, as some palms. Monolignols are synthesized in the cytosol as glucosides. The glucose is added to the monolignol to make them water soluble and to reduce their toxicity. The glucosides are transported through the cell membrane to the apoplast. The glucose is then removed and the monolignols are polymerized into lignin. Four main monolignols in the lignin structure are given in Fig. 2.4.

Wood and woody biomass also contain lesser amounts of tannins, simple sugars, starches, pectins and organic soluble extractives. Extractives are the organic substances which have low molecular weight and are soluble in neutral solvents. Extractives include terpenes, tall oil and the fatty acids, esters, and triglycerides, which contribute to paper mill pitch problems. Resin (combination of the following components: terpenes, lignans and other aromatics), fats, waxes, fatty acids and alcohols, terpentines, tannins and flavonoids are categorized as extractives.

Biomass has some advantageous chemical properties for use in current energy conversion systems. Compared to other carbon-based fuels, it has low ash content and high reactivity. Biomass combustion is a series of chemical reactions by which carbon is oxidized to carbon dioxide, and hydrogen is oxidized to water. Oxygen deficiency leads to incomplete combustion and the formation of many products of incomplete combustion. Excess air cools the system. The air requirements depend on the chemical and physical characteristics of the fuel. The combustion of the biomass relates to the fuel burn rate, the combustion products, the required excess air for complete combustion, and the fire temperatures.

The compositions of biomass among fuel types are considerably varied, especially with respect to inorganic constituents important to the critical problems of fouling and slagging. Alkali and alkaline earth metals, in combination with other fuel elements such as silica and sulfur, and facilitated by the presence of chlorine, are responsible for many undesirable reactions in combustion furnaces and power boilers.

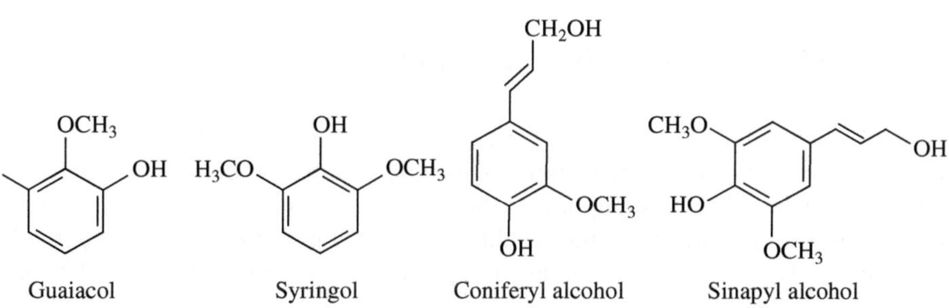

Fig. 2.4 Four main monolignols in the lignin structure

2.4 Production of Fuels from Wood Sources

Biomass includes 60% wood and 40% non-wood materials. The conversion of wood into biofuels and biochemicals is technically feasible. Wood valorization processes include fractionation, liquefaction, pyrolysis, hydrolysis, fermentation and gasification. Fast pyrolysis utilizes wood biomass to produce a product that is used both as an energy source and a feedstock for chemical production. The bio-oils from wood pyrolysis were composed of a range of cyclopentanone, methoxyphenol, acetic acid, methanol, acetone, furfural, phenol, formic acid, levoglucosan, guaiacol and their alkylated phenol derivatives. When wood is rapidly heated in a reduced oxygen environment, the feeding material does not combust but rather becomes a synthetic gas (syngas), a combination of hydrogen (H_2) and carbon monoxide (CO). Wood can be converted by hydrolysis into sugars and subsequently fermentation of sugars.

The wood pyrolysis is attractive because forest and industrial wood residues can be readily converted into liquid products. These liquids, as crude bio-oil or slurry of charcoal of water or oil, have advantages in transport, storage, combustion, retrofitting and flexibility in production and marketing (Demirbas, 2007). In the first step of pyrolysis of carbohydrates dehydration occurs and at low temperatures dehydration predominates. Dehydration is also known as a char-forming reaction. Between 550 and 675 K volatile products, tar, and char are formed. The volatile products are CO_2, CO, H_2O, acetals, furfural, aldehydes and ketones. Levoglucosan is the principle component in tar.

Gasification is carried out at high temperatures in order to optimize the gas production. Most wood gasification systems utilize air or oxygen in partial oxidation or combustion processes. The resulting gas, known as producer gas, is a mixture of carbon monoxide, hydrogen and methane, together with carbon dioxide and nitrogen. As mentioned above, when wood is rapidly heated in a reduced oxygen environment, the feeding material becomes a synthetic gas (syngas), a combination of H_2 and CO. Syngas can be combusted directly in a boiler or used in a gas turbine to generate electricity. The hydrogen from syngas can be isolated and used to power fuel cells. Syngas has the potential to produce a wide range of commercial fuels and chemicals, including synthetic diesel, methanol and lower carbon alcohols, acetic acid, dimethyl ether, etc. (Demirbas, 2000).

2.5 Production of Fuels from Crops

Biofuels such as liquid fuels, bio-oil and biochar, and biogas can be produced from crops. Agriculture-based liquid biofuels include bioethanol, biodiesel, biomethanol, methane, and bio-oil components. Various agricultural residues such as grain dust, crop residues, and fruit tree residues are available as the sources of agricultural energy. Bioenergy from biomass, both residues and energy crops, can be converted into modern energy carriers. Bioethanol is derived from renewable source feedstocks; typically plants such as wheat, sugar beet, corn, straw and wood. Biodiesel is a non-fossil fuel alternative to petrodiesel, which can be obtained from vegetable oil and animal fats by transesterification. Bio-oils are liquid or gaseous fuels made from biomass materials, such as agricultural crops, municipal wastes and agricultural and forestry byproducts via biochemical or thermochemical processes.

2.5.1 Fuels from Cereal Crops

Biofuels such as bioethanol and biodiesel originate from cereal crops such as plant oils, and sugar beets. Today the production cost of bioethanol cereal crops is still too high, which is the major reason why bioethanol has not made its breakthrough as a fuel source yet. When producing bioethanol from maize or sugar cane the raw material constitutes about 40–70% of the production cost.

Feedstock for bioethanol is essentially comprised of sugar cane and sugar beet. The two are produced in geographically distinct regions. Sugar cane is grown in tropical and subtropical countries, while sugar beet is only grown in temperate climate countries.

In European countries, beet molasses are the most utilized sucrose-containing feedstock. Sugar beet crops are grown in most of the EU-25 countries, and yield substantially more ethanol per hectare than wheat. The advantages with sugar beet are a lower cycle of crop production, higher yield, and high tolerance of a wide range of climatic variations, low water and fertilizer requirement. Sweet sorghum is one of the most drought resistant agricultural crops as it has the capability to remain dormant during the driest periods.

Another type of feedstock, which can be used for bioethanol production, are starch-based materials. Starch is a biopolymer and is defined as a homopolymer consisting of only one monomer, D-glucose. To produce bioethanol from starch it is necessary to break down the chains of this carbohydrate for obtaining glucose syrup, which can be converted into ethanol by yeasts. The single greatest cost in the production of bioethanol from corn, and the cost with the greatest variability, is the cost of the corn. Figure 2.5 shows the flow chart for the production of bioethanol from cereal grain or straw.

Bioethanol can be produced from a large variety of carbohydrates with a general formula of $(CH_2O)_n$. Chemical reaction is composed of enzymatic hydrolysis of sucrose followed by fermentation of simple sugars. Fermentation of sucrose is performed using commercial yeast such as *Saccharomyces cerevisiae*. First, invertase enzyme in the yeast catalyzes the hydrolysis of sucrose to convert it into glucose and fructose.

$$\underset{\text{Sucrose}}{C_{12}H_{22}O_{11}} \rightarrow \underset{\text{Glucose}}{C_6H_{12}O_6} + \underset{\text{Fructose}}{C_6H_{12}O_6} \tag{2.1}$$

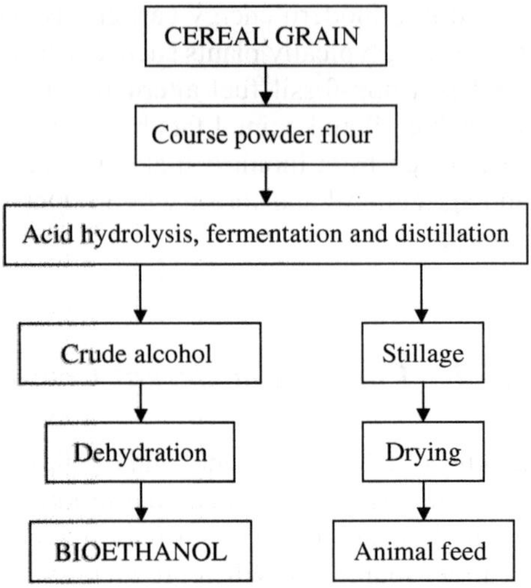

Fig. 2.5 Flow chart for the production bioethanol from cereal grain

Second, zymase, another enzyme also present in the yeast, converts the glucose and the fructose into ethanol.

$$C_6H_{12}O_6 \rightarrow 2\ C_2H_5OH + 2CO_2 \quad (2.2)$$

Gluco-amylase enzyme converts the starch into D-glucose. The enzymatic hydrolysis is then followed by fermentation, distillation and dehydration to yield anhydrous bioethanol. Corn (60–70% starch) is the dominant feedstock in the starch-to-bioethanol industry worldwide.

Lignocellulosic perennial crops (e.g., short-rotation coppices and grasses) are a promising feedstock because of high yields, low costs, good suitability for low-quality land (which is more easily available for energy crops), and the low environmental impacts.

2.5.2 *Fuels from Non-Cereal Crops*

Agricultural residues such as straws, nut shells, fruit shells, fruit seeds, plant stalks, green leaves and molasses are potential renewable energy resources.

A methane digester system, commonly referred to as an anaerobic digester is a device that promotes the decomposition of manure or digestion of the organics in manure to simple organics and gaseous biogas products. Several types of biodigesters have been developed including the floating drum, the fixed dome, the bag, the plastic tube (Beteta, 1995), the plug flow and the up-flow anaerobic sludge blanket digesters. Figure 2.6 shows an on-farm biogas system.

The following describes the process of manure and straw mixture digestion. For the first 3 days, the methane yield is almost 0% and carbon dioxide generation is roughly 100%. In this period, digestion occurs as aerobic fermentation to carbon

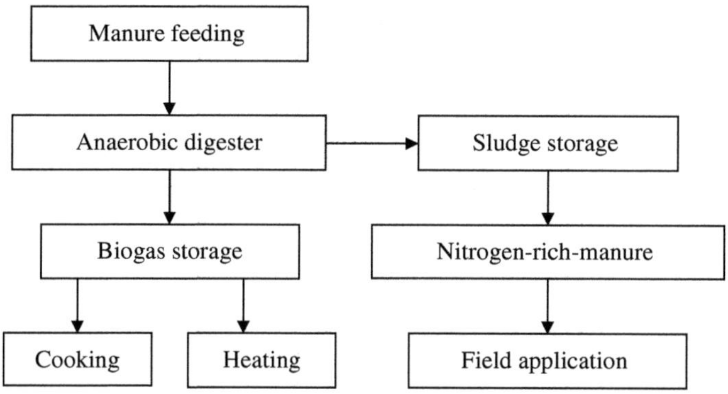

Fig. 2.6 An on-farm biogas system

dioxide. The yields of methane and carbon dioxide gases are then 50-50 on the eleventh day. At the end of the twentieth day, the digestion reaches the stationary phase. The methane content of the biogas is in the range of 73–79% for the runs, the remainder being principally carbon dioxide. During a 30-day digestion period, approximately 80–85% of the biogas is produced in the first 15–18 days. This implies that the digester retention time can be designed to be 15–18 days instead of 30 days.

Anaerobic digestion yields of the slurries are given in Table 2.5. Due to the higher proportion of lipids in manure, the methane production is higher than in wheat straw. The total amounts of lipid and protein, in particular, are lower in straw than in manure and thus the theoretical methane yield is significantly lower in wheat straw than in manure (Table 2.5). The average methane yields of manure and wheat straw were 14.7% and 10.4% of volatile solids, respectively (Demirbas and Ozturk, 2004).

Table 2.5 Anaerobic digestion yields of the slurries

Run	Inoculated solution (ml)	Manure (g)	Straw (g)	Volatile solids (wt.%)	Biogas yield (wt.% of VS)	Methane yield (wt.% of VS)
1	930	280	0	89.2	29.8	14.7
2	960	0	294	81.3	21.6	10.4
3	945	143	144	85.3	26.8	12.9
4	900	90	180	83.4	24.0	11.9
5	927	185	93	85.7	27.6	14.2
6	936	70	210	82.9	23.5	11.5
7	942	214	72	88.3	28.4	14.0

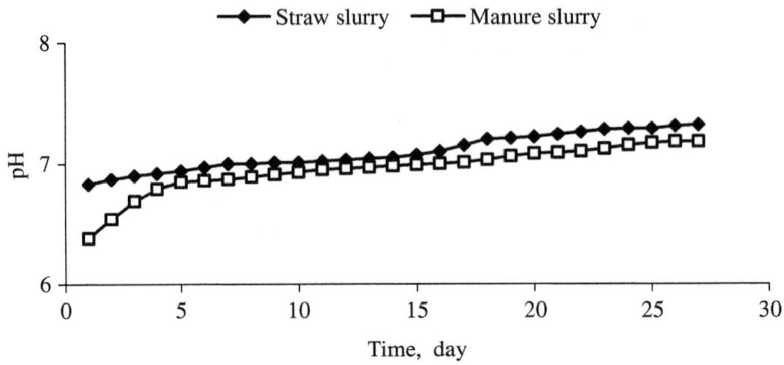

Fig. 2.7 The pH values plotted against time for wheat straw and manure slurries

Figure 2.7 shows pH values plotted against time for wheat straw and manure slurries. The pH value of slurry is a sensitive parameter for the digestion rate. During digestion, the volatile fatty acid concentration is lower and the pH higher. The pH of the slurry with manure increased from 6.4 initially, to 6.9–7.0 at the maximum methane production rate. The pH of the slurry with wheat straw was around 7.0–7.1 at the maximum methane production rate. Digester pH increased to its neutral value (6.9–7.1) due to the process stability and the optimal activity of methanogenic bacteria (Demirbas and Ozturk, 2004).

When the pyrolysis temperature increased the biochar yield decreased. The biochar yield increased with increasing particle size of the sample. The yield of bio-oil from pyrolysis of the samples increased with temperature. Anaerobic biogas production is an effective process for conversion of a broad variety of agricultural biomass to methane to substitute natural gas and medium calorific value gases.

Pyrolysis of agricultural residues can be used for the production of bio-oil if flash pyrolysis processes are used. Production of this type is currently in the pilot stage (EUREC Agency, 1996). Some problems in the conversion process and use of the oil need to be overcome; these include poor thermal stability and corrosivity of the oil. Upgrading by lowering the oxygen content and removing alkalis by means of hydrogenation and catalytic cracking of the oil may be required for certain applications. Table 2.6 shows the yields of pyrolysis liquid products from hazelnut shell, tea waste, and tobacco stalk samples at different temperatures. As seen in Table 2.6, the yields of liquid products for the samples increased with increasing temperature

Table 2.6 Yields of bio-oil by pyrolysis from agricultural residues at different temperatures (K)

Sample	675	725	775	825	875	925	975	1050
Hazelnut shell	38.0	39.5	40.4	41.9	42.2	41.0	39.2	38.5
Tea waste	34.9	35.8	36.0	36.2	38.0	37.0	35.5	33.4
Tobacco stalk	41.0	41.8	43.0	40.2	40.0	40.6	37.3	36.8

from 675 to 875 K and then decreased. The maximum liquid yields from the hazelnut shell, tea waste, tobacco stalk, and yellow pine samples were 42.2%, 38.0%, and 43.0%, respectively (Güllü, 2003).

2.5.3 Fuels from Energy Crops

The term "energy crop" can be used both for biomass crops that simply provide high output of biomass per hectare for low inputs, and for those that provide specific products that can be converted into other biofuels such as sugar or starch for bioethanol by fermentation, or into vegetable oil for biodiesel by transesterification.

Energy crops such as short-rotation woody crops, herbaceous woody crops, grasses, starch crops, sugar crops, forage crops, oilseed crops, switchgrass, and miscanthus can be converted to liquid biofuels by thermochemical and biochemical conversion processes.

Energy crops can be divided into short-rotation energy crops that are harvested on a cycle of anything from 2 to 20 years, depending on the crop and the system, and herbaceous energy crops that are harvested annually. Herbaceous energy crops such as the woody, perennial, rhizomatous grasses like miscanthus offer the advantage of an annual harvest, though typically establishment costs are high. All crops, including some biennial, permanent and multiannual crops, are produced for energy purposes. Energy is generated by burning plants grown for the purpose, often after the dry matter is pelletized. Energy crops are used for firing power plants, either alone or co-fired with other fuels. Alternatively they may be used for heat or combined heat and power (CHP) production.

Biogas plants can be directly supplemented with energy crops. European production of biodiesel from energy crops has grown steadily in the last decade, principally focused on rapeseed used for oil and energy. In the United States, biodiesel is being made from soybean oil because more soybean oil is produced than all other sources of fats and oil combined. There are many candidates for feedstocks, including recycled cooking oils, animal fats, and a variety of other oilseed crops. The European Union accounted for nearly 89% of all biodiesel production worldwide in 2005. By 2010, the United States is expected to become the world's largest single biodiesel market, accounting for roughly 18% of world biodiesel consumption, followed by Germany.

Switchgrass is a perennial grass native to North America. Switchgrass makes for a great energy crop because it grows fast, capturing lots of solar energy and turning it into chemical energy, which is stored as cellulose. Switchgrass reaches its full yield potential after the third year planted, producing approximately 6 to 8 tons per acre; that is 500 gallons of ethanol per acre.

Miscanthus is a tall perennial grass that has been evaluated in Europe during the past 5 to 10 years as a new bioenergy crop. Like other energy crops, the harvested stems of miscanthus may be used as fuel for production of heat and electric power, or for conversion to other useful products such as ethanol.

Sweet sorghum can be considered as an energy crop, because it can be grown in all continents, in tropical, subtropical, temperate regions as well as in poor quality soils. Sweet sorghum is a warm-season crop that matures earlier under high temperatures and short days. Sweet sorghum is an extraordinarily promising multifunctional crop not only for its high economic value but also for its capacity to provide a very wide range of renewable energy products, industrial commodities, food and animal feed products. Sweet sorghum biomass is rich in readily fermentable sugars and thus it can be considered an excellent raw material for fermentative hydrogen production. Integrated production of several crops (sweet sorghum/sugar cane/sweet sorghum corn/sweet potatoes, etc.) and simultaneous processing of the full crop components (starch, sugar, lignocellulosic) can improve the global economics of bioenergy schemes considerably in view of a viable production of bioethanol.

Summary

Biomass refers to the non-fossilized and biodegradable organic material that originates from plants, animals and microorganisms. Biomass is the world's fourth largest energy source, following coal, oil and natural gas. Biomass appears to be an attractive feedstock for three main reasons. First, it is a renewable resource that could be sustainably developed in the future. Second, it appears to have positive environmental properties, for example, reducing greenhouse gas emissions, possibly reducing NO_x and SO_x depending on the fossil fuels displaced. Third, it appears to have significant economic potential provided that fossil fuel prices increase in the future.

The components of biomass include cellulose, hemicelluloses, lignin, extractives, lipids, proteins, simple sugars, starches, water, hydrocarbons, ash, and other compounds. Two larger carbohydrate categories that have significant value are cellulose and hemicelluloses (holocellulose). The lignin fraction consists of non-sugar type macromolecules. Three structural components are cellulose, hemicelluloses and lignin, which have rough formulae as $CH_{1.67}O_{0.83}$, $CH_{1.64}O_{0.78}$, and $C_{10}H_{11}O_{3.5}$, respectively.

Biomass includes 60% wood and 40% non-wood materials. The conversion of wood into biofuels and biochemicals is technically feasible. Wood valorization processes include fractionation, liquefaction, pyrolysis, hydrolysis, fermentation and gasification.

References

Beteta, T. 1995. Experiences with plastic tube biodigesters in Colombia. Universidad Nacional Agraria, Managua, Nicaragua.

Demirbas, A. 2000. Recent advances in biomass conversion technologies. Energy Edu Sci Technol 6:19–41.

Demirbas, A. 2001. Biomass resource facilities and biomass conversion processing for fuel and chemicals. Energy Convers Manage 42:1357–1378.

Demirbas, A. 2003. Fuelwood characteristics of six indigenous wood species from Eastern Black Sea Region. Energy Sources 25:309–316.

Demirbas, A. 2004. Combustion characteristics of different biomass fuels. Prog Energy Combus Sci 30:219–230.

Demirbas, A. 2007. The influence of temperature on the yields of compounds existing in bio-oils obtained from biomass samples via pyrolysis. Fuel Proc Technol 88:591–598.

Demirbas, A. 2008a. Biodiesel: A realistic fuel alternative for diesel engines. Springer, London.

Demirbas, A. 2008b. Heavy metal adsorption onto agro-based waste materials: a review. J Hazard Mat 157:220–229.

Demirbas, A., Ozturk, T. 2004. Anaerobic digestion of agricultural solid residues. Int J Green Energy 1:483–494.

EUREC Agency. 1996. The future for renewable energy, prospects and directions. James and James Science, London.

Güllü, D. 2003. Effect of catalyst on yield of liquid products from biomass via pyrolysis. Energy Sources 25:753–765.

Jain, R.K. 1992. Fuelwood characteristics of certain hardwood and softwood tree species of India Biores Technol 41:129–133.

Jain, R.K., Singh, B. 1999. Fuelwood characteristics of selected indigenous tree species from central India. Biores Technol 68:305–308.

Sheehan, J., Dunahay, T., Benemann, J., Roessler, P. 1998. A look back at the US Department of Energy's Aquatic Species Program: biodiesel from algae. National Renewable Energy Laboratory (NREL) Report: NREL/TP-580-24190, Golden, Colorado.

Chapter 3
Biofuels

3.1 Introduction

The term biofuel refers to liquid, gas and solid fuels predominantly produced from biomass. Biofuels are important as they affect a wide array of topics, including energy security, environmental concerns, foreign exchange savings, and socioeconomic issues related to the rural sector. Biofuels include bioethanol, biomethanol, vegetable oils, biodiesel, biogas, biosynthetic gas (biosyngas), bio-oil, biochar, Fischer–Tropsch liquids, and biohydrogen.

There are two global biomass-based liquid transportation fuels that might replace gasoline and diesel fuel. These are bioethanol and biodiesel. World production of biofuel was about 68 billion liters in 2007. The primary feedstocks of bioethanol are sugarcane and corn. Bioethanol is a gasoline additive/substitute. Bioethanol is by far the most widely used biofuel for transportation worldwide. About 60% of global bioethanol production comes from sugarcane and 40% from other crops. Biodiesel refers to a diesel-equivalent mono-alkyl ester-based oxygenated fuel. Biodiesel production using inedible vegetable oil, waste oil and grease has become more attractive recently. The economic performance of a biodiesel plant can be determined once certain factors are identified, such as plant capacity, process technology, raw material cost and chemical costs. The central policy of biofuel concerns job creation, greater efficiency in the general business environment, and protection of the environment.

Biofuels offer a number of technical and environmental benefits over conventional fossil fuels, which make them attractive as alternatives for the transport sector. The benefits include greenhouse gas reductions including reduced carbon dioxide emissions, which will contribute to domestic and international targets, the diversification of the fuel sector, biodegradability, sustainability, and an additional market for agricultural products. Biofuels help to protect and create jobs. Table 3.1 shows the major benefits of biofuels.

Table 3.1 Major benefits of biofuels

Economic impacts	Sustainability
	Fuel diversity
	Increased number of rural manufacturing jobs
	Increased income taxes
	Increased investments in plant and equipment
	Agricultural development
	International
	Reducing the dependency on imported petroleum
Environmental impacts	Greenhouse gas reductions
	Reducing of air pollution
	Biodegradability
	Higher combustion efficiency
	Improved land and water use
	Carbon sequestration
Energy security	Domestic targets
	Supply reliability
	Reducing use of fossil fuels
	Ready availability
	Domestic distribution
	Renewability

According to a study by the European Commission, European production of biofuels equivalent to 1% of EU automotive fuel consumption would help to protect and/or create between 45 000 and 75 000 jobs. Biofuels can be used as an alternative fuel for transport, as can other alternatives such as liquid natural gas (LNG), compressed natural gas (CNG) and liquefied petroleum gas (LPG). In the longer term, significant use of biofuels could offer large carbon savings. Policy drivers for renewable liquid biofuels have attracted particularly high levels of assistance in some countries given their promise of benefits in several areas of interest to governments, including agricultural production, greenhouse gas emissions, energy security, trade balances, rural development and economic opportunities for developing countries (Demirbas, 2008a).

First generation biofuels refer to biofuels made from sugar, starch, vegetable oils, or animal fats using conventional technology. The basic feedstocks for the production of first generation biofuels are often seeds or grains such as wheat, which yields starch that is fermented into bioethanol, or sunflower seeds, which are pressed to yield vegetable oil that can be used in biodiesel. Table 3.2 shows the classification of renewable biofuels based on their production technologies.

Table 3.2 Classification of renewable biofuels based on their production technologies

Generation	Feedstock	Example
First generation biofuels	Sugar, starch, vegetable oils, or animal fats	Bioalcohols, vegetable oil, Biodiesel, biosyngas, biogas
Second generation biofuels	Non-food crops, wheat straw, corn, wood, solid waste, energy crop	Bioalcohols, bio-oil, bio-DMF, Biohydrogen, bio-Fischer–Tropsch diesel, wood diesel
Third generation biofuels	Algae	Vegetable oil, biodiesel
Fourth generation biofuels	Vegetable oil, biodiesel	Biogasoline

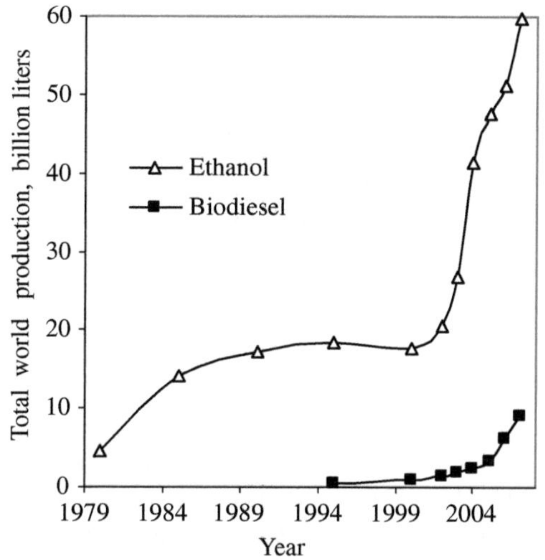

Fig. 3.1 World production of ethanol and biodiesel, 1980–2007

Second and third generation biofuels are also called advanced biofuels. Second generation biofuels made from non-food crops, wheat straw, corn, wood, energy crop using advanced technology. Algae fuel, also called oilgae or third generation biofuel, is a biofuel from algae. Algae are low-input/high-yield (30 times more energy per acre than land) feedstocks to produce biofuels using more advanced technology. On the other hand, an appearing fourth generation is based in the conversion of vegoil and biodiesel into biogasoline using the most advanced technology.

Figure 3.1 shows the world productions of ethanol and biodiesel between 1980 and 2007 (Demirbas, 2008a). Between 1991 and 2001, world ethanol production rose from around 16 billion liters a year to 18.5 billion liters. From 2001 to 2007, production is expected to have tripled, to almost 60 billion liters a year. Brazil was the world's leading ethanol producer until 2005 when U.S. production roughly

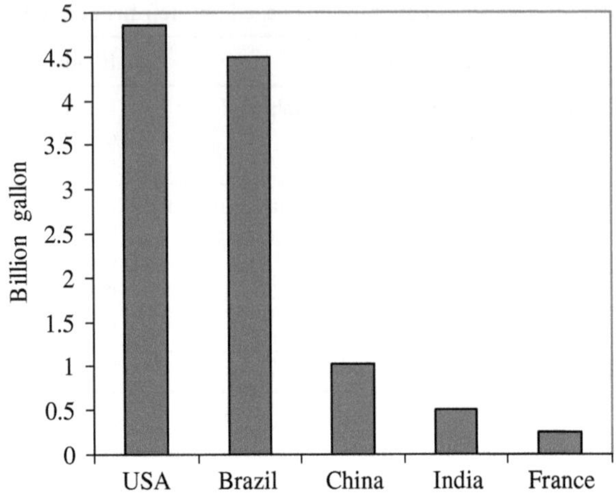

Fig. 3.2 The top five bioethanol producers (billion gallons) in 2006

equaled Brazil's. The United States became the world's leading ethanol producer in 2006. China holds a distant but important third place in world rankings, followed by India, France, Germany and Spain. Figure 3.2 shows the top five bioethanol producers in 2006 (RFA, 2007).

3.2 Bioethanol

Nowadays ethanol is the most popular fuel. Bioethanol is ethyl alcohol, grain alcohol, or chemically C_2H_5OH or EtOH. The use of ethanol as a motor fuel has as long a history as the car itself. Ethanol can be produced from cellulose feedstocks such as corn stalks, rice straw, sugar cane bagasse, pulpwood, switchgrass, and municipal solid waste is called bioethanol. Bioethanol is a renewable green fuel. Bioethanol is produced by fermentation of sugars, which can be obtained from natural sugars (e.g., sugar cane, sugar beet), starches (e.g., corn, wheat), or cellulosic biomass (e.g., corn stover, straw, grass, wood). The most common feedstock is sugar cane or sugar beet, and the second common feedstock is corn starch.

Currently, cellulosic biomass use is very limited due to the expensive pretreatment required for breaking the crystalline structure of cellulose. Bioethanol is already an established commodity due to its ongoing non-fuel uses in beverages, and in the manufacture of pharmaceuticals and cosmetics. In fact, ethanol is the oldest synthetic organic chemical used by mankind. Table 3.3 shows ethanol production in different continents (Demirbas, 2008b).

Bioethanol can be used as a 10% blend with gasoline without need for any engine modification. However, with some engine modification, bioethanol can be used at higher levels, for example, E85 (85% bioethanol).

Table 3.3 Ethanol production in different continents (billion liters/year) in 2006

America	Asia	Europe	Africa	Oceania
22.3	5.7	4.6	0.5	0.2

Bioethanol currently accounts for more than 94% of global biofuel production, with the majority coming from sugarcane. About 60% of global bioethanol production comes from sugarcane and 40% from other crops. Brazil and the United States are the world leaders, which together accounted for about 70% of the world bioethanol production exploiting sugarcane and corn respectively. Ethanol has been used in Germany and France as early as 1894 by the then incipient industry of internal combustion engines. Brazil has utilized ethanol as a fuel since 1925. Currently, ethanol is produced from sugar beets and from molasses. A typical yield is 72.5 l of ethanol per ton of sugar cane. Modern crops yield 60 tons of sugar cane *per* hectare of land. Production of ethanol from biomass is one way to reduce both the consumption of crude oil and environmental pollution. The use of gasohol (an ethanol and gasoline mixture) as an alternative motor fuel has been steadily increasing in the world for a number of reasons. Domestic production and use of ethanol for fuel can decrease dependence on foreign oil, reduce trade deficits, create jobs in rural areas, reduce air pollution, and reduce global climate change carbon dioxide build-up (Bala, 2005).

Bioethanol can be produced from a large variety of carbohydrates with a general formula of $(CH_2O)_n$. Fermentation of sucrose is performed using commercial yeast such as *Saccharomyces cerevisiae*. The chemical reaction is composed of enzymatic hydrolysis of sucrose followed by fermentation of simple sugars (Baltz et al., 1982; Sokhansanj et al., 2002; Kim and Dale, 2005). First, invertase enzyme in the yeast catalyzes the hydrolysis of sucrose to convert it into glucose and fructose. Second, *zymase*, another enzyme also present in the yeast, converts the glucose and the fructose into ethanol. A major processing step in an ethanol plant is enzymatic saccharification of cellulose to sugars through treatment by enzymes; this step requires lengthy processing and normally follows a short term pre-treatment step (Kumar et al., 2005).

3.3 Other Bioalcohols

The alcohols that can be used for motor fuels are methanol (CH_3OH), ethanol (C_2H_5OH), propanol (C_3H_7OH), and butanol (C_4H_9OH). However only the first two of the alcohols are technically and economically suitable as fuels for internal combustion engines (ICEs). Main commercial bioalcohols from renewable feedstocks are bioethanol and biomethanol in the world's energy market.

Methanol is known as "wood alcohol." Generally, methanol is easier to find than ethanol. Sustainable methods of methanol production are currently not economically viable. Methanol is produced from synthetic gas or biogas and evaluated as a fuel for internal combustion engines. The production of methanol is a cost-intensive chemical process.

Methanol is another possible replacement for conventional motor fuels. In fact, it has been considered as a possible large-volume motor fuel substitute at various times during gasoline shortages. In fact, it was often used in the early twentieth century to power automobiles before introduction of inexpensive gasoline. Later, synthetically produced methanol was widely used as a motor fuel in Germany during World War II. Again, during the oil crisis of 1970s, methanol blending with motor fuel received attention due to its availability and low cost. Methanol is poisonous and burns with an invisible flame. Similar to ethanol, methanol has a high octane rating and hence is suitable for the Otto engine. Today, methanol is commonly used in biodiesel production for its reactivity with vegetable oils. Methanol can be used as one possible replacement for conventional motor fuels. Many tests have shown promising results using 85–100% by volume methanol as a transportation fuel in automobiles, trucks and buses.

Before 1920s, methanol was obtained from wood as a co-product of charcoal production, hence the name wood alcohol. Methanol is currently manufactured worldwide from syngas, which is derived from natural gas, refinery off-gas, coal or petroleum, as:

$$2H_2 + CO \rightarrow CH_3OH \tag{3.1}$$

The above reaction can be carried out in the presence of a variety of catalysts including Ni, Cu/Zn, Cu/SiO_2, Pd/SiO_2, and Pd/ZnO. In the case of coal, it is first pulverized and cleaned, then fed to a gasifier bed where it is reacted with oxygen and steam to produce the syngas. A 2:1 mole ratio of hydrogen to carbon monoxide is fed to a fixed-catalyst bed reactor for methanol production. Also, the technology for making methanol from natural gas is already in place and in wide use. Current natural gas feedstocks are so inexpensive that even with tax incentives renewable methanol has not been able to compete economically.

The composition of biosyngas from biomass is shown in Table 3.4. The hydrogen to CO ratio in biosyngas is less than that from coal or natural gas, hence additional hydrogen is needed for full conversion to methanol. The gases produced from biomass can be steam reformed to produce hydrogen and followed by water–gas shift reaction to further enhance the hydrogen content. Wet biomass can be easily

Table 3.4 Composition of biosyngas from biomass gasification

Constituents	Percentage by volume (dry and nitrogen-free)
Carbon monoxide (CO)	28–36
Hydrogen (H_2)	22–32
Carbon dioxide (CO_2)	21–30
Methane (CH_4)	8–11
Ethene (C_2H_4)	2–4

Table 3.5 Main production facilities of methanol and biomethanol

Methanol	Biomethanol
Catalytic synthesis from CO and H_2	Catalytic synthesis from CO and H_2
Natural gas	Distillation of liquid from wood pyrolysis
Petroleum gas	Gaseous products from biomass gasification
Distillation of liquid from coal pyrolysis	Synthetic gas from biomass and coal

gasified using supercritical water conditions. Main production pathways for methanol and biomethanol are compared in Table 3.5.

Methanol can be produced from biomass, essentially any primary energy source. Thus, the choice of fuel in the transportation sector is to some extent determined by the availability of biomass. As regards to the difference between hydrogen and methanol production costs, conversion of natural gas, biomass and coal into hydrogen is generally more energy efficient and less expensive than the conversion into methanol.

3.4 Biorefinery

A biorefinery is a facility that integrates biomass conversion processes and equipment to produce fuels, power, and value-added chemicals from biomass. Biorefinery is the co-production of a spectrum of bio-based products and energy from biomass. The biorefinery concept is analogous to today's crude oil refinery. Biorefinery is a relatively new term referring to the conversion of biomass feedstock into a host of valuable chemicals and energy with minimal waste and emissions.

Biorefinery includes fractionation for separation of primary refinery products. The fractionation refers to the conversion of wood into its constituent components (cellulose, hemicelluloses and lignin). Processes include steam explosion, aqueous separation and hot water systems. Commercial products of biomass fractionation include levulinic acid, xylitol and alcohols. Figure 3.3 shows the fractionation of wood and chemicals from wood.

Main fractionation chemicals from wood ingredients are:

1. Dissociation of cell components → Lignin fragment + Oligosaccharides + Cellulose
2. Hydrolysis of cellulose (saccharification) → Glucose
3. Conversion of glucose (fermentation) → Ethanol + Lactic acid
4. Chemical degradation of cellulose → Levulinic acid + Xylitol
5. Chemical degradation of lignin → Phenolic products

Biorefinery is the process of extracting valuable chemicals and polymers from biomass. The main technologies to produce chemicals from biomass are: (a) biomass refining or pre-treatment, (b) thermochemical conversion (gasification, pyrolysis,

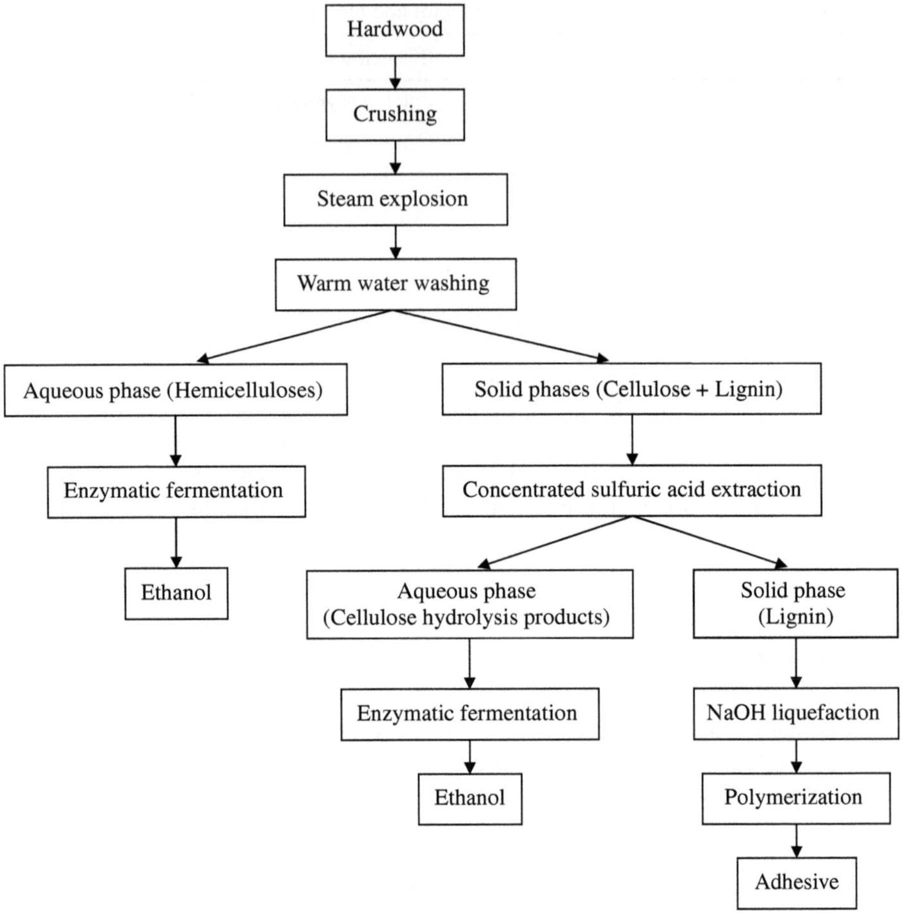

Fig. 3.3 Fractionation of wood and chemicals from wood

hydrothermal upgrading), (c) fermentation and bioconversion, and (d) product separation and upgrading.

There are four main biorefineries: biosyngas-based refinery, pyrolysis-based refinery, hydrothermal upgrading-based refinery, and fermentation-based refinery. Biosyngas is a multifunctional intermediate for the production of materials, chemicals, transportation fuels, power and/or heat from biomass. Figure 3.4 shows the gasification-based thermochemical biorefinery.

Pyrolysis oil (bio-oil) is produced in fast and flash pyrolysis processes and can be used for indirect co-firing for power production in conventional power plants and potentially as a high energy density intermediate for the final production of chemicals and/or transportation fuels. Gas chromatographic analysis of the liquid fraction of pyrolysis products from beech wood is given in Table 3.6 (Demirbas, 2007). Biocrude results from severe hydrothermal upgrading (HTU) of relatively wet biomass and potentially can be used for the production of materials, chemicals,

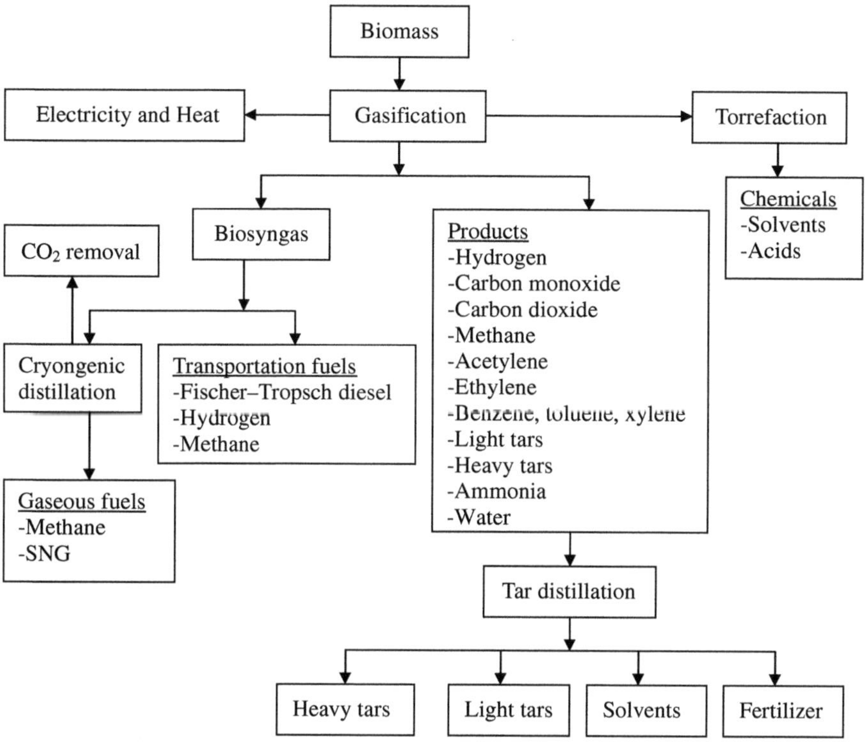

Fig. 3.4 Gasification-based thermochemical biorefinery

transportation fuels, power and/or heat. Mixed sugars resulting from fermentation (C_5 and C_6 sugars) are further refined substrates for chemical and bioconversion.

The bio-oil formed at 725 K contains high concentrations of compounds such as acetic acid, 1-hydroxy-2-butanone, 1-hydroxy-2-propanone, methanol, 2,6-dimethoxyphenol, 4-methyl-2,6-dimetoxyphenol and 2-Cyclopenten-1-one, etc. A significant characteristic of bio-oils was the high percentage of alkylated compounds especially methyl derivatives. The structural components of the biomass samples mainly have an effect on pyrolytic degradation products. A reaction mechanism is proposed which describes a possible reaction route for the formation of the characteristic compounds found in the oils. The supercritical water extraction and liquefaction partial reactions also occur during the pyrolysis. The phenol content increased to 52% and the yield of neutral oils increased from 18 to 33% with increasing temperature, while the methoxyl content decreased. The methoxyl content at 675 K amounted to 11.8%, at 875 K to only 5%. The coke formed could be briquetted without the addition of a binder. It has been reported that the first runs in the pyrolysis of the pyroligneous acid consists of about 50% methanol, 18% acetone, 7% esters, 6% aldehydes, 0.5% ethyl alcohol, 18.5% water, and small amounts of furfural (Demirbas, 2007).

Table 3.6 Gas chromatographic analysis of the liquid fraction of pyrolysis products from beech wood (wt.% dry basis)

Compound	Reaction temperature (K)					
	625	675	725	775	825	875
Acetic acid	16.8	16.5	15.9	12.6	8.42	5.30
Methyl acetate	0.47	0.35	0.21	0.16	0.14	0.11
1-Hydroxy-2-propanone	6.32	6.84	7.26	7.66	8.21	8.46
Methanol	4.16	4.63	5.08	5.34	5.63	5.82
1-Hydroxy-2-butanone	3.40	3.62	3.82	3.88	3.96	4.11
1-Hydroxy-2-propane acetate	1.06	0.97	0.88	0.83	0.78	0.75
Levoglucosan	2.59	2.10	1.62	1.30	1.09	0.38
1-Hydroxy-2-butanone acetate	0.97	0.78	0.62	0.54	0.48	0.45
Formic acid	1.18	1.04	0.84	0.72	0.60	0.48
Guaiacol	0.74	0.78	0.82	0.86	0.89	0.93
Crotonic acid	0.96	0.74	0.62	0.41	0.30	0.18
Butyrolactone	0.74	0.68	0.66	0.67	0.62	0.63
Propionic acid	0.96	0.81	0.60	0.49	0.41	0.34
Acetone	0.62	0.78	0.93	1.08	1.22	1.28
2,3-Butanedione	0.46	0.50	0.56	0.56	0.58	0.61
2,3-Pentanedione	0.34	0.42	0.50	0.53	0.59	0.64
Valeric acid	0.72	0.62	0.55	0.46	0.38	0.30
Isovaleric acid	0.68	0.59	0.51	0.42	0.35	0.26
Furfural	2.52	2.26	2.09	1.84	1.72	1.58
5-Methyl-furfural	0.65	0.51	0.42	0.44	0.40	0.36
Butyric acid	0.56	0.50	0.46	0.39	0.31	0.23
Isobutyric acid	0.49	0.44	0.38	0.30	0.25	0.18
Valerolactone	0.51	0.45	0.38	0.32	0.34	0.35
Propanone	0.41	0.35	0.28	0.25	0.26	0.21
2-Butanone	0.18	0.17	0.32	0.38	0.45	0.43
Crotonolactone	0.12	0.19	0.29	0.36	0.40	0.44
Acrylic acid	0.44	0.39	0.33	0.25	0.19	0.15
2-Cyclopenten-1-one	1.48	1.65	1.86	1.96	2.05	2.13
2-Methyl-2-cyclopenten-1-one	0.40	0.31	0.24	0.17	0.13	0.14
2-Methyl-cyclopentenone	0.20	0.18	0.17	0.22	0.25	0.29
Cyclopentenone	0.10	0.14	0.16	0.23	0.27	0.31
Methyl-2-furancarboxaldehyde	0.73	0.65	0.58	0.50	0.44	0.38
Phenol	0.24	0.30	0.36	0.43	0.54	0.66
2,6-Dimethoxyphenol	2.28	2.09	1.98	1.88	1.81	1.76

Table 3.6 (*continued*)

Compound	Reaction temperature (K)					
	625	675	725	775	825	875
Dimethyl phenol	0.08	0.13	0.18	0.42	0.64	0.90
Methyl phenol	0.32	0.38	0.44	0.50	0.66	0.87
4-Methyl-2,6-dimetoxyphenol	2.24	2.05	1.84	1.74	1.69	1.58

3.5 Biodiesel

Fossil fuel resources are decreasing daily. Biodiesel is attracting increasing attention worldwide as blending components or direct replacements for diesel fuel in vehicle engines. Biodiesel typically comprises lower alkyl fatty acid (chain length C_{14}–C_{22}), esters of short-chain alcohols, primarily, methanol or ethanol. Various methods have been reported for the production of biodiesel from vegetable oil, such as direct use and blending, microemulsification, pyrolysis, and transesterification. Among these, transesterification is an attractive and widely accepted technique. The purpose of the transesterification process is to lower the viscosity of the oil. The most important variables affecting methyl ester yield during the transesterification reaction are the molar ratio of alcohol to vegetable oil and the reaction temperature. Methanol is the commonly used alcohol in this process, due in part to its low cost. Methyl esters of vegetable oils have several outstanding advantages over other new renewable and clean engine fuel alternatives.

The interest in fatty acid methyl esters based on biodiesel has grown considerably worldwide. A rapid expansion in production is being observed not only in the United States but also in many other developed and developing countries around the world. Increasing the share of biodiesel in transportation fuels may decrease the net emission of CO_2 and alleviate the global warming problem. Biodiesel fuel is a renewable substitute fuel for petroleum diesel or petrodiesel fuel made from vegetable or animal fats; it can be used in any mixture with petrodiesel fuel, as it has very similar characteristics, but it has lower exhaust emissions. Biodiesel fuel has better properties than petrodiesel fuel; it is renewable, biodegradable, non-toxic, and essentially free of sulfur and aromatics. Biodiesel seems to be a realistic fuel for the future, and has become more attractive recently because of its environmental benefits. Biodiesel is an environmentally friendly fuel that can be used in any diesel engine without modification.

Typical raw materials of biodiesel are rapeseed oil, soybean oil, sunflower oil and palm oil. Beef and sheep tallow and poultry oil from animal sources and cooking oil are also sources of raw materials. There are various other biodiesel sources: almond, andiroba (*Carapa guianensis*), babassu (*Orbignia* sp.), barley, camelina (*Camelina sativa*), coconut, copra, cumaru (*Dipteryx odorata*), *Cynara cardunculus*, fish oil, groundnut, *Jatropha curcas*, karanja (*Pongamia glabra*), laurel, *Lesquerella fendleri*, *Madhuca indica*, microalgae (*Chlorella vulgaris*), oat, piqui (*Caryocar* sp.), poppy seed, rice, rubber seed, sesame, sorghum, tobacco seed, and wheat (Pinto et al., 2005).

The biggest advantage of biodiesel is the environmentally friendliness that it has over gasoline and petroleum diesel. The advantages of biodiesel as a diesel fuel are its portability, ready availability, renewability, higher combustion efficiency, lower sulfur and aromatic content, higher cetane number, and higher biodegradability (Ma and Hanna, 1999). The main advantages of biodiesel given in the literature include its domestic origin, its potential for reducing a given economy's dependency on imported petroleum, biodegradability, high flash point, and inherent lubricity in the neat form.

Biodiesel can be made from domestically produced, renewable oilseed crops such as soybean, rapeseed, and sunflower. The risks of handling, transporting, and storing biodiesel are much lower than those associated with petrodiesel. Biodiesel is the only alternative fuel in which low-concentration biodiesel–diesel blends run on conventional unmodified engines. It can be stored anywhere that petroleum diesel fuel is stored. Biodiesel is safe to handle and transport because it is as biodegradable as sugar and has a high flash point compared to petroleum diesel fuel. Biodiesel can be used alone or mixed in any ratio with petroleum diesel fuel. The most common blend is a mix of 20% biodiesel with 80% petroleum diesel, or B20 in recent scientific investigations; however, in Europe the current regulation foresees a maximum 5.75% of biodiesel.

The commercial biodiesel fuel significantly reduced PM exhaust emissions (75–83%) compared to the petrodiesel base fuel. However, NO_x exhaust emissions increased slightly with commercial biodiesel compared to the base fuel. The chain length of the compounds had little effect on NO_x and PM exhaust emissions, while the influence was greater on HC and CO, the latter being reduced with decreasing chain length. Non-saturation in the fatty compounds causes an increase in NO_x exhaust emissions.

Many studies on the performances and emissions of compression ignition engines, fueled with pure biodiesel and blends with diesel fuel, have been conducted and are reported in the literature (Laforgia and Ardito, 1994). Fuel characterization data show some similarities and differences between biodiesel and petrodiesel fuels. The sulfur content of petrodiesel is 20 to 50 times that of biodiesel. Biodiesel has demonstrated a number of promising characteristics, including reduction of exhaust emissions.

For soybean-based biodiesel at this concentration, the estimated emission impacts for percent change in emissions of NO_x, particular matter (PM), HC, and CO were +20%, –10.1%, –21.1%, and –11.0%, respectively (EPA, 2002). The use of blends of biodiesel and diesel oil are preferred in engines in order to avoid some problems related to the decrease of power and torque, and to the increase of NO_x emissions (a contributing factor in the localized formation of smog and ozone) that occurs with an increase in the content of pure biodiesel in a blend. Emissions of all pollutants except NO_x appear to decrease when biodiesel is used. The use of biodiesel in a conventional diesel engine dramatically reduces the emissions of unburned hydrocarbons, carbon dioxide, carbon monoxide, sulfates, polycyclic aromatic hydrocarbons, nitrated polycyclic aromatic hydrocarbons, ozone-forming hydrocarbons, and particulate matter. The net contribution of carbon dioxide from biomass combustion is small.

Biodegradability of biodiesel has been proposed as a solution for the waste problem. Biodegradable fuels such as biodiesels have an expanding range of potential applications and they are environmentally friendly. Therefore, there is growing interest in degradable diesel fuels that degrade more rapidly than conventional disposable fuels. Biodiesel is non-toxic and degrades about four times faster than petrodiesel. Its oxygen content improves the biodegradation process, leading to a decreased level of quick biodegradation. About 90 to 98% of biodiesel is mineralized in 21 to 28 days under aerobic as well as anaerobic conditions. The biodegradabilities of several biodiesels in the aquatic environment show that all biodiesel fuels are readily biodegradable. In one study, after 28 days all biodiesel fuels were 77 to 89% biodegraded; diesel fuel was only 18% biodegraded. The enzymes responsible for the dehydrogenation/oxidation reactions that occur in the process of degradation recognize oxygen atoms and attack them immediately.

Biodiesel methyl esters improve the lubrication properties of the diesel fuel blend. Fuel injectors and some types of fuel pumps rely on fuel for lubrication. Biodiesel reduced long-term engine wear in test diesel engines to less than half of what was observed in engines running on current low-sulfur diesel fuel. Lubricity properties of fuel are important for reducing friction wear in engine components normally lubricated by the fuel rather than crankcase oil (Ma and Hanna, 1999; Demirbas, 2003). Biodiesel provides significant lubricity improvement over petroleum diesel fuel. Lubricity results of biodiesel and petroleum diesel using industry test methods indicate that there is a marked improvement in lubricity when biodiesel is added to conventional diesel fuel. Even biodiesel levels below 1% can provide up to a 30% increase in lubricity. Lubricity results of biodiesel and petroleum diesel using industry test methods indicate that there is a marked improvement in lubricity when biodiesel is added to conventional diesel fuel.

Biodiesels are mono-alkyl esters containing approximately 10% oxygen by weight. The oxygen improves the efficiency of combustion, but it takes up space in the blend and therefore slightly increases the apparent fuel consumption rate observed while operating an engine with biodiesel. The high combustion temperature at high engine speed becomes the dominant factor, making both heated and unheated fuel to acquire the same temperature before fuel injection. Brake thermal efficiency is better in the dual fuel operation and with the methyl ester of vegetable oil as compared to the blend. In an earlier study (Canakci et al., 2006), the tests were performed with commercial diesel fuel and biodiesel. The maximum brake power values of biodiesel and diesel were 4.390 and 5.208 kW obtained at 2750 and 2500 rpm, respectively. According to these values, the commercial diesel fuel has the greatest brake power.

The major disadvantages of biodiesel are its higher viscosity, lower energy content, higher cloud point and pour point, higher nitrogen oxide (NO_x) emissions, lower engine speed and power, injector coking, engine compatibility, and high price (Demirbas, 2008c). The biodiesels on the average decrease power by 5% compared to that of diesel at the rated load. The maximum torque values are about 21.0 Nm at 1500 rpm for the diesel fuel, and 19.7 Nm at 1500 rpm for the biodiesel. The torque values of commercial diesel fuel are greater than those of biodiesel. Peak torque applies less to biodiesel fuels than it does to No. 2 diesel fuel but occurs at lower engine speed and generally its torque curves are flatter.

The specific fuel consumption values of biodiesel are greater than those of commercial diesel fuel. The effective efficiency and effective pressure values of commercial diesel fuel are greater than those of biodiesel (Canakci et al., 2006). Important operating disadvantages of biodiesel in comparison with petrodiesel are cold-start problems, the lower energy content, higher copper strip corrosion and fuel pumping difficulty from higher viscosity. Fuel consumption at full load condition and low speeds generally is high. Fuel consumption first decreases and then increases with increasing speed. The reason is that, the produced power in low speeds is low and the main part of fuel is consumed to overcome the engine friction.

The effects of oxidative degradation caused by contact with ambient air (autoxidation) during long-term storage present a legitimate concern in terms of maintaining the fuel quality of biodiesel. Transunsaturated fatty acids, or *transfats*, are solid fats produced artificially by heating liquid vegetable oils in the presence of metal catalysts and hydrogen. This process, partial hydrogenation, causes carbon atoms to bond in a straight configuration and remain in a solid state at room temperature. Physical properties that are sensitive to the effects of fatty oil oxidation include viscosity, refractive index, and dielectric constant. In oxidative instability, the methylene group ($-CH_2-$) carbons between the olefinic carbons are the sites of first attack.

Oxidation to CO_2 of biodiesel results in the formation of hydroperoxides. The formation of a hydroperoxide follows a well-known peroxidation chain mechanism. Oxidative lipid modifications occur through lipid peroxidation mechanisms in which free radicals and reactive oxygen species abstract a methylene hydrogen atom from polyunsaturated fatty acids, producing a carbon-centered lipid radical. Spontaneous rearrangement of the 1,4-pentadiene yields a conjugated diene, which reacts with molecular oxygen to form a lipid peroxyl radical.

3.6 Biogas

Biogas is an environment friendly, clean, cheap and versatile gaseous fuel. It is mainly a mixture of methane and carbon dioxide obtained by anaerobic digestion of biomass, sewage sludge, animal wastes, and industrial effluents. Anaerobic digestion occurs in the absence of air and is typically carried out for a few weeks. Typical compositions of biogas and landfill gas are given in Table 3.7. CH_4 and CO_2 make up around 90% of the gas volume produced, both of which are greenhouse gases. However, CO_2 is recycled back by the plants, but methane can contribute to global warming. Hence, the capture and fuel use of biogas is beneficial in two ways: (a) fuel value, and (b) conversion of CH_4 into CO_2, a plant-recyclable carbon.

Anaerobic digestion (AD) is the conversion of organic material directly to a gas, termed biogas, a mixture of mainly methane and carbon dioxide with small quantities of other gases such as hydrogen sulfide. Methane is the major component of the biogas used in many homes for cooking and heating. The biogas has a chemical composition close to that of natural gas. A biodigester, or a biogas plant,

Table 3.7 Composition of biogas and landfill gas

Component	Chemical formula	Landfill gas	Biogas
Methane (% by vol.)	CH_4	40–60	55–70
Carbon dioxide (% by vol.)	CO_2	20–40	30–45
Nitrogen (% by vol.)	N_2	2–20	0–2
Oxygen (% by vol.)	O_2	< 1	< 1
Heavier hydrocarbons (% by vol.)	C_nH_{2n+2}	< 1	0
Hydrogen sulfide (ppm)	H_2S	20–200	100–500
Ammonia (NH_3) (ppm)	NH_3	0	80–100
Volatile organic compounds (% by vol.)	–	0.25–0.50	0

is a physical structure used to provide an anaerobic condition which stimulates various chemical and microbiological reactions resulting in the decomposition of input slurries and the production of biogas—mainly methane (Demirbas and Ozturk, 2004).

Biogas can be used after appropriate gas cleanup as a fuel for engines, gas turbines, fuel cells, boilers, industrial heaters, other processes, or for the manufacturing of chemicals. Before landfilling, treatment or stabilization of biodegradable materials can be accomplished by a combination of anaerobic digestion followed by aerobic composting.

The same types of anaerobic bacteria that produced natural gas also produce methane today. Anaerobic bacteria are some of the oldest forms of life on earth. They evolved before the photosynthesis of green plants released large quantities of oxygen into the atmosphere. Anaerobic bacteria break down or digest organic material in the absence of oxygen and produce biogas as a waste product.

The first methane digester plant was built at a leper colony in Bombay, India, in 1859 (Meynell, 1976). Most of the biogas plants utilize animal dung or sewage. A schematic of a biogas plant utilizing cow dung is illustrated in Fig. 3.5 (Balat, 2008). Anaerobic digestion is a commercially proven technology and is widely used for treating high moisture content organic wastes including +80–90% moisture. Biogas can be used directly in spark ignition gas engines (SIGEs) and gas turbines. Used as a fuel in SIGE to produce electricity only, the overall conversion efficiency from biomass to electricity is about 10–16% (Demirbas, 2006).

Aerobic conversion includes most commercial composting and activated sludge wastewater treatment processes. Aerobic conversion uses air or oxygen to support the metabolism of the aerobic microorganisms degrading the substrate. Nutritional considerations are also important to the proper functioning of aerobic processes. Aerobic processes operate at much higher reaction rates than anaerobic processes and produce more cell mass, but generally do not produce useful fuel gases. Aerobic decomposition can occur from as low as near freezing to about 344 K.

Respiration refers to those biochemical processes in which organisms oxidize organic matter and extract the stored chemical energy needed for growth and repro-

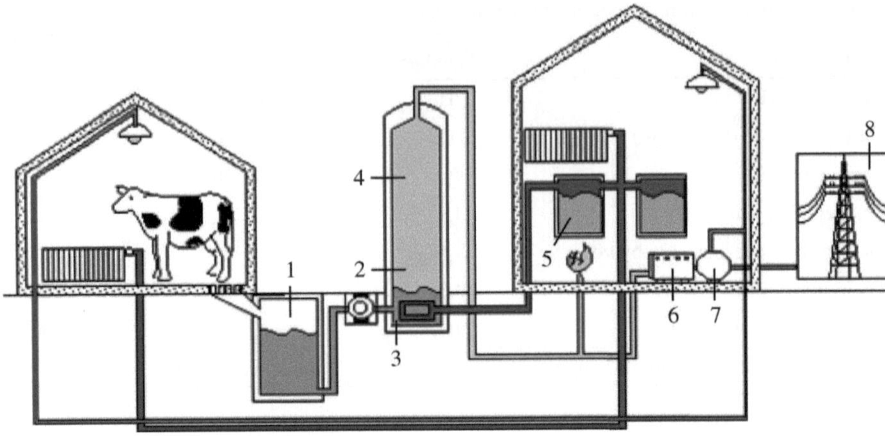

Fig. 3.5 Schematic of biogas plant utilizing cow dung: *1* Compost storage, *2* pump, *3* internal heater, *4* digester, *5* combustor, *6–8* power generator

duction. Respiration patterns may be subdivided into two major groups, based on the nature of the ultimate election acceptor. Although alternative pathways exist for the oxidation of various organic substrates, it is convenient to consider only the degradation of glucose. The breakdown of glucose is via the Embden–Meyerhof–Parnas glycolytic pathway, which yields 2 moles each of pyruvate, ATP, and reduced nicotinamide adenine dinucleotide (NAD) per mole of glucose.

Under aerobic conditions, the pyruvate is oxidized to CO_2 and H_2O via the tricarboxylic acid or Krebs cycle and the electron transport system. The net yield for glycolysis followed by complete oxidation is 38 moles ATP per mole glucose, although there is evidence that the yield for bacteria is 16 moles ATP per mole glucose (Aiba et al., 1973). Thus, 673 kcal are liberated per mole glucose, much of which is stored as ATP.

For anaerobic systems, methane gas is an important product. Depending on the type and nature of the biological components, different yields can be obtained for different biodegradable wastes. For pure cellulose, for example, the biogas product is 50% methane and 50% carbon dioxide. Mixed waste feedstocks yield biogas with methane concentrations of 40–60% (by volume). Fats and oils can yield biogas with 70% methane content.

Anaerobic digestion functions over a wide temperature range from the so-called psychrophilic temperature near 283 K to extreme thermophilic temperatures above 344 K. The temperature of the reaction has a very strong influence on the anaerobic activity, but there are two optimal temperature ranges in which microbial activity and biogas production rate are highest, the so-called mesophilic and thermophilic ranges. The mesophilic regime is associated with temperatures of about 308 K, the thermophilic regime of about 328 K. Operation at thermophilic temperature allows for shorter retention time and a higher biogas production rate; however, maintaining the high temperature generally requires an outside heat source because anaero-

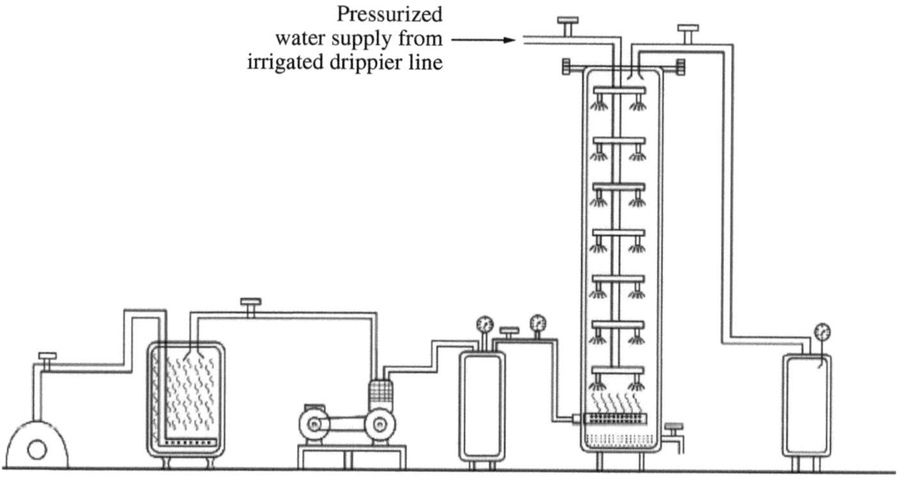

Fig. 3.6 Biogas purification plant

bic bacteria do not generate sufficient heat. Aerobic composting can achieve relatively high temperatures (up to 344 K) without heat addition because reaction rates for aerobic systems are much higher than those for anaerobic systems. If heat is not conducted away from the hot center of a compost pile, then thermochemical reactions can initiate which can lead to spontaneous combustion if sufficient oxygen reaches the hot areas. Managed compost operations use aeration to provide oxygen to the bacteria but also to transport heat out of the pile. The anaerobic digestion of lignocellulosic waste occurs in a three-step process often termed hydrolysis, acetogenesis, and methanogenesis. The molecular structure of the biodegradable portion of the waste that contains proteins and carbohydrates is first broken down through hydrolysis. The lipids are converted to volatile fatty acids and amino acids. Carbohydrates and proteins are hydrolyzed to sugars and amino acids. In acetogenesis, acid-forming bacteria use these byproducts to generate intermediary products such as propionate and butyrate. Further microbial action results in the degradation of these intermediary products into hydrogen and acetate. Methanogenic bacteria consume the hydrogen and acetate to produce methane and carbon dioxide. Figure 3.6 shows a biogas purification plant.

Under anaerobic conditions, various pathways exist for pyruvate metabolism which serve to re-oxidize the reduced hydrogen carriers formed during glycolysis. The ultimate acceptor builds up as a waste product in the culture medium. The end-products of the pathways are: (1) CO_2, ATP, and acetate; (2) CO_2 and ethanol; (3) H_2 and CO_2; (4) CO_2 and 2,3-butylene glycol; (5) CO_2, H_2, acetone, ATP, and butanol; (6) succinate; and (7) lactate. The pathway that occurs depends on the microorganism cultivated and the culture.

3.7 Landfill Gas

Main biorenewable gaseous fuels are biogas, landfill gas, gaseous fuels from pyrolysis and gasification of biomass, gaseous fuels from Fischer–Tropsch synthesis and biohydrogen. There are a number of processes for converting of biomass into gaseous fuels such as methane or hydrogen. One pathway uses plant and animal wastes in a fermentation process leading to biogas from which the desired fuels can be isolated. This technology is established and in widespread use for waste treatment. Anaerobic digestion of biowastes occurs in the absence of air, the resulting gas called a biogas is a mixture consisting mainly of methane and carbon dioxide. Biogas is a valuable fuel which is produced in digesters filled with the feedstock like dung or sewage. The digestion is allowed to continue for a period of ten days to a few weeks. A second pathway uses algae and bacteria that have been genetically modified to produce hydrogen directly instead of the conventional biological energy carriers. Finally, high-temperature gasification supplies a crude gas, which may be transformed into hydrogen by a second reaction step. This pathway may offer the highest overall efficiency.

Contents of domestic solid waste are given in Table 3.8.

Landfill leachate treatment has received significant attention in recent years, especially in municipal areas (Uygur and Kargi, 2004). The generation of municipal solid wastes (MSW) has increased in parallel to rapid industrialization. Approximately 16% of all discarded MSW is incinerated (EPA, 1994); the remainder is disposed of in landfills. Effective management of these wastes has become a major social and environmental concern (Erses and Onay, 2003). Disposal of MSW in sanitary landfills is usually associated with soil, surface water and groundwater contamination when the landfill is not properly constructed. The flow rate and composition of leachate vary from site to site, seasonally at each site and depending on the age of the landfill. Young leachate normally contains high amounts of volatile fatty acids (Timur and Ozturk, 1999). MSW statistics and management practices including waste recovery and recycling initiatives have been evaluated (Metin et al., 2003). The organic MSW was chemically and biologically characterized, in order to study its behavior during anaerobic digestion, and its pH, biogas production, alkalinity, and volatile fatty acid production was determined by Plaza et al. (1996). Anaerobic digestion of the organic food fraction of MSW, on its own or co-digested with primary sewage sludge, produces high-quality biogas, suitable as renewable energy (Kiely et al., 1997). The processing of MSW (i.e., landfill, incineration, aerobic composting) secures many advantages and limitations (Braber, 1995). The greenhouse gas emissions produced by the uncontrolled releasing of methane from improperly disposed organic waste in a large landfill can be reduced (Al-Dabbas, 1998).

The decomposition in landfills occurs in a series of stages, each of which is characterized by the increase or decrease of specific bacterial populations and the formation and utilization of certain metabolic products. The first stage of decomposition, which usually lasts less than a week, is characterized by the removal of

Table 3.8 Contents of domestic solid waste (wt.% of total)

Component	Lower limit	Upper limit
Paper waste	33.2	50.7
Food waste	18.3	21.2
Plastic matter	7.8	11.2
Metal	7.3	10.5
Glass	8.6	10.2
Textile	2.0	2.8
Wood	1.8	2.9
Leather and rubber	0.6	1.0
Miscellaneous	1.2	1.8

oxygen from the waste by aerobic bacteria (Augenstein and Pacey, 1991). In the second stage, which has been termed the anaerobic acid stage, a diverse population of hydrolytic and fermentative bacteria hydrolyzes polymers, such as cellulose, hemicellulose, proteins, and lipids, into soluble sugars, amino acids, long-chain carboxylic acids, and glycerol (Micales and Skog, 1997).

3.8 Fischer–Tropsch Liquids from Biorenewable Feedstocks

Fischer–Tropsch synthesis (FTS) is a process for producing mainly straight-chain hydrocarbons from a synthesis gas rich in CO and H_2. Catalysts are usually employed. Typical operating conditions for FT synthesis are temperatures of 475–625 K and very high pressure depending on the desired products. The product range includes light hydrocarbons such as methane (CH_4) and ethane (C_2H_6), propane (C_3H_8) butane (C_4H_{10}), gasoline (C_5–C_{12}), diesel (C_{13}–C_{22}), and light and waxes (C_{23}–C_{33}). The distribution of the products depends on the catalyst and the process conditions (temperature, pressure, and residence time). The synthesis gas must have very low tar and particulate matter content. Biomass-derived synthesis gas for FT liquid production is feasible and is not limited due to gas cleaning issues. FT from natural gas and ethanol from biomass may become widespread. The FT liquids will penetrate if there are large amounts of stranded natural gas selling for very low prices at the same time that petroleum is expensive or extremely low sulfur is required in diesel fuel (Demirbas, 2005).

The FTS was established in 1923 by German scientists Franz Fischer and Hans Tropsch. The main aim of FTS is the synthesis of long-chain hydrocarbons from CO and H_2 gas mixture. The FTS is described by the set of equations (Schulz, 1999):

$$nCO + (n + m/2)\,H_2 \rightarrow C_nH_m + nH_2O \tag{3.2}$$

where n is the average length of the hydrocarbon chain and m is the number of hydrogen atoms per carbon. All reactions are exothermic and the product is a mixture of different hydrocarbons in that paraffin and olefins are the main parts.

Fischer–Tropsch (FT) liquid fuel can be used as an alternate to diesel fuel. The FT catalytic conversion process can be used to synthesize diesel fuels from a variety of feedstocks, including coal, natural gas and biomass. The alternate fuel source is coal, indirectly converted to diesel fuel. The FT process uses various catalysts to produce linear hydrocarbons and oxygenates, including unrefined gasoline, diesel, and wax ranges. Gasification followed by FTS is currently the most promising method for upgrading low-value coal and biomass to high-value liquid fuels and chemicals. The total biomass produced each year as waste material from agriculture and forest operations could be converted into roughly 40 billion gallon per year liquid fuels, roughly 25% of the current US gasoline usage).

Synthetic FT diesel fuels can have excellent autoignition characteristics. The FT diesel is composed of only straight-chain hydrocarbons and has no aromatics or sulfur. Reaction parameters are temperature, pressure and H_2/CO ratio. FT product composition is strongly influenced by catalyst composition: the yield of paraffins is higher with cobalt catalytic run and the yield of olefins and oxygenates is higher with ironcatalytic run.

Basic FT reactions are:

$$nCO + 2nH_2 \rightarrow n(\text{-}CH_2\text{-}) + nH_2O \tag{3.3}$$

$$nCO + (2n+1)H_2 \rightarrow C_nH_{2n+2} + nH_2O \tag{3.4}$$

Catalysts and reactors have been extensively investigated for liquid phase Fischer–Tropsch synthesis (Davis, 2002). The synthetic Fischer–Tropsch diesel fuel can provide benefits in terms of both PM and NO_x emissions (May, 2003). Properties of FT and No. 2 diesel fuels are given in Table 3.9.

FT is most compatible with existing distribution for conventional diesel and only minimal adjustments are required to obtain optimal performance from existing diesel engines. Physical properties of FT are very similar to No. 2 diesel fuel, and its chemical properties are superior in that the FT process yields middle distillates that, if correctly processed (as through a cobalt-based catalyst), contain no aromatics or sulfur compounds.

Table 3.9 Properties of Fischer–Tropsch (FT) diesel and No. 2 diesel fuels

Property	Fischer–Tropsch diesel	No. 2 petroleum diesel
Density, g/cm³	0.7836	0.8320
Higher heating value, MJ/kg	47.1	46.2
Aromatics, %	0–0.1	8–16
Cetane number	76–80	40–44
Sulfur content, ppm	0–0.1	25–125

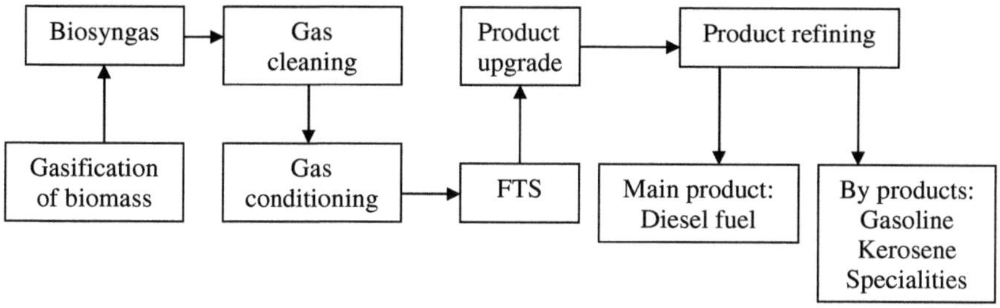

Fig. 3.7 Production of diesel fuel from biosyngas by Fischer–Tropsch synthesis (FTS)

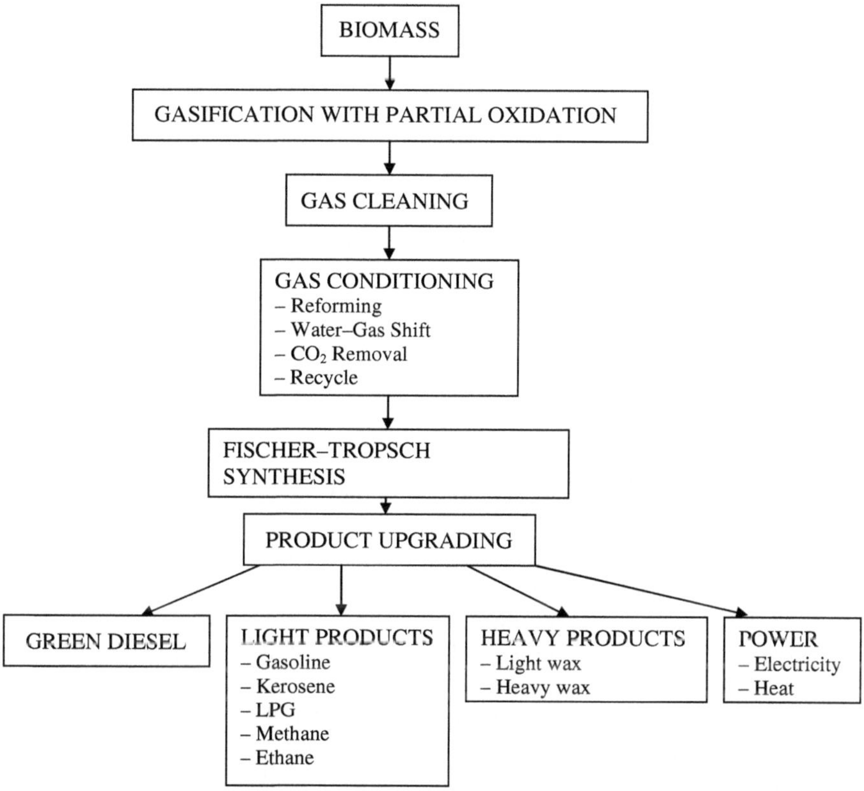

Fig. 3.8 Green diesel and other products from biomass via Fischer–Tropsch synthesis

Figure 3.7 shows the production of diesel fuel from biosyngas by FTS. The design of a biomass gasifier integrated with a FTS reactor must be aimed at achieving a high yield of liquid hydrocarbons. For the gasifier, it is important to avoid methane formation as much as possible, and convert all carbon in the biomass to mainly carbon monoxide and carbon dioxide (Prins et al., 2004).

To produce biosyngas from a biomass fuel the following procedures are necessary: (a) gasification of the fuel, (b) cleaning of the product gas, and (c) usage of the synthesis gas to produce chemicals, (d) usage of the synthesis gas as energy carrier in fuel cells. Figure 3.8 shows the green diesel and other products from biomass via Fischer–Tropsch synthesis.

Supercritical fluids (SCFs) offer several advantages as reaction media for catalytic reactions. These advantages include the ability to manipulate the reaction environment through simple changes in pressure to enhance solubility of reactants and products, to eliminate interphase transport limitations, and to integrate reaction and separation unit operations. Benefits derived from the SCF phase Fischer–Tropsch synthesis (SCF–FTS) involve the gas-like diffusivities and liquid-like solubilities, which together combine the desirable features of the gas- and liquid-phase FT synthesis routes.

Steady-state operation was quickly achieved under SCF conditions and the SCF–FT process has a marked effect on the hydrocarbon product distribution with a shift to higher carbon number products owing to enhanced heat and mass transfer from the catalyst surface. In addition, an obvious difference in the olefin content was observed where the 1-olefin content in the SCF phase was always higher than in the gas phase. Based on the experimental observations, a mechanistic explanation is provided for the difference of the reaction behavior under supercritical and gas-phase environments.

Summary

The term biofuel refers to liquid, gas and solid fuels predominantly produced from biomass. Biofuels affect a wide range of topics, including energy security, environmental concerns, foreign exchange savings, and socioeconomic issues related to the rural sector. Biofuels include bioethanol, biomethanol, vegetable oils, biodiesel, biogas, biosynthetic gas (biosyngas), bio-oil, biochar, Fischer–Tropsch liquids, and biohydrogen. Biofuels offer a number of technical and environmental benefits over conventional fossil fuels, which make them attractive as alternatives for the transport sector. The benefits include greenhouse gas reductions including reduced carbon dioxide emissions, which will contribute to domestic and international targets, the diversification of the fuel sector, biodegradability, sustainability, and an additional market for agricultural products.

Nowadays ethanol is the most popular fuel. The use of ethanol as a motor fuel has as long a history as the car itself. Ethanol can be produced from cellulose feedstocks such as corn stalks, rice straw, sugar cane bagasse, pulpwood, switchgrass; ethanol produced from municipal solid waste is called bioethanol. Methanol is known as wood alcohol. Generally, methanol is easier to find than ethanol. Sustainable methods of methanol production are currently not economically viable. Methanol is produced from synthetic gas or biogas and evaluated as a fuel for internal combustion engines.

Biorefinery is the co-production of a spectrum of bio-based products and energy from biomass. The biorefinery concept is analogous to today's crude oil refinery. Biorefinery is a relatively new term referring to the conversion of biomass feedstock into a host of valuable chemicals and energy with minimal waste and emissions.

Biodiesel is attracting increasing attention worldwide as blending components or direct replacements for diesel fuel in vehicle engines. Biodiesel typically comprises lower alkyl fatty acid (chain length C_{14}–C_{22}), esters of short-chain alcohols, primarily, methanol or ethanol.

Main biorenewable gaseous fuels are biogas, landfill gas, gaseous fuels from pyrolysis and gasification of biomass, gaseous fuels from Fischer–Tropsch synthesis and biohydrogen. There are a number of processes for converting biomass into gaseous fuels such as methane or hydrogen. One pathway uses plant and animal wastes in a fermentation process leading to biogas from which the desired fuels can be isolated.

Biogas is considered a clean, cheap and versatile gaseous fuel that is also environmentally friendly. It is mainly a mixture of methane and carbon dioxide obtained by anaerobic digestion of biomass, sewage sludge, animal wastes, and industrial effluents. Anaerobic digestion occurs in the absence of air and is typically carried out for a few weeks.

References

Aiba, S., Humphrey, A.E., Milis, N.F. 1973. Biochemical Engineering, 2nd ed.. Academic, New York.

AICHE. 1997. Alternative transportation fuels: a comparative analysis. Report by AICHE Government

Relations Committee. AICHE, New York.

Al-Dabbas, M.A.F. 1998. Reduction of methane emissions and utilization of municipal waste for energy in Amman. Renew Energy 14:427–434.

Augenstein, D., Pacey, J. 1991. Landfill methane models. Proceedings from the Technical Sessions of SWANA's 29th Annual International Solid Waste Exposition, SWANA, Silver Spring, MD.

Bala, B.K. 2005. Studies on biodiesels from transformation of vegetable oils for diesel engines. Energy Edu Sci Technol 15:1–45.

Balat, M. 2008. Progress in biogas production processes. Energy Edu Sci Technol 22:15–35.

Baltz, R.A., Burcham, A.F., Sitton, O.C., Book, N.L. 1982. The recycle of sulfuric acid and xylose in the prehydrolysis of corn stover. Energy 7:259–265.

Braber, K. 1995. Anaerobic digestion of municipal solid waste: A modern waste disposal option on the verge of breakthrough. Biomass Bioenergy 9:365–376.

Canakci, M., Erdil, A., Arcaklioglu, E. 2006. Performance and exhaust emissions of a biodiesel engine. Appl Energy 83:594–605.

Chigier, N.A. 1981. Energy, combustion and the environment. McGraw Hill, New York.

Davis, B.H. 2002. Overview of reactors for liquid phase Fischer–Tropsch synthesis. Catal Today 71:249–300.

Demirbas, A. 2003. Biodiesel fuels from vegetable oils via catalytic and non-catalytic supercritical alcohol transesterifications and other methods: a survey. Energy Convers Manage 44:2093–2109.

Demirbas, A. 2005. Potential applications of renewable energy sources, biomass combustion problems in boiler power systems and combustion related environmental issues. Progress Energy Combus Sci 31:171–192.
Demirbas, A. 2006. Biogas potential of manure and straw mixtures. Energy Sources Part A 28:71–78.
Demirbas, A. 2007. The influence of temperature on the yields of compounds existing in bio-oils obtaining from biomass samples via pyrolysis. Fuel Proc Technol 88: 591–597.
Demirbas, A. 2008a. Economic and environmental impacts of the liquid biofuels. Energy Edu Sci Technol 22:37–58.
Demirbas, A. 2008b. Biodiesel: A realistic fuel alternative for diesel engines. Springer, London.
Demirbas, A. 2008c. New liquid biofuels from vegetable oils via catalytic pyrolysis. Energy Edu Sci Technol 21:1–59.
Demirbas, A., Ozturk, T. 2004. Anaerobic digestion of agricultural solid residues. Int J Gren Energy 1:483–494.
EPA (US Environmental Protection Agency). 2002. A comprehensive analysis of biodiesel impacts on exhaust emissions. Draft Technical Report, EPA420-P-02-001, October 2002. EPA, Washington, DC.
EPA. 1994. Characterization of potential of municipal solid waste (MSW) components. Municipal Solid Waste in the United States: 1992 Update. EPA/530-R-94-042, NTS #PB 95-147690. Solid Waste and Emergency Response (5305). EPA, Washington, DC.
Erses, A.S., Onay, T.T. 2003. In situ heavy metal attenuation in landfills under methanogenic conditions. J Hazard Mat B99:159–175.
Kiely, G., Tayfur, G., Dolan, C., Tanji, K. 1997. Physical and mathematical modelling of anaerobic digestion of organic wastes. War Res 31:534–540.
Kim, S., Dale, B.E. 2005. Life cycle assessment of various cropping systems utilized for producing biofuels: bioethanol and biodiesel. Biomass Bioenergy 29:426–439.
Kumar, A., Cameron, J.B., Flynn, P.C. 2005. Pipeline transport and simultaneous saccharification of corn stover. Biores Technol 96:819–829.
Laforgia, D., Ardito, V. 1994. Biodiesel fueled IDI engines: performances, emissions and heat release investigation. Biores Technol 51:53–59.
Ma, F., Hanna, M.A. 1999. Biodiesel production: a review. Biores Technol 70:1–15.
May, M. 2003. Development and demonstration of Fischer–Tropsch fueled heavy-duty vehicles with control technologies for reduced diesel exhaust emissions. 9th Diesel Engine Emissions Reduction Conference. Newport, Rhode Island, 24–28 Aug 2003.
Metin, E., Erozturk, A., Neyim, C. 2003. Solid waste management practices and review of recovery and recycling operations in Turkey. Waste Manage 23:425–432.
Meynell, P.-J. 1976. Methane: planning a digester. Schocken, New York.
Micales, J.A., Skog, K.E. 1997. The decomposition of forest products in landfills. Int Biodeterioration Biodegradation 39:145–158.
Piel, W.J. 2001. Transportation fuels of the future? Fuel Proces Technol 71:167–179.
Pinto, A.C., Guarieiro, L.L.N., Rezende, M.J.C., Ribeiro, N.M., Torres, E.A., Lopes, W.A., Pereira, P.A.P., Andrade, J.B. 2005. Biodiesel: an overview. J Brazil Chem Soc 16:1313–1330.
Plaza, G., Robredo, P., Pacheco, O., Toledo, A.S. 1996. Anaerobic treatment of municipal solid waste. Water Sci Technol 33:169–175.
Prins, M.J., Ptasinski, K.J., Janssen, F.J.J.G. 2004. Exergetic optimisation of a production process of Fischer–Tropsch fuels from biomass. Fuel Proc Technol 86:375–389.
RFA (Renewable Fuels Association). 2007. Ethanol Industry Statistics. RFA, Washington, DC.
Schulz, H. 1999. Short history and present trends of FT synthesis. Appl Catal A General 186:1–16.
Sokhansanj, S., Turhollow, A., Cushman, J., Cundiff, J. 2002. Engineering aspects of collecting corn stover for bioenergy. Biomass Bioenergy 23:347–355.
Timur, H., Ozturk, I. 1999. Anaerobic sequencing batch reactor treatment of landfill leachate. Water Res 33:3225–3230.
Uygur, A., Kargi, F. 2004. Biological nutrient removal from pre-treated landfill leachate in a sequencing batch reactor. J Environ Manage 71: 9–14.

Chapter 4
Transportation Fuels

4.1 Introduction

Today, most of the energy we use comes from fossil fuels: petroleum, coal, and natural gas. Fossil fuels are used mainly in transportation and electric industries. The combustion of fossil fuels is the greatest source of atmospheric pollution. All fossil fuels release greenhouse gases and other pollutants into the atmosphere. There is also a growing concern that the world may run out of petroleum-based fuel resources. The diesel engine is used mainly for heavy vehicles. The diesel process means that air is compressed in the engine and then the fuel is injected and ignited by the hot, compressed air. In the Otto cycle engine a mixture of fuel and air is compressed and is then ignited by a spark plug.

Almost all motor vehicles today are powered by either diesel or gasoline. Both fuels are mainly derived from petroleum. Diesel fuel consists of straight-chain hydrocarbons with between C_9 and C_{27} carbon atoms. Gasoline is a blend of hydrocarbons with some contaminants, including sulfur, nitrogen, oxygen and certain metals. Gasoline has four major constituent groups: olefins, aromatics, paraffins, and naphthenes. The main alternate fuels include alcohol, liquefied petroleum gas, compressed natural gas, hydrogen and electricity for the operation of gasoline-type vehicles.

Known petroleum reserves are estimated to last only about 50 years at the present rate of consumption. In developed countries there is a growing trend towards employing modern technologies and efficient bioenergy conversion using a range of biofuels, which are becoming cost-wise competitive with fossil fuels. For this reason, alternate transportation fuels such as bioethanol, biodiesel, and biohydrogen from biomass will play an important role in the world's future. There are four alternative fuels that can be relatively easily used in conventional compression ignition (CI) engines: vegetable oil, biodiesel, Fischer–Tropsch (FT) diesel, and dimethyl ether (DME). The vegetable oils, such as palm, soybean, sunflower, peanut and olive oils can be used as alternative fuels for diesel engines. Both the FT and DME fuels can be manufactured from natural gas and are therefore not limited by feedstock availability. Table 4.1 presents availabilities of modern transport fuels. FT product composition is strongly influenced by the catalyst composition: the product from a cobalt catalyst is higher in paraffins and the product from an iron catalyst is higher in olefins and oxygenates. Vegetable oil (m)ethyl esters, commonly referred to as biodiesel, are prominent candidates as alternative diesel fuels. Biodiesel is technically competitive with conventional petroleum diesel fuel, and offers technical advantages over conventional fuels. The vegetable oils as alternative engine fuels are all extremely viscous with viscosities ranging from 10 to 20 times greater than petroleum diesel fuel. To refine these oils, the process of transesterification is employed, whose purpose is to lower the viscosity of the oil.

Gasoline is a blend of hydrocarbons, which also includes some contaminants: sulfur, nitrogen, oxygen and certain other metals. Gasoline is comprised of four major constituent groups, which are olefins, aromatics, paraffins, and napthenes. The important characteristics of gasoline are density, vapor pressure, distillation range, octane, and chemical composition. To be attractive, a motor gasoline must have (a) desirable volatility, (b) anti-knock resistance (related to octane rating), (c) good fuel economy, (d) minimal deposition on engine component surfaces, and (e) complete combustion and low pollutant emissions (Chigier, 1981).

Alternative fuels are substitute fuel sources to petroleum. These fuels are important because they replace petroleum fuels; however, some still include a small amount of petroleum in the mixture. By replacing petroleum-based fuels, we will no longer need to rely on fossil fuel, which is a non-renewable resource. There are many benefits in using alternative fuels as well. The population of automobiles is currently dominated by vehicles burning gasoline, where the main alternative fuels for these types of vehicles are alcohol, liquefied petroleum gas, compressed natural gas, hydrogen and electricity.

Figure 4.1 shows the whole sale prices of a number of possible alternative fuels on an energy equivalent basis compared to conventional gasoline (AICHE, 1997). Only compressed natural gas (CNG) and liquid petroleum gas (LPG) appear to have some economic advantage relative to gasoline while ethanol, methanol and electricity are at a severe economic disadvantage (Piel, 2001).

Table 4.1 Availability of modern transportation fuels

Fuel type	Availability	
	Current	Future
Gasoline	Excellent	Moderate-poor
Bioethanol	Moderate	Excellent
Biodiesel	Moderate	Excellent
Compressed natural gas (CNG)	Excellent	Moderate
Hydrogen for fuel cells	Poor	Excellent

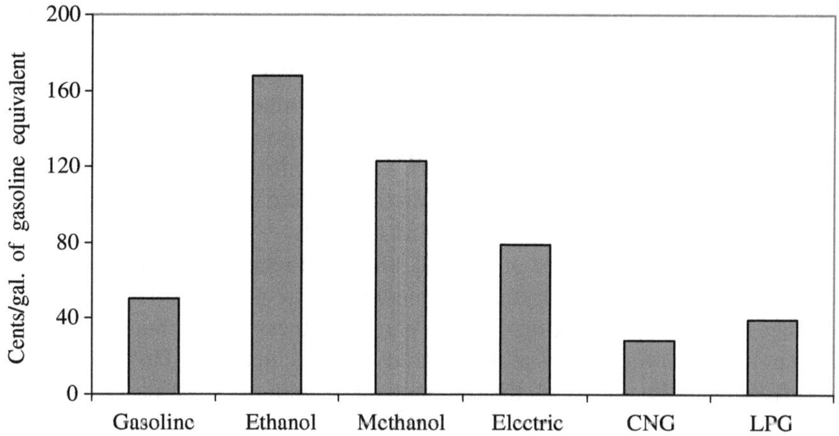

Fig. 4.1 Economics of gasoline alternative fuels

Diesel fuel is produced by distilling raw oil which is extracted from bedrock. Diesel is a fossil fuel, which consists of hydrocarbons with between 9 and 27 carbon atoms in a chain as well as a smaller amount of sulfur, nitrogen, oxygen and metal compounds. It is a general property of hydrocarbons that the autoignition temperature is higher for more volatile hydrocarbons. The hydrocarbons present in the diesel fuels include alkanes, naphthenes, olefins and aromatics. In addition, other substances are added to improve the characteristics of diesel fuel. Its boiling point is between 445 and 640 K. A good diesel fuel is characterized by low sulfur and aromatic content, good ignition quality, the right cold weather properties, low content of pollutants and also the right density, viscosity and boiling point.

The main advantage of the diesel engine is that the level of efficiency is greater than in the Otto cycle engine. This means that a greater part of the energy content of the fuel is exploited. The efficiency of a diesel engine is at best 45%, compared with 30% for the Otto engine.

Diesel emissions contain low concentrations of carbon monoxide and hydrocarbons. The major problem with diesel emissions are nitric oxides and particles as these are the most difficult to reduce.

The alternative fuels FT and DME fuels can be manufactured from natural gas and are therefore not limited by feedstock availability. Biodiesel on the other hand, is produced from vegetable (and some waste animal) oils whose supply for non-nutritional uses is presently quite limited.

Dimethyl ether (DME, CH_3–O–CH_3), is a new fuel which has attracted much attention recently. Today DME is made from natural gas, but DME can also be produced by gasifying biomass. DME can be stored in liquid form at pressures from 5 to 10 bars at normal temperatures. A major advantage of DME is its naturally high cetane number, which means that self-ignition will be easier. The high cetane rating makes DME most suitable for use in diesel engines, which implies that the high level of efficiency of the diesel engine is retained when using DME. The energy content of DME is lower than in diesel.

DME can be produced effectively from syngas in a single-stage, liquid phase (LPDME) process. The origin of syngas includes a wide spectrum of feedstocks such as coal, natural gas, biomass, and others. Non-toxic, high-density liquid DME fuel can be easily stored at modest pressures. The production of DME is very similar to that of methanol. DME conversion to hydrocarbons, lower olefins in particular, is studied using ZSM-5 catalysts with varying SiO_2/Al_2O_3 ratios, whereas the DME carbonylation reaction to produce methyl acetate is studied over a variety of group VIII metal substituted heteropolyacid catalysts.

Syngas (typically a mixture of CO, H_2, and CO_2) reacts over the active catalyst ($Cu/Zn/Al_2O_3$) dispersed in an inert oil medium. This process offers considerable advantages over the conventional vapor phase synthesis of methanol in the areas of heat transfer, exothermicity, and selectivity toward methanol. However, this process suffers from the drawback that the methanol synthesis reaction is a thermodynamically governed equilibrium reaction.

The single-stage LPDME process is very significant from both scientific and commercial perspectives. DME can be effectively converted to gasoline-range hydrocarbons, lower olefins, and other oxygenates. It may be used directly as a transportation fuel in admixture with methanol or as a fuel additive. In particular, dimethyl ether shows promise as an ultraclean alternative fuel for diesel engines. The advantages of using DME are ultralow emissions of nitrogen oxides (NO_x), reduced engine noise or quiet combustion, practically soot-free or smokeless operation and hence no exhaust after treatment, and high diesel thermal efficiency.

The Fischer–Tropsch (FT) catalytic conversion process can be used to synthesize diesel fuels from a variety of feedstocks, including coal, natural gas and biomass. Synthetic diesel fuels can have excellent autoignition characteristics. The Fischer–Tropsch diesel is composed of only straight-chain hydrocarbons and has no aromatics or sulfur. The synthetic Fischer–Tropsch diesel fuel can provide benefits in terms of both PM and NO_x emissions.

FT is most compatible with the existing distribution for conventional diesel and only minimal adjustments are required to obtain optimal performance from existing CI engines. Physical properties of FT fuels are very similar to No. 2 diesel fuel, and its chemical properties are superior in that the FT process yields middle distillates that, if correctly processed (as through a cobalt-based catalyst), contain no aromatics or sulfur compounds. Thus, only FT and DME were considered feasible near-term substitutes in our analysis.

The vegetable oils, such as palm, soybean, sunflower, peanut and olive oils can be used as alternative fuels for diesel engines. Due to the rapid decline in crude oil reserves, the use of vegetable oils as diesel fuels is again promoted in many countries. The effect of using a coconut oil as a diesel fuel alternative or as a direct fuel blend are investigated using a single-cylinder, direct-injection diesel engine (Machacon et al., 2001).

Vegetable oils have the potential to substitute a fraction of petroleum distillates and petroleum-based petrochemicals in the near future. Vegetable oil fuels are not petroleum-competitive fuels because they are more expensive than petroleum fuels. However, with recent increases in petroleum prices and uncertainties concerning petroleum availability, there is renewed interest in using vegetable oils, called biodiesel, in diesel engines.

4.2 Liquefied Petroleum Gas

Liquefied petroleum gas (LPG), also known as autogas, mainly contains propane (C_3H_8) and butane (C_4H_{10}). Both propane and butane are easily liquefied and stored in pressure containers. These properties make the fuel highly portable, and hence, can be easily transported in cylinders or tanks to end-users. LPG is a gas blend of gases produced commercially in petroleum refineries and stored under pressure below critical temperatures in order to keep it in a liquid state. LPG can also be produced from natural gas. The normal boiling point of LPG varies from 229 to 273 K so that the pressure required liquefying.

LPG is in widespread use in homes, industry and agriculture. Among the many uses of these gases are heating and refrigeration, as a supplement for natural gas, as fuel for industrial equipment and mobile homes, in the manufacture of ethylene, and as a solvent. Worldwide LPG production is limited to about 10% of total gasoline and diesel fuel consumption and is used to a great extent for domestic and industrial purposes. Since LPG burns cleaner with less carbon build-up and oil contamination, engine wear is reduced and the life of some components such as rings and bearings is much longer than with gasoline. The high octane of LPG also minimizes wear from engine knock. The rapid development of LPG technology including ease of vehicular conversion and estimations of increased LPG availability suggest LPG will soon be recognized as a premium automotive fuel. The composition of LPG components and their higher heating values are given in Table 4.2 (Demirbas, 2002).

When prepared as fuel, LPG is largely propane; common uses are for powering automotive vehicles, for cooking and heating, and sometimes for lighting in rural areas. LPG is an attractive fuel for internal combustion engines because it burns with little air pollution and little solid residue, it does not dilute lubricants, and it has a high octane rating.

Unlike gasoline, LPG is used as a dry gas and contains no fuel additives. Without lead as a lubricating agent, some early LPG conversions resulted in increased wear rates in the engine cylinder valves. LPG has higher ignition temperatures than gasoline, increasing the importance of maintaining proper ignition system opera-

Table 4.2 Composition of LPG components and their higher heating values (HHVs)

Component	Percentage by volume	HHV (MJ/kg)
Propane (C_3H_8)	60–85	50.4
Butane (C_4H_{10})	14–38	49.6
Pentane (C_5H_{12})	0–6	49.1
Isopentane (C_5H_{12})	0–0.2	49.0
Cyclopentane (C_5H_{12})	0–0.1	45.5
Ethylene (C_2H_4)	0–1.5	49.6
Others	0–2	–

tion. If not maintained, this will result in improper combustion and sluggish vehicle performance.

The LPG fuel/air mixture burns well, so engine starting problems are less common than with liquid fuels. Since LPG has a lower energy density but higher octane rating than gasoline, LPG produces fewer miles per gallon but the octane value allows higher compression ratios and higher power and fuel efficiency. For the same compression ratio, the amount of gaseous LPG air/fuel mixture that can be burned is about 5–10% less than in gasoline engines.

Total HC, CO and NO_x emissions are a function of propane concentration in the LPG blend. It is observed that upon changing the propane concentration from 60 to 85% the HC emissions are reduced 30–44%, CO 73–95%, while NO_x emissions are increased seven to nine times for various vehicles using LPG blends (Diaz et al., 2000).

LPG is a good substitute for petrol in spark ignition engines. Its clean-burning properties, in a properly tuned engine, give reduced exhaust emissions, extended lubricant and spark plug life. LPG is one of the cleanest alternative fuels for motor vehicles. LPG presents a useful combination of combustion and storage properties that makes it an attractive vehicular fuel.

Propane and butane are the most common liquefied natural gases, and are extremely hazardous. Highly flammable, these gases easily ignite and burn, and will burst their containers with explosive force. Because of the large amount of LPG in transportation, storage and use, it is important that an active safety training programs be set up for workers who handle these gases. It is also important to set up an emergency response program to handle incidents involving these hazardous materials. The fuel tanks of LPG vehicles are designed to withstand severe impact without rupture and can survive vehicle fire. Overseas experience shows that LPG vehicles are at least as safe as diesel and petrol vehicles.

Among alternative road-transport fuels, autogas and biodiesel are probably the most secure. In heating and electricity markets, LPG's security advantage over other fuels is less obvious (Johnson, 2003). A risk assessment consultant commissioned by Demirbas and co-workers carried out a survey of previous accident experience with gas-fueled vehicles. The consultant found that there had been very few accidents involving the fuel systems of gas vehicles. There were 4 reported LPG vehicle accidents on the road. It should be noted that all 4 cases involved converted LPG vehicles, which are not allowed in Hong Kong. Only custom-made LPG vehicles produced by original vehicle manufacturers were allowed, which are subject to very high safety and performance standards. The risk consultant confirmed that there have not been any accidents relating to the fuel system of custom-made LPG vehicles.

The risk of explosion or fire associated with the use of mobile telephones in a LPG vehicle is extremely low. First, LPG vehicle fuel systems are closed systems with safety features to prevent accidental release of LPG. The risk of fuel leakage is less than that of a petrol or diesel vehicle. Second, LPG will only burn when mixed with air in proportions within the flammable limits and when there is an ignition source. Working with higher-pressure fuel systems requires special tools and safety precautions. Mechanics should receive special training and certification. The training should include the fundamentals of LPG combustion and storage, working with high-pressure conduit, connectors, regulators and cylinders, safety codes and industry standards, cylinder inspections, conversion system calibration, and troubleshooting. The vapors of many liquid fuels such as gasoline and LPG are heavier than air. As they evaporate, gasoline and LPG vapors tend to accumulate around the source and pose an explosion hazard.

4.3 Compressed Natural Gas

Natural gas (NG) consists of 85–95% methane (CH_4), which is the simplest hydrocarbon. NG is the cleanest burning alternative fuel. Exhaust emissions from NG vehicles are much lower than those from gasoline-powered vehicles. Combustion of NG reduces carbon dioxide content compared with diesel, but the lost efficiency when the Otto process is used means that carbon dioxide emissions increase. All in all combustion of NG in a gasoline engine gives rise to about as much carbon dioxide as the combustion of diesel in a diesel engine, measured in units of energy.

NG is a cleaner burning fuel than either gasoline or diesel. However, NG also contains active compounds, such as sulfur, and inert compounds such as nitrogen and CO_2. NG has a high octane number (110–130) and therefore potential for use in a high-compression engine. NG is a fossil fuel and is extracted from gas sources in bedrock. It is lighter than air and a carries a certain risk of explosion. Natural gas can be stored in compressed form and is then known as compressed natural gas (CNG), or at a temperature (−162°C) in liquid form, known as liquefied natural gas (LNG). LNG has been considered as an option for fueling automobiles.

The composition of pipeline natural gas (NG) varies depending on the source and processing of the gas. Typically, it is over 90% methane and small amounts of ethane and other hydrocarbons. NG may also contain nitrogen, carbon dioxide, and trace amounts of water vapor. NG at ambient temperatures and pressures is a gas. As is typical of gases, it has a very low energy density compared to other fuels. On average, it takes 0.921 cubic meters of NG to equal the same energy content as 1 l of gasoline. This makes use of NG as a transportation fuel at ambient temperatures and pressures unfeasible. To use NG as a transportation fuel it must either be compressed or liquefied to increase its volumetric energy density (Demirbas, 2002).

As mentioned above, the vapors of many liquid fuels, including gasoline and LPG, are heavier than air and as they evaporate the vapors tend to accumulate around the source, creating an explosion hazard. NG requires a greater concentration in air and a higher temperature to ignite than gasoline. In the event of a leak or accident, natural gas will ignite at a temperature of 811 K and between 5 and 15% concentration in air. Gasoline will ignite at only 533 K and 1.5% concentration in air.

Ventilation must be provided in CNG vehicle maintenance garages and vehicle storage buildings through ceiling exhaust systems to prevent hazardous CNG accumulations. In vehicle operation, as the pressure regulators reduce CNG pressure, the temperature will drop, causing water vapor in the NG to condense. The condensed water will restrict or block fuel flow. CNG vehicle fueling stations normally dehydrate the NG to prevent water condensation.

Use of CNG as a transport sector fuel requires investment in a dramatically different fuel supply system. Gas has to be taken from a gas distribution system, usually at a pressure of 0.3 to 1 MPa, and compressed to 20 MPa into vehicle CNG tanks. Without intermediate storage at the CNG vehicle filling station, a slow fill system is used that only has compressors. About 8 hours are required for refueling the vehicles. For a more rapid fill but usually no faster than 5 min, larger compressors and intermediate storage are required.

Typically, to use CNG as a transportation fuel, NG is taken from a gas distribution system at pressures ranging from 0.3 to 1 MPa, compressed at a fueling station, and stored in cylinders on vehicles at pressures of about 20 MPa. Currently, the international NG vehicle industry has not set a standardized CNG pressure. At 20 MPa, the volumetric energy density of CNG is about one-fourth that of gasoline. Thus, with all efficiencies being equal, a CNG vehicle requires four times the size of a gasoline tank for the same driving range. Liquefying NG by cooling it to approximately 400 K increases its energy density. Liquefied natural gas (LNG) is typically stored at low pressures between 0.07 and 0.34 MPa to maintain its liquid state, but LNG is not available in most markets. Increasing the travel distance between refueling increases the investment, weight of the storage tank, and space given up for making it possible to use CNG. Natural gas has an octane value of 130, which is considerably higher than gasoline, which usually ranges between 84 and 97. The higher octane of CNG generally gives very good engine performance characteristics (Demirbas, 2002).

Despite the large difference in volumetric energy density between gasoline and CNG, the impact of CNG energy density on engine performance is less dramatic. As a gas, it has few cold-start problems. Its higher octane value allows for higher engine compression ratios than can be used with gasoline alone. Higher compression ratios allow for higher power and fuel efficiency. However, for the same compression ratio, the amount of natural gas air/fuel mixture that can be burned in each piston stroke is 10–15% less than for gasoline. Thus, there is a 10–15% loss of engine output power (Demirbas, 2002).

To use CNG safely, technicians and drivers need to know what the differences are and how to work with them. Other subjects that must be understood are NG combustion and storage, working with high-pressure conduits, connectors, regulators, and cylinders, safety codes and industry standards, and recommended CNG cylinder inspections. Detailed hands-on training for installation and maintenance technicians should normally be provided by the conversion kit manufacturers (Demirbas, 2002).

In general, the use of NG results in cleaner and longer-lasting engines. Less carbon builds up on spark plugs, engine oil, and in the combustion chamber. With higher ignition temperatures than gasoline, the use of NG increases the importance of maintaining proper ignition system operation.

The toxic emissions with CNG, without exception, are lower than for any other hydrocarbon fuel. This is a direct result of the fact that CNG is a single hydrocarbon, 90% methane, whereas all of the other fuels are a mix of hydrocarbons. LPG is a relatively simple mix of propane, butane, and pentane compared to CNG and the complex mix that makes up the gasoline and diesel typically purchased at the service station. Gasoline and diesel emit compounds into the air: methanol, formaldehyde, aldehydes, acrolein, benzene, toluene, xylene, etc., some of which are not yet part of any established emission standard but certainly are not beneficial to people's health (Demirbas, 2002).

Soot emission from hydrocarbon flames is an important subject since it plays an important role in relation to both heat transfer by radiation and air pollution (Shahad and Mohammed, 2000). The use of CNG in internal combustion (IC) engines permits operation with decreased advances and decreases NO_x without increasing soot formation.

The production, processing, transportation, and compression of NG to the CNG fuel that is used by vehicles results in less environmental impact than the production, transportation, and processing of crude oil and the transportation of gasoline or diesel to the service stations. CO_2 emission of NG is lower than both diesel fuel and gasoline, which makes natural gas engines favorable also in terms of the greenhouse effect.

4.4 Hydrogen

Hydrogen has been suggested as a convenient, clean-burning fuel. Hydrogen gas may be stored as a compressed gas or as a liquid, and it also has properties that make it suitable as a fuel for internal combustion engines in automobiles.

The use of hydrogen as a fuel for transportation and stationary applications is receiving much favorable attention as a technical and policy issue. Hydrogen gas is being explored for use in combustion engines and fuel-cell electric vehicles. It is a gas at normal temperatures and pressures, which is more challenging to transport and store than liquid fuels. Several hydrogen technologies are under development; the most promising of these is the fuel cell. Fuel cells use hydrogen, oxygen, catalyst, and electrolytic solution to produce energy in the form of heat and electricity. A primary advantage of hydrogen over other fuels is that its only major oxidation product is water vapor; its use produces no CO_2. Combustion of hydrogen in air can result in the formation of NO_x, but it may be reduced to low levels by proper design. Hydrogen is non-toxic and not carcinogenic.

Hydrogen can be stored as a gas, a cryogenic liquid, or, in addition, solid-state storage is also possible. A particular problem with liquid hydrogen is boiling off. As the liquid warms, boil off gas is released which must be vented from the storage tank. In confined spaces there is a risk of fire or explosion if contacted by a flame.

4.5 Electricity

Electricity can be used as a transportation fuel to power battery electric and fuel cell vehicles. Electric fuel is electricity that is used to directly power the vehicle. In a sustainable energy future, electricity will become the prime energy carrier. We now have to focus our research on electricity storage, electric cars and the modernization of the existing electricity infrastructure.

The power is created by batteries and other electricity sources. Batteries are energy storage devices, but unlike batteries, fuel cells convert chemical energy to electricity. Fuel cell vehicles use electricity produced from an electrochemical reaction that takes place when hydrogen and oxygen are combined in the fuel cell stack. The production of electricity using fuel cells takes place without combustion or pollution and leaves only two byproducts, heat and water. Benefits include no emissions and fewer parts to be serviced and replaced. Electricity is also cheaper than gasoline.

4.6 Solar

Solar energy technologies use sunlight to produce heat and electricity. Solar fuel is light from the Sun transformed into energy. Electricity produced by solar energy through photovoltaic technologies can be used in conventional electric vehicles. Solar energy cannot be used directly in cars except as a publicity stunt. Solar power may be used to run certain auxiliary systems in the vehicle. In order to collect the energy derived from the Sun and use it to fuel a vehicle, photovoltaic cells are used. Pure solar energy is 100% renewable and a vehicle run on this fuel emits no emissions.

4.7 Biorenewable Liquid Fuels

Alcohols such as methanol, ethanol, propanol and butanol can be used as alternate motor fuels. Practically, any of the organic molecules of the alcohol family can be used as a fuel. However, only methanol and ethanol fuels are technically and economically suitable for internal combustion engines (ICEs).

Ethanol has a higher octane number (107), broader flammability limits, higher flame speeds and higher heats of vaporization than gasoline. These properties allow for a higher compression ratio, shorter burn time and leaner burn engine, which lead to theoretical efficiency advantages over gasoline in an ICE. Disadvantages of ethanol include its lower energy density than gasoline, its corrosiveness, low flame luminosity, lower vapor pressure, miscibility with water, and toxicity to ecosystems.

Currently, ethanol is produced from sugar beets and from molasses. A typical yield is 72.5 liters of ethanol per ton of sugar cane. Modern crops yield 60 tons of sugar cane per hector of land. Production of ethanol from biomass is one way to reduce both the consumption of crude oil and environmental pollution. Domestic production and use of ethanol for fuel can decrease dependence on foreign oil, reduce trade deficits, create jobs in rural areas, reduce air pollution, and reduce global climate change carbon dioxide build-up.

Bioethanol is a liquid fuel made from plant material and recycled elements of the food chain. Bioethanol is a fuel derived from renewable sources of feedstock; typically plants such as wheat, sugar beet, corn, straw, and wood. Bioethanol is currently made by large-scale yeast fermentation of sugars that are extracted or prepared from crops followed by separation of the ethanol by distillation. Bioethanol is renewable and sustainable. By contrast, petrol, diesel and the road fuel gases liquefied petroleum gas (LPG) and compressed natural gas (CNG), are fossil fuels in finite supply. Bioethanol is a petrol additive/substitute. It is possible that wood, straw and even household wastes may be economically converted to bioethanol. Bioethanol can be used as a 5% blend with petrol under the EU quality standard EN 228. This blend requires no engine modification and is covered by vehicle warranties. With engine modification, bioethanol can be used at higher levels, for example, E85 (85% bioethanol) is common in the USA. Ethanol is also a safer alternative to methyl tertiary butyl ether (MTBE), the most common additive to gasoline used to provide cleaner combustion. MTBE is a toxic chemical compound and has been found to contaminate groundwater.

The reduced CO_2 emissions mean that bioethanol is good for the environment. Using ethanol-blended fuel for automobiles can significantly reduce petroleum use and exhaust greenhouse gas emission. Reducing the use of non-renewable fossil energy reserves together with improving the environment are two important reasons that drive interest in the use of bioethanol as an automotive fuel. While more expensive to produce than other fuel types, bioethanol production boosts the farm economy.

Methanol can be used as one possible replacement for conventional motor fuels. Methanol has been seen as a possible large volume motor fuel substitute at various times during gasoline shortages. It was often used in the early part of the century to power automobiles before inexpensive gasoline was widely introduced. Methanol is poisonous and burns with an invisible flame. Methanol has just like ethyl alcohol a high octane rating and hence an Otto engine is preferable. If an ignition booster is used, methanol can be used in a diesel engine.

Methanol is mainly manufactured from natural gas, but biomass can also be gasified to methanol. Methanol can be made with any renewable resource containing carbon such as seaweed, waste wood and garbage. This is a promising alternative, with a diversity of fuel applications with proven environmental, economic and consumer benefits. Problems occurred early in the development of gasoline-methanol blends. As a result of its low price some gasoline marketers over blended. Many tests have shown promising results using 85–100% by volume methanol as a transportation fuel in automobiles, trucks and buses.

Ethanol is produced as a more environmentally benign fuel. The systematic effect of ethyl alcohol differs from that of methyl alcohol. Ethyl alcohol is rapidly oxidized in the body to carbon dioxide and water, and in contrast to methyl alcohol no cumulative effect occurs. Ethanol is also a preferred alcohol in the transportation sector compared to methanol because it is derived from agricultural products and is renewable and biologically less objectionable in the environment.

Methanol use in current-technology vehicles has some distinct advantages and disadvantages. On the plus side, methanol has a higher octane rating than gasoline. Methanol has a high heat of vaporization that results in lower peak flame temperatures than gasoline and lower nitrogen oxide emissions. Its greater tolerance to lean combustion higher air-to-fuel equivalence ratio results in generally lower overall emissions and higher energy efficiency. However, several disadvantages must be studied and overcome before neat methanol is considered a viable alternative to gasoline. The energy density of methanol is about half that of gasoline, reducing the range a vehicle can travel on an equivalent tank of fuel.

There are some important differences in the combustion characteristics of alcohols and hydrocarbons. Alcohols have higher flame speeds and extended flammability limits. Pure methanol is very flammable and its flame is colorless when ignited. The alcohols mix in all proportions with water due to the polar nature of the OH group. Low volatility is indicated by high boiling point and high flash point. Combustion of alcohol in presence of air can be initiated by an intensive source of localized energy, such as a flame or a spark and also, the mixture can be ignited by application of energy by means of heat and pressure, such as happens in the compression stroke of a piston engine. The high latent heat of vaporization of alcohols cools the air entering the combustion chamber of the engine, thereby increasing the air density and mass flow. This leads to increased volumetric efficiency and reduced compression temperatures. The oxygen contents of alcohols depress the heating value of the fuel in comparison with hydrocarbon fuels. The heat of combustion per unit volume of alcohol is approximately half that of isooctane (Bala, 2005).

Methanol is not miscible with hydrocarbons and separation ensues readily in the presence of small quantities of water, particularly with reduction in temperature. On the other hand, anhydrous ethanol is completely miscible in all proportions with gasoline, although separation may be effected by water addition or by cooling. If water is already present, the water tolerance is higher for ethanol than for methanol, and can be improved by the addition of higher alcohols, such as butanol. Also benzene or acetone can be used. The wear problem is believed to be caused by formic acid attack, when methanol is used or acetic acid attack when ethanol is used.

Dry methanol is very corrosive to some aluminum alloys, but additional water at 1% almost completely inhibits corrosion. It must be noted that methanol with additional water at more than 2% becomes corrosive again. Ethanol always contains some acetic acid and is particularly corrosive to aluminum alloys.

Since alcohols, especially methanol, can be readily ignited by hot surfaces, pre-ignition can occur. It must be emphasized here that pre-ignition and knocking in alcohol engine is a much more dangerous condition than gasoline engines. Other properties, however, are favorable to the increase of power and reduction of fuel consumption. Such properties are as follows: (1) number of molecules or products

Table 4.3 Some properties of transportation fuels

Fuel property	Gasoline	Diesel No. 2	Isooctane	Methanol	Ethanol
Cetane number	–	50	–	5	8
Octane number	96	–	100	112	107
Autoignition temperature (K)	644	588	530	737	606
Latent heat of vaporization (MJ/Kg)	0.35	0.22	0.26	1.18	0.91
Lower heating value (MJ/Kg)	44.0	42.6	45.0	19.9	26.7

Table 4.4 Potential and available motor fuels

Fuel type	Available motor fuel
Traditional fuels	Diesel and Gasoline
Oxygenated fuels	Ethanol 10% (E10), Methanol, Methyl tertiary butyl ether (MTBE), Ethyl tertiary butyl ether (ETBE), Tertiary butyl alcohol (TBA), and Tertiary amyl methyl ether (TAME)
Alternative fuels	Liquefied petroleum gases (LPG), Ethanol, 85% (E85), Ethanol, 95% (E95), Methanol, 85% (M85), Methanol, neat (M100), Compressed natural gas (CNG), Liquefied natural gas (LNG), Biodiesel (BD), Hydrogen, and Electricity

is more than that of reactants, (2) extended limits of flammability, (3) high octane number, (4) high latent heat of vaporization, (5) constant boiling temperature, and (6) high density.

When diesel engines are converted to alcohols, some properties of gasoline, diesel and alcohol should be taken into account. Table 4.3 shows the properties of the transportation fuels. There are several methods for converting a diesel engine to alcohol.

Production of motor fuel alternatives from biomass materials is an important application area of biotechnological methods. Table 4.4 shows the potential and available motor fuels. Biorenewable motor fuel alternatives are:

1. Gasoline-alcohol mixtures
2. Alcohol substituting gasoline
3. Gasoline-vegetable oil mixtures
4. Diesel fuel-vegetable oil mixtures
5. Vegetable oil substituting diesel fuel

The application of alcohol and gasoline-alcohol mixtures in the gasoline (Otto) engines began in the first half of the twentieth century. It is possible to find information about various studies on the change in octane numbers of gasoline-alcohol mixtures, composition of the exhaust gases, motor tests and about the materials used in constructing the engines. Between 1980 and 1985 studies were intensified on solving the phase separation problem of gasoline-alcohol mixtures.

In gasoline-alcohol mixtures ethanol and methanol are generally used and in gasoline engine mixtures containing 20% or less alcohol by volume can be used without altering the construction of the engine. Because of the hygroscopic properties of ethanol and methanol the gasoline-alcohol mixtures are in fact ternary mixtures composed of gasoline-alcohol and water. In the evaluation of such mixtures as motor fuel there is the phase separation problem, which depends on several factors.

In gasoline-methanol mixtures containing 0.1% water i-propanol is added to the environment (medium) in order to decrease the phase separation temperature. Fuels containing different ratios of gasoline-methanol-i-propanol and water are also composed, which are proved to be stable in the climatic conditions. An increase in the aromatic character of the gasoline, a decrease in the water content of the mixture, an increase in the amount of the additive used results in a decrease in the phase separation temperature of the mixture. In gasoline-ethanol mixtures the additive used is also i-propanol. In gasoline-alcohol mixtures various additives like i-propanol, n-butanol, i-butanol, and i-amylalcohol are used.

P-series fuel is a unique blend of natural gas liquids, ethanol, hydrocarbons methyltetrahydrofuran (MeTHF). MeTHF is a biomass-derived co-solvent. P-series is made primarily from renewable resources and provides significant emissions benefits over reformulated gasoline. P-series fuels are clear, colorless, 89–93 octane number, liquid blends that are formulated to be used in flexible fuel vehicles. Like gasoline, low vapor pressure formulations are produced to prevent excessive evaporation during summer months, and high vapor pressure formulations are used for easy starting in the winter. P-series is at least 60% non-petroleum. Because a majority of the components that make up P-series fuels come from domestically produced renewable resources, this alternative fuel promotes both energy security and environmental quality. P-series could be 96% derived from domestic resources. P-series fuels could reduce fossil energy use by 49% to 57% and petroleum use by 80% relative to gasoline. Greenhouse gas emissions of the P-series fuels are 45% to 50% below those of reformulated gasoline. P-series would be used in severe cold-weather conditions to meet cold-start requirements.

P-series also has many environmental benefits. Emissions from the production and use of P-series are substantially better than those from gasoline. Each unit of P-series fuel emits approx 50% less carbon dioxide, 35% less hydrocarbons, 15% less carbon monoxide than gasoline. It also has 40% less ozone-forming potential.

Another biorenewable fuel is dimethyl ether (DME or CH_3–O–CH_3). DME is made from natural gas, but it can also be produced by gasifying biomass. DME can be stored in liquid form at 5–10 bars of pressure at normal temperatures. A major advantage of DME is its naturally high cetane number. This makes self-ignition easier. This high cetane rating makes DME most suitable for use in diesel engines, as this fuel maintains the high level of efficiency of the diesel engine. The energy content of DME is also lower than in diesel.

DME can be produced effectively from biosyngas in a single-stage, liquid phase (LPDME) process. The origin of syngas includes a wide spectrum of feedstocks such as coal, natural gas, biomass, and others. Non-toxic, high-density, liquid DME fuel can be easily stored at modest pressures. DME and methanol are produced similarly. The DME conversion to hydrocarbons is studied using ZSM-5 catalysts with varying SiO_2/Al_2O_3 ratios, whereas the DME carbonization reaction to produce methyl acetate is studied over a variety of group VIII metal substituted heteropoly-acid catalysts.

Vegetable oils have the potential to substitute a fraction of petroleum distillates and petroleum-based petrochemicals in the near future. Possible acceptable converting processes of vegetable oils into reusable products are transesterification, solvent extraction, cracking and pyrolysis. Pyrolysis has received a significant amount of interest as this gives products of better quality compared to any other thermochemical process. The liquid fuel produced from vegetable oil pyrolysis has similar chemical components to conventional petroleum diesel fuel.

Vegetable oils are biorenewable and potentially inexhaustible sources of energy with energetic contents close to diesel fuel. There are more than 350 oil-bearing crops identified, among which only sunflower, safflower, soybean, cottonseed, rapeseed, and peanut oils are considered as potential alternative fuels for diesel engines. The major problem associated with the use of pure vegetable oils as fuels, for diesel engines are caused by high fuel viscosity in compression ignition.

Limitations to vegetable oil use are costs and potential production. Production of vegetable oil is limited by available land area. Vegetable oil fuels are not as competitive as petroleum-based fuels because they are more expensive than petroleum fuels. However, with recent increases in petroleum prices and uncertainties concerning petroleum availability, there is renewed interest in using vegetable oils in diesel engines.

Biodiesel is a fuel consisting of long-chain fatty acid alkyl esters made from renewable vegetable oils, recycled cooking greases, or animal fats. Vegetable oils are generally converted to their methyl esters by the transesterification reaction in the presence of a catalyst. Methyl esters of vegetable oils (biodiesels) have several outstanding advantages among other new-renewable and clean engine fuel alternatives. Compared to No. 2 diesel fuel, all of the vegetable oils are much more viscous, while methyl esters of vegetable oils are slightly more viscous. The methyl esters are more volatile than those of their vegetable oils.

The soaps obtained from the vegetable oils can be pyrolyzed into hydrocarbon-rich products. The yield of conversion of the sunflower oil reached the maximum 78.3% at 660 K over $ZnCl_2$ catalyst (Demirbas, 2003).

Palm oil has been cracked at atmospheric pressure and a reaction temperature of 723 K to produce biofuel in a fixed-bed microreactor. The reaction was carried out over microporous HZSM-5 zeolite, mesoporous MCM-41, and composite micromesoporous zeolite as catalysts. The products obtained were gas, organic liquid product, water, and coke. The organic liquid product was composed of hydrocarbons corresponding to gasoline, kerosene, and diesel boiling point range. The maximum conversion of palm oil, 99 wt.%, and gasoline yield of 48 wt.% was

Table 4.5 Catalytic cracking of palm oil over HZSM-5 with different Si/Al ratios

Catalyst ID	HZSM-5(50)	HZSM-5(240)	HZSM-5(400)
Conversion (wt.%)	96.9	96.0	94.0
Gas yield (wt.%)	17.5	14.0	8.2
Water yield (wt.%)	6.8	4.6	6.1
OLP (wt.%)			
Total organic liquid yield	70.9	76.0	78.0
Gasoline (wt.%)	44.6	45.9	49.3
Kerosene (wt.%)	19.6	24.6	26.1
Diesel (wt.%)	6.7	5.5	2.6
Coke (wt.%)	1.7	1.4	1.7

obtained with composite micromesoporous zeolite (Sang et al., 2003). Table 4.5 presents the conversion of the palm oil over HZSM-5 with different Si/Al ratios catalyst by catalytic cracking. The gasoline yield increased with the increase in the Si/Al ratio due to the decrease in the secondary cracking reactions and the drop in the yield of gaseous products (Sang et al., 2003).

Some 26–34% by weight of fractions obtained from the fractional distillation of the vegetable oils were within the ASTM limits. When the ASTM D86 procedure was used to distill the vegetable oils, they cleaved into a two-phase distillate. Preliminary data indicate a complex mixture of products including alkanes, alkenes and carboxylic compounds (Demirbas, 1998). Typically, it was not possible to distill all of the vegetable oil and some brownish residue remained in the distillation flask. However, the soaps obtained from the vegetable oils can be distilled into hydrocarbon-rich products with higher yields. These findings are in general agreement with results given in the literature (Barsic and Humke, 1981).

The replacement of petroleum with biorenewables for fuels and chemicals requires the identification of feedstock, intermediate, and product species that replace their fossil fuel counterparts. The properties of biorenewable feedstocks are compared to petroleum as shown in Table 4.6.

The biomass pyrolysis is attractive because solid biomass and wastes can be readily converted into liquid products. These liquids, as crude bio-oil or slurry of charcoal of water or oil, have advantages in transport, storage, combustion, retrofitting and flexibility in production and marketing.

The pyrolysis of biomass is a thermal treatment that results in the production of charcoal, liquid, and gaseous products. Among the liquid products, methanol is one of the most valuable products. The liquid fraction of the pyrolysis products consists of two phases: an aqueous phase containing a wide variety of organo-oxygen compounds of low molecular weight and a non-aqueous phase containing insoluble organics of high molecular weight. This phase is called tar and is the product of

Table 4.6 Typical properties of petroleum and biorenewable feedstocks and biodiesel

Property	Petroleum	Biorenewable		
	Crude oil	Soybean oil	Soybean oil methyl ester (biodiesel)	Pyrolysis oil
%C	83–86	77.7	78.1	60.8
%H	11–14	11.7	12.0	6.6
%O	–	10.5	9.8	32.5
%N	0.1–1	0.0011	0.001	0.3
%S	0.1–4	0.0006	0.0004	0.0008
Density (kg/L at 293 K)	0.86	0.92	0.87	1.23
Viscosity (mm^2/s at 313 K)	1.9–3.4	34.4	4.1	175
Higher heating value (MJ/kg)	45.3	39.6	41.2	26.7

greatest interest. The ratios of acetic acid, methanol, and acetone of the aqueous phase were higher than those of the non-aqueous phase. The point where the cost of producing energy from fossil fuels exceeds the cost of biomass fuels has been reached. With a few exceptions, energy from fossil fuels will cost more money than the same amount of energy supplied through biomass conversion.

Table 4.7 shows the fuel properties of diesel, biodiesel and biomass pyrolysis oil. The kinematic viscosity of pyrolysis oil varies from as low as 11 cSt to as high as 115 mm^2/s (measured at 313 K) depending on the nature of the feedstock, temperature of pyrolysis process, thermal degradation degree and catalytic cracking, the water content of the pyrolysis oil, the amount of light ends that have collected, and the pyrolysis process used. The pyrolysis oils have water contents of typically 15–30 wt.% of the oil mass, which cannot be removed by conventional methods like distillation. Phase separation may partially occur above certain water contents. The water content of pyrolysis oils contributes to their low energy density, lowers the flame temperature of the oils, leads to ignition difficulties, and, when preheating the oil, can lead to premature evaporation of the oil and resultant injection difficulties. The higher heating value (HHV) of pyrolysis oils is below 26 MJ/kg (compared to 42–45 MJ/kg for conventional petroleum fuel oils). In contrast to petroleum oils, which are non-polar and in which water is insoluble, biomass oils are highly polar and can readily absorb over 35% water (Demirbas, 2007).

The pyrolysis oil (bio-oil) from wood is typically a liquid, almost black through dark red brown. The density of the liquid is about 1200 kg/m^3, which is higher than that of fuel oil and significantly higher than that of the original biomass. The bio-oils have water contents of typically 14–33 wt.%, which cannot be removed by

Table 4.7 Fuel properties of diesel, biodiesel and biomass pyrolysis oil

Property	Test method	ASTM D975 (diesel)	ASTM D6751 (biodiesel, B100)	Pyrolysis oil (bio-oil)
Flash point	D 93	325 K min	403 K	–
Water and sediment	D 2709	0.05 Max % vol.	0.05 Max % vol.	0.01–0.04
Kinematic viscosity (at 313 K)	D 445	1.3–4.1 mm²/s	1.9–6.0 mm²/s	25–1000
Sulfated ash	D 874	–	0.02 Max wt.%	–
Ash	D 482	0.01 Max wt.%	–	0.05–0.01 wt.%
Sulfur	D 5453	0.05 Max wt.%	–	–
Sulfur	D 2622/129	–	0.05 Max wt.%	0.001–0.02 wt.%
Copper strip corrosion	D 130	No. 3 max	No. 3 max	–
Cetane number	D 613	40 min	47 min	–
Aromaticity	D 1319	–	35 Max % vol.	–
Carbon residue	D 4530	–	0.05 Max mass%	0.001–0.02 wt.%
Carbon residue	D 524	0.35 Max mass%	–	–
Distillation temperature	D 1160	555 K min –611 K max	–	–

conventional methods like distillation. Phase separation may occur above certain water contents. The higher heating value (HHV) is below 27 MJ/kg (compared to 43–46 MJ/kg for conventional fuel oils).

The bio-oil formed at 725 K contained high concentrations of compounds such as acetic acid, 1-hydroxy-2-butanone, 1-hydroxy-2-propanone, methanol, 2,6-dimethoxyphenol, 4-methyl-2,6-dimetoxyphenol and 2-Cyclopenten-1-one, etc. A significant characteristic of the bio-oils was the high percentage of alkylated compounds especially methyl derivatives. As the temperature increased, some of these compounds were transformed via hydrolysis (Kuhlmann et al., 1994). The formation of unsaturated compounds from biomass materials generally involves a variety of reaction pathways such as dehydration, cyclization, Diels–Alder cycloaddition reactions and ring rearrangement. For example, 2,5-hexandione can undergo cyclization under hydrothermal conditions to produce 3-methyl-2-cyclopenten-1-one with very high selectivity of up to 81% (An et al., 1997).

The mechanism of pyrolysis reactions of biomass was extensively discussed in an earlier study (Demirbas, 2000). Water is formed by dehydration. In the pyrolysis reactions, methanol arises from the breakdown of methyl esters and/or ethers from decomposition of pectin-like plant materials. Methanol also arises from methoxyl groups of uronic acid (Demirbas and Güllü, 1998). Acetic acid is formed in the thermal decomposition of all three main components of wood. When the yield of acetic acid originating from the cellulose, hemicelluloses, and lignin is taken into account, the total is considerably less than the yield from the wood itself. Acetic acid comes from the elimination of acetyl groups originally linked to the xylose unit.

Summary

Almost all motor vehicles today are powered by either diesel or gasoline. Both fuels are mainly derived from petroleum. Diesel fuel consists of straight-chain hydrocarbons with between C_9 and C_{27} carbon atoms. Gasoline is a blend of hydrocarbons with some contaminants, including sulfur, nitrogen, oxygen and certain metals. The four major constituent groups of gasoline are olefins, aromatics, paraffins, and napthenes. The important characteristics of gasoline are density, vapor pressure, distillation range, octane, and chemical composition. The population of automobiles is currently dominated by vehicles burning gasoline. The main alternative fuels include alcohol, liquefied petroleum gas, compressed natural gas, hydrogen and electricity for the operation of gasoline-type vehicles.

Diesel fuel is produced by distilling raw oil, which is extracted from bedrock. Diesel is a fossil fuel, consisting of hydrocarbons with between 9 and 27 carbon atoms in a chain, as well as a smaller amount of sulfur, nitrogen, oxygen and metal compounds. It is a general property of hydrocarbons that the autoignition temperature is higher for more volatile hydrocarbons. The hydrocarbons present in the diesel fuels include alkanes, naphthenes, olefins and aromatics.

There are four alternative fuels that can be relatively easily used in conventional compression ignition (CI) engines: vegetable oil, biodiesel, Fischer–Tropsch (FT), and dimethyl ether (DME). Both FT and DME can be manufactured from natural gas and are therefore not limited by feedstock availability. Biodiesel on the other hand, is produced from vegetable (and some waste animal) oils whose supply for non-nutritional uses is presently quite limited.

Hydrogen has been recognized as a convenient, clean-burning fuel. Hydrogen gas may be stored as a compressed gas or as a liquid, and has certain properties that make it suitable for internal combustion engines in automobiles. The use of hydrogen as a fuel for transportation and stationary applications is receiving much favorable attention as a technical and policy issue. Hydrogen gas is being explored for use in combustion engines and fuel cell electric vehicles. Since hydrogen is a gas at normal temperatures and pressures, it is more challenging to transport and store than liquid fuels. Several hydrogen technologies are under development; the most promising of these is the fuel cell. Fuel cells use hydrogen, oxygen, catalysts, and electrolytic solution to produce energy in the form of heat and electricity. A primary advantage of hydrogen over other fuels is that its only major oxidation product is water vapor; its use produces no CO_2. Combustion of hydrogen in air can result in the formation of NO_x, but it may be reduced to low levels by proper design. Hydrogen itself is non-toxic and not carcinogenic.

Electricity can be used as a transportation fuel to power electric and fuel cell vehicles. Electric fuel is electricity that is used to directly power the vehicle. In a sustainable energy future, electricity will become the prime energy carrier. We now have to focus our research on electricity storage, electric cars and the modernization of the existing electricity infrastructure.

Biorenewable liquid fuels such as are methanol, ethanol, propanol and butanol can be used as alternate motor fuels. Practically, any of the organic molecules of the alcohol family can be used as a fuel. However, only methanol and ethanol fuels are technically and economically suitable for internal combustion engines (ICEs).

References

An, J., Bagnell, L., Cablewski, T., Strauss, C.R., Trainor, R.W. 1997. Applications of high temperature aqueous media for synthetic organic reactions. J Org Chem 62:2505–2511.

Bala, B.K. 2005. Studies on biodiesels from transformation of vegetable oils for diesel engines. Energy Edu Sci Technol 15:1–43.

Barsic, N.J., Humke, A.L. 1981. Performance and emissions characteristics of a naturally aspirated diesel engine with vegetable oil fuels. SAE paper no. 810262. Society of Automotive Engineers, Warrendale, PA.

Demirbas, A. 1998. Fuel properties and calculation of higher heating values of vegetable oils. Fuel 77:1117–1120.

Demirbas, A. 2000. Mechanism of liquefaction and pyrolysis reactions of biomass. Energy Convers Manage 41:633–46.

Demirbas, A. 2002. Fuel properties of hydrogen, liquefied petroleum gas (LPG), and compressed natural gas (CNG) for transportation. Energy Sources 24:601–610.

Demirbas, A. 2003. Fuel conversional aspects of palm oil and sunflower oil. Energy Sources 25:457–466.

Demirbas, A. 2007. The influence of temperature on the yields of compounds existing in bio-oils obtaining from biomass samples via pyrolysis. Fuel Proc Technol 88:591–597.

Demirbas, A., Güllü, D. 1998. Acetic acid, methanol and acetone from lignocellulosics by pyrolysis. Energy Edu Sci Technol 2:102–110.

Diaz, L., Schifter, I., Lopez-Salinas, E., Gamas, E., Rodriguez, R., Avalos, S. 2000. Optimizing automotive LPG blend for Mexico City. Fuel 79:79–88.

Johnson, E. 2003. LPG: a secure, cleaner transport fuel? A policy recommendation for Europe. Energy Policy 31:1573–1577.

Kuhlmann, B., Arnett, E.M., Siskin, M. 1994. Classical organic reactions in pure superheated water. J Org Chem 59:3098–3101.

Machacon, H.T.C., Matsumoto, Y., Ohkawara, C., Shiga, S., Karasawa, T., Nakamura, H. 2001. The effect of coconut oil and diesel fuel blends on diesel engine performance and exhaust emissions. JSAE Rev 22:349–355.

Sang, O.Y., Twaiq, F., Zakaria, R., Mohamed, A., Bhatia, S. 2003. Biofuel production from catalytic cracking of palm oil. Energy Sources 25:859–869.

Shahad, H.A.K., Mohammed, Y.K.A. 2000. Investigation of soot formation and temperature field in laminar diffusion flames of LPG-air mixture. Energy Convers Manage 41:1897–1916.

Chapter 5
Hydrogen

5.1 Introduction

Hydrogen energy is a clean or inexhaustible energy like renewable energy and nuclear energy.

Today's energy supply has a considerable impact on the environment. Hydrogen energy is a promising alternative solution because it is clean and environmentally safe. It also produces negligible levels of greenhouse gases and other pollutants when compared with the fossil energy sources they replace. It is well-known that hydrogen is a clean and renewable fuel. Hydrogen is a secondary source of energy—an energy carrier—which is used to move, store and deliver energy in an easily usable form.

Future energy technology will utilize hydrogen with an increasing trend in steady as well as unsteady combustion processes. Hydrogen is the most common element in the universe and can be made from water. Converting hydrogen into energy is compatible with existing energy technologies, such as fuel cells, engines, and combustion turbines.

Hydrogen is produced from fossil fuels, hydrocarbon polymers, biomass, and water by electrolysis, and the biological processes of photolysis or thermolysis. Hydrogen gas may be stored as a gas or as a liquid. Hydrogen possesses properties that make it an ideal fuel for internal combustion engines. Hydrogen could be used advantageously as a clean energy carrier for heat supply and transportation purposes. However, the very high cost of hydrogen fuel and hydrogen vehicles is thought to outweigh the environmental advantages. It serves as an agent, through which a primary energy source (nuclear, solar, etc.) can be stored, transmitted and utilized to fulfill present and future energy needs. In the future, hydrogen may be used in furnaces and as a transportation fuel for automobiles, buses, trains, ships and airplanes. Hydrogen could also be converted directly into electricity by fuel cells (Caglar and Ozmen, 2000). Fuel cells can produce electricity and heat from hydrogen, natural gas, and petroleum fuels, and fuel gases derived from coal and biomass. What makes fuel cells unique is that they can use fuels without combustion, simply by chemical reactions, making them extremely clean and efficient.

Hydrogen is considered one of the most promising fuels for generalized use in the future, mainly because it is an energy-efficient, low-polluting and renewable fuel. Much of the hydrogen produced in the world, and especially for the petrochemical industry, is obtained from natural gas, which is mostly made up of methane. The following energy carriers methanol, dimethyl ether, ethanol, gasoline, diesel, and propane can be utilized for hydrogen production in a reformer with autothermal reforming, that is a partial oxidation reformer above a temperature of 825 K (Höhlein et al., 2000).

Hydrogen can be generated from renewable energy sources such as from biomass, hydropower, solar thermal energy, solar energy using photovoltaics for direct conversion, and wind power. In the future, hydrogen could also be produced by gasification of biomass (Miranda, 2004). The only primary hydrogen combustion product is water. There is no CO_2, no CO, no sulfuric acid, no soot, no unburned hydrocarbons, and only very small amounts of NO_x are produced (Koroneos et al., 2005). Because hydrogen can be easily generated from renewable energy sources and water it has great potential as an energy source. However, storage of hydrogen in liquid form is difficult, as very low temperatures are required to liquefy hydrogen (Demirbas, 2002).

5.2 History

In 1766, hydrogen was first identified as a distinct element by British scientist Henry Cavendish after he evolved hydrogen gas by reacting zinc metal with hydrochloric acid. In a demonstration to the Royal Society of London, Cavendish applied a spark to hydrogen gas yielding water. This discovery led to his later finding that water is made of hydrogen and oxygen. In 1788, building on the discoveries of Cavendish, French chemist Antoine Lavoisier gave hydrogen its name, which was derived from the Greek words *hydro* and *genes*, meaning *water* and *born of* or *forming*. In 1800, English scientists William Nicholson and Sir Anthony Carlisle discovered that applying electric current to water produced hydrogen and oxygen gases by electrolysis. In 1838 the fuel cell effect, combining hydrogen and oxygen gases to produce water and an electric current, was discovered by Swiss chemist Christian Friedrich Schoenbein. In 1889, Ludwig Mond and Charles Langer attempted to build the first fuel cell device using air and industrial coal gas. They named the device a fuel cell (Caglar and Ozmen, 2000).

In the 1920s, German engineer Rudolf Erren, converted the internal combustion engines of trucks, buses, and submarines to use hydrogen or hydrogen mixtures. British scientist and Marxist writer, J.B.S. Haldane, introduced the concept of renewable hydrogen in his paper "Science and the Future" by proposing that "there will be great power stations where during windy weather the surplus power will be used for the electrolytic decomposition of water into oxygen and hydrogen."

In 1958, The United States formed the National Aeronautics and Space Administration (NASA). NASA's space program currently uses the most liquid hydrogen worldwide, primarily for rocket propulsion and as a fuel for fuel cells. In 1959, Francis T. Bacon of Cambridge University in England built the first practical hydrogen-air fuel cell. The 5-kilowatt (kW) system powered a welding machine. In 1970, electrochemist John O'M. Bockris coined the term "hydrogen economy" during a discussion at the General Motors (GM) Technical Center in Warren, Michigan. In 1973, after the OPEC oil embargo and the resulting supply shock suggested that the era of cheap petroleum had ended and that the world needed alternative fuels, the development of hydrogen fuel cells for conventional commercial applications began. In 1974, the National Science Foundation transferred the Federal Hydrogen R&D Program to the US Department of Energy. Professor T. Nejat Veziroglu of the University of Miami organized The Hydrogen Economy Miami Energy Conference (THEME), the first international conference held to discuss hydrogen energy. In 1990, the world's first solar-powered hydrogen production plant at Solar-Wasserstoff-Bayern, a research and testing facility in southern Germany, became operational. In 1994, Daimler Benz demonstrated its first NECAR I (New Electric CAR) fuel cell vehicle at a press conference in Ulm, Germany. In 2000, Ballard Power Systems presented the world's first production-ready PEM fuel cell for automotive applications at the Detroit Auto Show. In 2004, the world's first fuel cell-powered submarine underwent deepwater trials (Caglar and Ozmen, 2000).

5.3 Properties of Hydrogen

Hydrogen is a colorless, tasteless gas and it is the lightest chemical substance. The hydrogen molecule is non-polar; the weak nature of the intermolecular forces of attraction is indicated by the low normal boiling point (15.5 K), the normal melting point (14.1 K) and the critical temperature (30.2 K at a critical pressure of 12.8 atm). Hydrogen is virtually insoluble in water; approximately 2 ml of hydrogen will dissolve in 1 l of water at room temperature and atmospheric pressure. There are three isotopes of hydrogen. The most abundant isotope ${}_1H^1$, constitutes 99.985% of naturally occurring hydrogen; deuterium ${}_1H^2$ (also indicated as ${}_1D^2$), constitutes 0.015%; and the radioactive tritium ${}_1H^3$ (also indicated as ${}_1T^3$) occurs only in trace amounts.

Hydrogen consists of a mixture of two types of molecules, *ortho*-hydrogen, in which the two proton spins are parallel, and *para*-hydrogen, in which they are antiparallel. There are a number of possible methods of measuring the proportion of *ortho*- to *para*-molecules in a sample of hydrogen gas (Farkas, 1935; Stewart and Squires, 1955; Noganow, 1992; Silvera and Pravica, 1998). Figure 5.1 shows the schematic model of the *ortho*/*para*-H_2 conversion on solid surfaces (Buntkowsky et al., 2006).

Figure 5.2 shows the structure of the hydrogen molecule. Two hydrogen atoms, each with one electron, combine to form a hydrogen molecule, in which the two

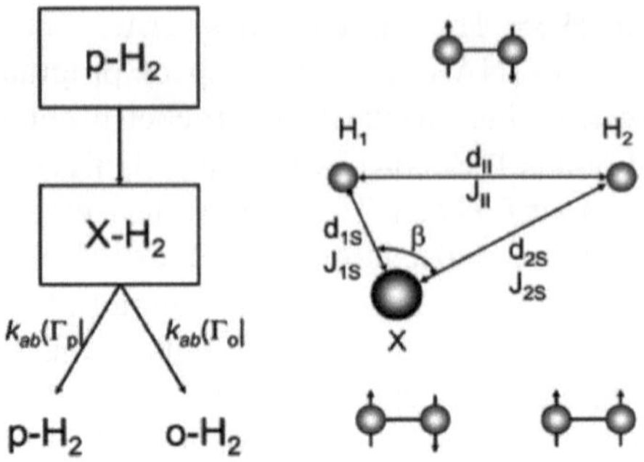

Fig. 5.1 Schematic model of the *ortho/para*-H_2 conversion on solid surfaces

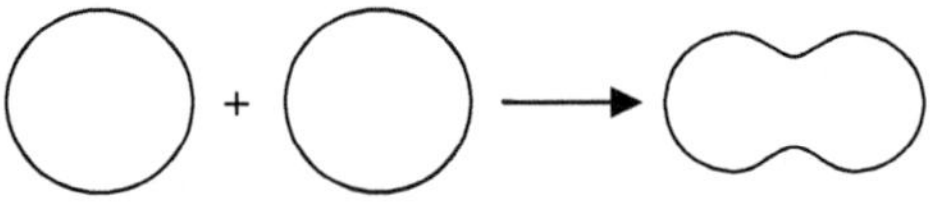

Fig. 5.2 Structure of the hydrogen molecule

electrons are shared between the atoms and serve to give each atom a filled valence shell. The simplest chemical bond is that formed between two hydrogen atoms. Each hydrogen atom has one electron. As the two atoms approach each other, the nucleus of one atom attracts the electron of the other. Eventually the two orbitals overlap, becoming a single orbital containing two electrons.

5.4 Fuel Properties of Hydrogen

Hydrogen is a synthetic energy carrier. It carries energy generated by some other process. The combustion products of hydrogen when it is burned completely with air consist of water, oxygen, and nitrogen—although it has been suggested that hydrogen is too valuable to burn. Laboratory tests conducted on internal combustion engines burning hydrogen demonstrate good performance (Berry et al., 1996). In comparison with an engine burning gasoline, the emission of nitrogen oxides is far less for the engine fueled by hydrogen. The product of hydrogen combustion with air is water vapor and negligible pollution when the peak temperature is limited. Some oxides of nitrogen (NO_x) are formed at very high combustion temperatures (< 2300 K); fortunately, the autoignition temperature of hydrogen is only 858 K.

Hydrogen is emerging as the favorite alternative to fossil fuels as an energy carrier, since it has good properties as a motor fuel and it can be used in internal combustion engines of automobiles. Some of the characteristic properties of a hydrogen-air mixture that can definitely influence the engine design and performance are low ignition energy, low density, a wide range of ignition limits, high diffusion speed, and high flame speed (Plass et al., 1990).

The bond energy of the H-H bond is 431 kJ/mol. Electron affinity of hydrogen is low (–72 kJ/mol), where hydrogen reacts in this manner only with the most reactive metals. The ionization potential of hydrogen is relatively high (1312 kJ/mol). Hydrogen burns with a very hot flame and explosion in air:

$$2H_2 + O_2 \rightarrow 2H_2O + Q \tag{5.1}$$

The higher heating value (HHV) of hydrogen is $Q_{HHV} = 142.324$ kJ/kg or $Q_{HHV} = 284.648$ kJ/kg mol. The lower heating value (LHV) of hydrogen is $Q_{LHV} = 118.976$ kJ/kg mol.

The catch fire energy of hydrogen is the lowest compared to other fuels. While the ignition energy of a gasoline-air mixture is 0.2 and 0.4 MJ under atmospheric pressure and ambient temperature, it is about 0.02 MJ for hydrogen-air mixtures. The ignition barrier, as a percentage of hydrogen by volume in a hydrogen-air mixture, is found between the lowest limit of 4–10% and the highest limit of 60–75% (Caglar and Ozmen, 2000).

Hydrogen reacts with the reactive metals to produce metal hydrides:

$$2Na\ (s) + H_2\ (g) \rightarrow 2NaH \tag{5.2}$$

Hydrogen reacts with many metal oxides to produce water and the free metal:

$$CuO\ (s) + H_2\ (g) \rightarrow Cu\ (s) + H_2O\ (g) \tag{5.3}$$

$$WO_3\ (s) + 3H_2\ (g) \rightarrow W\ (s) + 3H_2O\ (g) \tag{5.4}$$

Some of these reactions are employed in the metallurgy of oxide ores, for example, in the commercial production of tungsten metal. WO_3 in Eq. 5.4 is reduced to the free metal by hydrogen.

CO and H_2 react at high temperatures in the presence of a catalyst to produce methanol:

$$CO\ (g) + 2H_2\ (g) \rightarrow CH_3OH\ (g) \tag{5.5}$$

Hydrogen reacts with nitrogen at high pressures (300 to 1000 atm) and at high temperatures (675 to 875 K), in the presence of a catalyst to produce ammonia:

$$3H_2\ (g) + N_2\ (g) \rightarrow 2NH_3\ (g) \tag{5.6}$$

The advantages of hydrogen as a universal energy medium are: (a) the combustion of hydrogen results in the formation of steam and liquid water; (b) it is nontoxic and, as an energy carrier, extremely environmentally benign since water is the only exhaust product when hydrogen is converted into energy; (c) it is possible to produce hydrogen from the most abundant chemical on earth: water; (d) hydrogen can be used as a feedstock for numerous chemicals; (e) it is the most suitable fuel for use in fuel cells; and (f) transmission of energy in the form of hydrogen is more economical than through high-voltage AC lines for large distances. Table 5.1 shows the fuel properties of hydrogen (Midilli et al., 2005).

Table 5.1 Fuel properties of hydrogen

Property	Unit	Value
Boiling point	K	20.41
Freezing point	K	13.97
Density (liquid)	kg/m^3	70.8
Specific heat at constant pressure	kJ/kg K	14.89
Explosion limits in air	% (vol.)	4–75
Ignition energy in air	mJ (millijoule)	0.02
Ignition temperature	K	585
Flame temperature in air	K	2318
Flame emissivity	%	17–25
Stoichiometric mixture in air	%	29.53
Stoichiometric air/fuel	kg/kg	34.3/1
Flame velocity	cm/s	2.75
HHV and LHV	MJ/kg	141.9–119.90
HHV and LHV	MJ/m^3	11.89–10.05

5.5 Hydrogen Production Processes

Today the two most common methods used to produce hydrogen are: (1) steam reforming of natural gas, and (2) electrolysis of water. The predominant method for producing synthesis gas is steam reforming of natural gas, although other hydrocarbons can be used as feedstocks. For example, hydrogen can be produced from the biomass reforming process.

Hydrogen is produced commercially in almost a dozen processes. Most of them involve the extraction of the "hydro" part from hydrocarbons. The most widely used, least costly process is steam reforming, in which natural gas is made to react with steam, releasing hydrogen (Hohhmann, 2002). While using steam to reform natural gas has proven the cheapest way to produce commercial hydrogen, natural gas is still a hydrocarbon and emits CO_2 in the conversion process (Rifkin, 2002; Arni, 2004). Global hydrogen is produced from natural gas, oil, coal, and water by electrolysis with percentages of 48, 30, 18, and 4%, respectively. The main hydrogen production processes are given in Table 5.2.

Table 5.2 Main hydrogen production processes

Method	Process	Feedstock
Thermal	Steam reformation	Natural gas
	Thermochemical water splitting	Water
	Pyrolysis	Biomass
	Gasification	Coal, biomass
Electrochemical	Electrolysis	Water
	Photoelectrochemical	Water
Biological	Photobiological	Water and algae strains
	Fermentative microorganisms	Biomass
	Anaerobic digestion	Biomass

5.5.1 Hydrogen from Natural Gas by Steam Reforming

Hydrogen can also be extracted or "reformed" from natural gas. The process of steam reforming of hydrocarbons, developed in 1924, is the main industrial method for production of hydrogen. In the steam reforming reaction, steam reacts with hydrocarbons in the feed to predominantly produce carbon monoxide and hydrogen, commonly called synthesis gas. Natural gas or methane reforming is widely used in industry to obtain hydrogen or syngas (H_2 + CO), which are utilized in industry.

Steam reforming of natural gas is currently the least expensive method of producing hydrogen, and used for about half of the world's production of hydrogen. Steam reforming of methane has been carried out at high temperatures (Rostrup-Nielsen, 1984). The steam reforming is a highly endothermic reaction of methane and steam whereas the partial oxidation is slightly exothermic:

$$CH_4 + H_2O \rightarrow CO + 3H_2 \qquad \Delta H^{\circ} = 206 \text{ kJ/mol} \qquad (5.7)$$

$$CH_4 + 0.5O_2 \rightarrow CO + 2H_2 \qquad \Delta H^{\circ} = -36 \text{ kJ/mol} \qquad (5.8)$$

CH_4 is a very stable molecule and has to be processed under very severe conditions. Although its conversion to synthesis gas can be conducted at temperatures even below 700 K, high yields to syngas need substantially higher temperatures, typically 1100 K. The reaction products H_2, CO, and H_2O are for all practical purposes stable at reaction conditions.

5.5.2 Hydrogen Production from Hydrocarbons

Hydrogen production by catalytic steam reforming and partial oxidation of hydrocarbons has been the most efficient, economically and widely used process for the production of hydrogen and hydrogen/carbon monoxide mixtures (synthesis gas or syngas). The process basically involves a catalytic conversion of the hydrocarbon and steam to hydrogen and carbon oxides. Since the process works only with light hydrocarbons, which can be vaporized completely without carbon formation, the feedstock used range from methane, to naphtha, to No. 2 fuel oil. The basic steps in conventional steam reforming are as follows:

1. Synthesis gas generation:
   A desulfurized hydrocarbon is mixed with process steam over a nickel-based catalyst in the reformer, where the endothermic reforming reaction occurs at 1175 K:

$$C_nH_m + nH_2O \rightarrow nCO + (n + m/2)H_2 \tag{5.9}$$

2. Supplemental hydrogen generation (water–gas shift):
   The synthesis gas enters the shift converter, where the exothermic water–gas shift reaction occurs at 475 to 675 K as follows:

$$CO + H_2O \rightarrow CO_2 + H_2 \tag{5.10}$$

3. Gas purification:
   CO_2 is removed in a scrubbing unit. Residual CO is converted in the methanator, where the following exothermic methanation reaction occurs:

$$CO + 3H_2 \rightarrow CH_4 + H_2O \tag{5.11}$$

The product hydrogen typically has a purity of 97–98%; 50% of the hydrogen is derived from H_2O when the hydrocarbon feedstock is methane and 64.5% when it is naphtha.

The typical composition of a synthesis gas at 0.7 MPa leaving a steam methane reformer is given in Table 5.3 (Caglar and Ozmen, 2000).

Hydrogen production by partial oxidation is similar to production by catalytic steam reforming. The process basically involves the conversion of steam, oxygen and hydrocarbons to hydrogen and carbon oxides. The process proceeds at moderately high pressures with or without a catalyst depending on the feedstock and process selected. The catalytic POX, which occurs at about 865 K, will work with feedstock ranging from methane to naphtha. The non-catalytic POX, which occurs

Table 5.3 Typical composition of synthesis gas at 0.7 MPa from a steam methane reformer

Component	Vol. (%)
H_2	34
CO	48
CO_2	17
N_2 and Ar	1
Total	100

at 1425–1590 K, can operate on hydrocarbons including methane, heavy oil, and coal. The synthesis gas step is slightly modified. A hydrocarbon feedstock is mixed with process steam and oxygen in the partial oxidation unit, and the following exothermic partial oxidation reactions occur:

$$C_nH_m + (n/2)O_2 \rightarrow nCO + (m/2)H_2 \quad (5.12)$$

$$C_nH_m + nH_2O \rightarrow nCO + (m/2)H_2 \quad (5.13)$$

$$CO + H_2O \rightarrow CO_2 + H_2$$

The product hydrogen typically has a purity of 93–98%.

There are three main steps:

1. Synthesis gas generation: Process in which light hydrocarbons are partially oxidized over a catalyst at about 875 K with oxygen and the carbon monoxide is shifted with steam to produce CO_2 and H_2.
2. Water–gas shift reaction: Processes in which heavy hydrocarbons such as heavy oil are partially oxidized without a catalyst present at about 1675 K, a temperature at which catalysts are not required to promote the reforming reaction.
3. Gas purification: Solid hydrocarbons such as coal are partially oxidized without a catalyst in the process.

The typical composition of the synthesis gas at 5.4 MPa leaving a POX reactor with heavy oil as feedstock is given in Table 5.4 (Caglar and Ozmen, 2000).

5.5.3 *Hydrogen from Coal*

Hydrogen production from carbonaceous feedstocks requires multiple catalytic reaction steps: For the production of high-purity hydrogen, the reforming of fuels is followed by two water–gas shift reaction steps, a final carbon monoxide purification and carbon dioxide removal. Steam reforming, partial oxidation and autothermal reforming of methane are well-developed processes for the production of hydro-

Table 5.4 Typical composition of synthesis gas at 5.4 MPa leaving POX reactor

Component	Vol. (%)
H_2	46
CO	46
CO_2	6
CH_4	1
N_2 and Ar	1
Total	100

gen. Stepwise steam reforming of methane for production of carbon monoxide-free hydrogen has been investigated at various process conditions by Choudhary and Goodman (2000). The process consists of two steps involving the decomposition of methane to carbon monoxide-free hydrogen and surface carbon in the first step followed by steam gasification of this surface carbon in the second step. The amount of carbon monoxide-free hydrogen formed in the first step hydrogen is produced in the second step of the reaction. The mixture of gases (primarily CH_4 and CO_2) can be separated or simply methanated and returned to the first step (Choudhary and Goodman, 1999).

Hydrogen derived from fossil fuels is not a clean, renewable resource. It may be easiest to sustain a transition to hydrogen by expanding commercially available, relatively inexpensive production processes, such as the manufacture of H_2 from coal. Coal is relatively abundant and could provide a low-cost feedstock for hydrogen. Coal represents a major energy source that can be transformed into transportation fuels and chemical feedstocks. A gasifier converts coal feedstock into gaseous components by applying heat and pressure to the coal in the presence of steam and oxygen.

A gasifier differs from a combustor in that the amount of oxygen inside the gasifier is carefully controlled such that only a relatively small portion of the fuel burns completely, minimizing the formation of carbon dioxide. The actual composition depends on the conditions in the gasifier and the type of feedstock. Typical coal syngas H_2 to CO ratios are in the 0.4:1 to 0.9:1 range. The gasification of coal is generally represented by the following reaction:

$$C + H_2O \rightarrow CO_2 + H_2 \tag{5.14}$$

The second step in the process consists of cleaning and shifting the CO to form hydrogen and CO_2 by reacting with steam. The overall reaction for converting the CO is the chemical shift reaction:

$$CO + H_2O \rightarrow CO_2 + H_2 \tag{5.15}$$

There are commercial or near commercial technologies for coal gasification processes used for hydrogen production. These are Koppers–Totzek and Texaco gasification processes.

In the Koppers–Totzek process, the polarized coal is rapidly partially oxidized with oxygen and steam at essentially atmospheric pressure under slogging con-

Table 5.5 Typical composition of synthesis gas from Koppers–Totzek gasifier

Component	Vol. (%)
H_2	29
CO	60
CO_2	10
N_2 and Ar	1
Total	100

Table 5.6 Typical composition of synthesis gas leaving the Texaco gasifier

Component	Vol. (%)
H_2	34
CO	48
CO_2	17
N_2 and Ar	1
Total	100

ditions. The typical composition of synthesis gas leaving the gasifier is given in Table 5.5 (Caglar and Ozmen, 2000).

The product hydrogen is about 2.8 MPa, with a purity higher than 97.5%. It is more efficient to gasify coal at elevated pressures.

The Texaco gasifier is a higher-pressure gasifier operating at around 5.5 MPa. By operating with a direct quench at this pressure, high steam content in the synthesis gas is desirable for use in the shift reaction for hydrogen production. The typical composition of synthesis gas leaving the Texaco gasifier is given in Table 5.6. The product hydrogen is about 4 MPa, with a purity higher than 97%.

5.5.4 Hydrogen from Water

Hydrogen can be obtained by direct electrolysis, by direct thermal conversion, thermochemically, photochemically, photoelectrochemically, and biochemically from water.

5.5.4.1 Direct Electrolysis of Water

Hydrogen production by electrolysis using electricity from fossil, nuclear, hydro, biomass, solar, and wind energy resources can be evaluated. But the hydrogen produced from electrolysis will in no way help reduce the pollution of the atmosphere if the electricity needed for the reaction is obtained through fossil fuels. Although presently not economically competitive with hydrogen produced from hydrocarbons, electrolytic hydrogen has several advantages such as high product purity and flexibility of operation. Non-polluting hydrogen would come from renewable sources like hydro, solar, wind and photocatalysis. Electrolysis produces extremely pure hydrogen, which is necessary for some types of fuel cells. But a significant amount of electricity is required to produce a usable amount of hydrogen from electrolysis.

Water electrolysis involves passing an electric current through water to separate it into hydrogen (H_2) and oxygen (O_2). Hydrogen gas rises from the cathode electrode and oxygen gas collects at the anode electrode. The reactions involved in the electrolysis of water are:

Reduction cathode electrode:

$$4H_2O + 4e \rightarrow 2H_2 + 4OH^- \quad (5.16)$$

Oxidation anode electrode:

$$4OH^- \rightarrow 2H_2O + O_2 + 4e \quad (5.17)$$

Complete cell reaction:

$$H_2O + \text{Electricity} \rightarrow H_2 + 1/2\ O_2 \quad (5.18)$$

The values of the cathode and anode half-cell potentials are known to be 0.401 V and –0.828 V, respectively, at 298 K at a pH of 14. If the activities of water and the gaseous species are considered to be equal to unity, the cathode (E_c) and anode (E_a) potentials required according to Nernst equation will be:

$$E_c = -0.828 \quad - 0.059 \log a_{OH^-}$$
$$E_a = 0.401 \quad - 0.059 \log a_{OH^-}$$

The potential required to split water into H_2 and O_2, i.e., ($E_a - E_c$) is equal to 1.229 V. Though the theoretical potential is 1.23 V for water electrolysis, in practice the actual water decomposition will occur only above 1.7 V. The extra potential, which is essential for the water decomposition, is called overpotential. Overvoltages are composed of activation or charge transfer overvoltage, concentration or diffusion or mass transfer overvoltage and resistance overvoltage. Overvoltage is evaluated mainly as a function of current and temperature (Viswanathan, 2006).

In general, an aqueous solution of caustic potash or soda is used as the electrolyte for water electrolysis. The nature of anode and cathode is decided based on their hydrogen and oxygen overvoltages in the electrolytic medium in addition to their stability in the particular medium. The cathode and anode are separated by a diaphragm, which prevents the mixing of hydrogen and oxygen gases produced at the cathode and anode surfaces respectively. The diaphragm should be stable in the electrolyte and minimizes the diffusion of gas molecules without affecting the conductivity of the medium. For ideal electrolyzers, heat is added to make up for the difference between the minimum electrical and total energy requirements. For practical electrolyzers enough heat is evolved due to cell inefficiencies to more than make up for the difference (Caglar and Ozmen, 2000; Viswanathan, 2006). When the temperature increases the reversible voltage decreases, whereas the thermoneutral voltage slightly increases with temperature. In general, the commercial industrial electrolytic cells operate between 333 and 353 K.

5.5.4.2 Direct Thermal Conversion of Water

Water can be decomposed thermally:

$$H_2O + \text{Heat} \rightarrow aH_2O + bH_2 \quad (5.19)$$

In Eq. 5.19, a and b are the mole fractions. At temperatures above 2500 K, the water molecules start dissociating into hydrogen and oxygen. This method (a) processes high thermal efficiency, (b) offers no or little environmental hazards, and (c) needs no immediate chemicals (Caglar and Ozmen, 2000). However, there are serious difficulties with direct thermal decomposition methods. Due to the increase of the kinetic energy of H_2O molecules with increasing temperature, this process requires more intensive energy.

5.5.4.3 Thermochemical Conversion of Water

The decomposition of water into hydrogen and oxygen can be achieved when energy is supplied in the form of heat and work. The positive value of ΔG^o decreases with increase in temperature, but rather slowly because of the nearly constant enthalpy change, as a function of temperature and ΔG^o becomes zero around 4700 K. This means that even the highest temperature available from a nuclear reactor, in the range of 1300 K, is not sufficient to decompose water. Therefore, single-step thermal decomposition of water is difficult unless other methods like electrolysis are then utilized. There is a two-step decomposition of water wherein a metal oxide, metal hydride or hydrogen halide is involved according to the equations:

$$H_2O + M \rightarrow MO + H_2 \tag{5.20}$$

$$MO \rightarrow M + 0.5O_2 \tag{5.21}$$

or

$$H_2O + M \rightarrow MH_2 + 0.5O_2 \tag{5.22}$$

$$MH_2 \rightarrow M + H_2 \tag{5.23}$$

However, even these two-step routes require temperatures on the order of 1273 K or more. Water cannot be decomposed in one or two thermochemical steps when the available temperature is below 1273 K.

High temperature (often exceeding 1000 K) drives the endothermic chemical reactions. Multistep cycles for water splitting are used because very high temperatures are required before an appreciable amount of water decomposes in single-step cycles. Thus, in one or more subsequent chemical reactions, the intermediary compounds can be recovered to the original substance, which is used repeatedly. The thermochemical water decomposition steps involve the following five principal reactions:

$$2H_2O + SO_2 + I_2 + 4NH_3 \rightarrow 2NH_4I + (NH_4)_2SO_4 \tag{5.24}$$

$$2NH_4I \rightarrow 2NH_3 + H_2 + I_2 \tag{5.25}$$

$$(NH_4)_2SO_4 + Na_2SO_4 \rightarrow Na_2S_2O_7 + H_2O + 2NH_3 \tag{5.26}$$

$$Na_2S_2O_7 \rightarrow SO_3 + Na_2SO_4 \tag{5.27}$$

$$SO_3 \rightarrow SO_2 + 0.5O_2 \tag{5.28}$$

These reactions take place at a range of temperatures: 325, 900, 675, 825, and 1140 K, respectively. In order to have this process viable 99.90 to 99.99% recovery must be achieved in each step to avoid the high cost of intermediary reagent replacement and to avoid the environmental problem of spoiling reagents, which cause pollution.

5.5.4.4 Photolytic Hydrogen Production

Photolytic hydrogen production processes use the energy in sunlight to separate water into hydrogen and oxygen. These processes are in the very early stages of research but offer long-term potential for sustainable hydrogen production with low environmental impact.

When water molecules adsorb energy at a rate of 68.3 kcal/mol of water from ultraviolet radiation, hydrogen in principle can be released. Photolysis, the splitting of water by light with the aid of photochemical electron transfer reagents analogous to chlorophyll has been described as "the most reagent solution" to the hydrogen production problem (Caglar and Ozmen, 2000).

5.5.4.5 Photochemical Hydrogen Production

A photochemical hydrogen production system is similar to a thermochemical system, in that it also employs a system of chemical reactants, which carry out the splitting of water. However, the driving force is not thermal energy but light, generally solar light. In this sense, this system is similar to the photosynthetic system present in green plants. One can effectively utilize photochemical means to promote endergonic (energy requiring) reactions. The sensitized oxidation of water by Ce^{4+} using irradiation of 254 nm light by the following reaction is known (Viswanathan, 2006):

$$Ce^{4+} + 0.5H_2O \rightarrow Ce^{3+} + 0.25O^2 + H^+ \qquad \Delta H = 3.8 \text{ kcal/mol} \tag{5.29}$$

Ce^{3+} can be used with light of shorter wavelength to promote the hydrogen generation reaction:

$$Ce^{3+} + H_2O \rightarrow Ce^{4+} + 0.5H_2 + OH^- \tag{5.30}$$

5.5.4.6 Photoelectrochemical Hydrogen Production

Photoelectrochemical (PEC) hydrogen production replaces one electrode of an electrolyzer with photovoltaic (PV) semiconductor material to generate the electricity needed for the water-splitting reaction. The efficiency loss of separate steps is done away with, as is the cost of the other components of a solar cell. PEC is elegantly simple, but finding PV materials both strong enough to drive the water split and stable in a liquid system presents great challenges for researchers. In its simplest

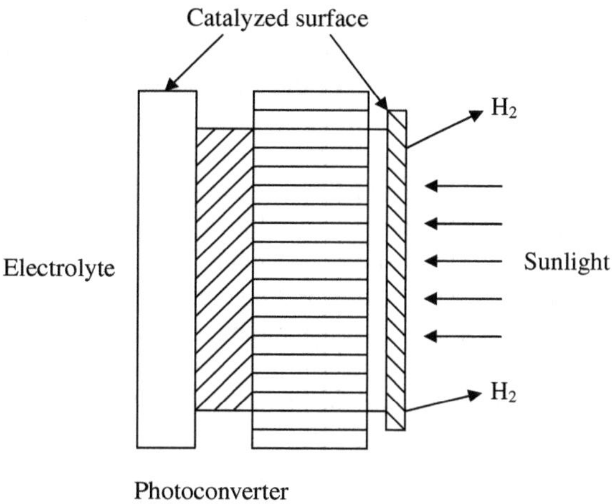

Fig. 5.3 Photoelectrochemical hydrogen production

form, a PEC hydrogen production cell consists of a semiconductor electrode and a metal counter electrode immersed in an aqueous electrolyte. When light is incident on the semiconductor electrode, it absorbs part of the light and generates electricity. This electricity is then used for the electrolysis of water.

In the illustration of a generic hydrogen photoelectrode shown in Fig. 5.3, sunlight shining on photoactive regions of the electrode produces an electric current to drive the hydrogen and oxygen evolution reactions (HER, OER) at opposite surfaces. Hydrogen photoelectrode operation represents a complex interaction of photovoltaic, optical, and electrochemical effects, and an important part of this research has been the development of integrated models combining these effects (Rocheleau and Vierthaler, 1994).

Iron oxide (Fe_2O_3) and tungsten oxide (WO_3) films have been studied and developed as candidate semiconductor materials for the PEC junction (photoanode). High-temperature synthesis methods, as reported for some high-performance metal oxides, have been found incompatible with multijunction device fabrication. A low-temperature reactive sputtering process has been developed instead. In the parameter space investigated so far, the optoelectronic properties of WO_3 films were superior to those of Fe_2O_3 films, which showed high recombination of photogenerated carriers (Miller et al., 2004).

5.5.5 Photocatalytic Hydrogen Production

Essentially, photocatalyzed reactions have generated considerable interest after the photocatalytic splitting of water on TiO_2 electrodes was first demonstrated by Fujishima and Honda in 1972. Subsequently, various kinds of photocatalysts have been employed for hydrogen production and remediation of pollutants from water.

Dispersed heterogeneous semiconductor surfaces provide a fixed environment that influences the chemical reactivity. Simultaneous oxidation and reduction reaction occurs on the surface of the catalyst during photoexcitation. The other advantages of this process are: the easy separation of catalyst after the reaction by centrifugation, availability of large surface area, low cost and stability. In heterogeneous photocatalytic systems, absorption of the light is an essential requirement for successful photocatalysis. In addition, it should be stable at the reaction conditions employed and it should be chemically inert. Among the available materials like metals, semiconductors and insulators, the semiconductors have been used because of their optimal band-gap, the band-edge positions are suitable for oxidation/reduction of water, and one can possibly use sunlight as an energy source to excite the electron from the valence band (Viswanathan, 2006).

Hydrogen has the potential to play an important role in future energy systems because of the possibility of producing hydrogen by splitting water using photocatalysts and solar energy. Photocatalytic water splitting with the use of solar energy is an attractive route for energy conversion because it converts water directly into hydrogen, the critical energy carrier of the future.

The method of catalytic photolysis of water, leading to obtainment of hydrogen, by using the oxide semiconductors, is an economic and clean technology; there are unlimited stores of material (water) and the needed energy (Sun). Therefore this method is very attractive. However, presently, renewable energy contributes only about 5% of the commercial hydrogen production primarily via water electrolysis, while the other 95% of hydrogen is mainly derived from fossil fuels (Ni et al., 2004). Photovoltaic water electrolysis may become more competitive as the cost continues to decrease with the technology advancement. Figure 5.4 shows the mechanism of dye-sensitized photocatalytic hydrogen production under visible light irradiation (Ni et al., 2007).

Some photocatalysts are used to adsorb visible light and then transmit the energy of appropriate wavelength and intensity to water molecules to liberate the constituent gases. The photolysis with a photo catalyst "X" can be expressed as follows:

$$H_2O + X \quad + \text{Light} \rightarrow \text{Reduced X} + 2H^+ + 0.5O_2 \tag{5.31}$$

$$\text{Reduced X} \quad + 2H^+ \quad \rightarrow X + H_2 \tag{5.32}$$

$$\text{Overall reaction:} \quad H_2O + X + \text{Light} \rightarrow H_2 + 0.5O_2 + X \tag{5.33}$$

The photocatalysts "X" are some compound salts, compound semiconductors, photosynthetic dyes, some intact cells of some species of blue-green, green and red algae, or some photosynthetic bacteria (Caglar and Ozmen, 2000). Unfortunately, these systems typically are very inefficient, utilizing less than 1% of the incoming energy, and thus are very costly; although recent developments suggest that improvements are possible.

Production of hydrogen from an inexhaustible source, water, by a cheaper route has been under extensive investigation in recent years (Koca and Sahin, 2003). The requirement for the photoproduction of hydrogen using a semiconductor is the need for a hydrogen evolution catalyst on a semiconductor surface as reported by many

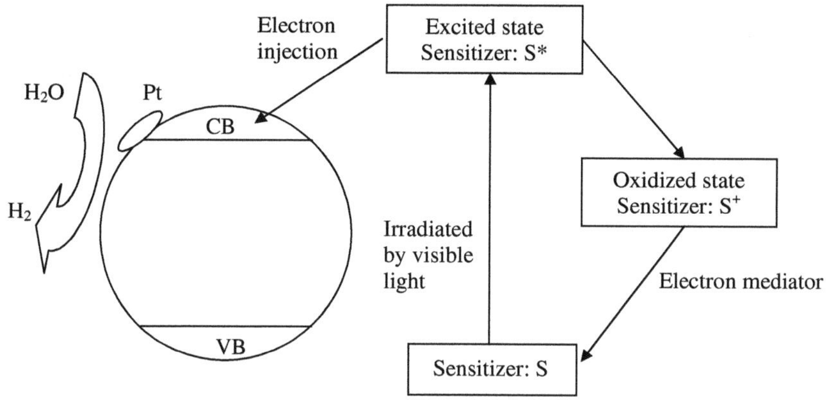

Fig. 5.4 Mechanism of dye-sensitized photocatalytic hydrogen production under visible light irradiation

research groups (Domen et al., 1982; Gurunathan, 2000). In view of the current importance of hydrogen as a clean chemical source, the discovery of new kinds of the photocatalysts for water decomposition is a very important issue (Sato et al., 2001). Selection of a suitable photocatalyst is an important criterion to establish the process as workable for the maximum quantity of hydrogen production. Extensive research has so far been performed, but stable photocatalysts found in the past two decades for the overall splitting of pure water to produce hydrogen and oxygen are confined to the transition metals involving Ti^{4+}, Zr^{4+}, Nb^{5+}, and Ta^{5+}; $SrTiO_3$ (Domen and Kudo, 1986), $A_2Ti_6O_{13}$ (A=Na, K, Rb) (Inoue et al., 1991; Ogura et al., 1997), $BaTi_4O_9$ (Kohno et al., 1998), $A_2La_2Ti_3O_{10}$ (A=K, Rb, Cs) (Takata et al., 1997a, 1997b), ZrO_2 (Sayama and Arakawa, 1993), $A_4Nb_6O_{17}$ (A=K, Rb) (Kudo et al., 1988), $Sr_2Nb_2O_7$ (Kudo et al., 2000), $ATaO_3$ (A=Na, K) (Kato and Kudo, 1999a; Ishihara et al., 1999), MTa_2O_6 (M=Ca, Sr, Ba) (Kato and Kudo, 1998, 1999b), and $Sr_2Ta_2O_7$ (Kudo et al., 2000) were active when combined with NiO or RuO_2 as a promoter (Sato et al., 2001). It will be greatly beneficial to demonstrate that metal oxides having other electronic structures are useful for photoassisted water decomposition. The development of visible light-active photocatalysts, therefore, has become one of the most important topics in photocatalysis research today (Jang et al., 2005).

Fujishima and Honda (1972) reported the photocurrent generation using semiconductor electrodes. Since then, scientific and engineering interests in semiconductor photocatalysis have grown significantly. In this system, electrons and holes are produced in a photocatalyst by incident light if the light energy is greater than the band-gap of the photocatalyst. Then, photogenerated electrons and holes react with water to produce H_2 and O_2, respectively, theoretically provided that the conduction band potential of the photocatalyst is more negative than the redox potential of H^+/H_2 (0 V vs. NHE; pH=0). The valence band potential is more positive that the redox potential of O_2/H_2O (1.23 V vs. NHE; pH=0) (Bak et al., 2002; Kudo, 2001;

Kida et al., 2004). Semiconductors have been utilized for this purpose in the form of electrodes (Desilvestro and Neumannspallart, 1985; Mackor and Blasse, 1981; Gringue et al., 1987; Ashokkumar et al., 1994), colloids (Kennedy and Dunnwald, 1983; Lee et al., 1984; Kamat and Fox, 1983; Kamat, 1989), powders (Ashokkumar and Maruthamuthu, 1989; Herrmann et al., 1986; Okamoto et al., 1985; Oosawa, 1984) and thin films (Fonash, 1981; Green, 1982; Faherburch and Bube, 1983).

The performance of semiconducting inorganic materials as a photocatalyst depends much on the physical properties of the particles such as crystallinity, size and morphology. The effect can be even more pronounced when using two or more materials together, since an intimate interaction between particles is more important in the multicomponent system (So et al., 2004).

Many chemical reactions have been tested using heterogeneous photocatalysts. These catalytic systems adsorb sunlight, transfer electrons and promote redox reactions (Cerveramarch, et al., 1992; Gurunathan, 2000; Moon et al., 2000; Tryk et al., 2000; Bamwenda and Arakawa, 2000; Namon et al., 1986; Gurunathan et al., 1997; Fletcher et al., 1984). Many photocatalyst systems have been tried for the production of hydrogen from water. Most of the photocatalyst systems studied recently include materials such as TiO_2 and B/TiO_2 (Moon et al., 2000), CeO_2–Ce^{+4}–Fe^{+3} (Bamwenda and Arakawa, 2000), CdS (Cerveramarch et al., 1992; De et al., 1996) or V_2O_4 (Namon et al., 1986). Gurunathan (2000) sensitized TiO_2 with CdS, Cu(II) ions, $Ru(bpy)_3^{+2}$, organic dyes and other photosensitizers and recorded 0.0225 ml/h g as the maximum hydrogen evolution rate. These are the ones that yielded maximum rates.

Table 5.7 Effect of the composition of CdS/ZnS photocatalyst system to the rate of hydrogen evaluation

CdS:ZnS	Wt.% CdS (mg)	Wt.% ZnS (mg)	Rate of H_2 (ml/h g)
1:0	100.0	00.0	2.80
4:1	80.0	20.0	4.20
3:1	75.0	25.0	8.30
2:1	66.6	33.4	15.60
3:2	60.0	40.0	12.50
1:1	50.0	50.0	5.40
2:3	40.0	60.0	4.50
1:2	33.4	66.6	3.80
1:3	25.0	75.0	3.30
1:4	20.0	80.0	1.50
0:1	00.0	100.0	0.00

5.5.5.1 Photocatalytic H_2 by Direct Sunlight from Sulfide/Sulfite Solution

The photocatalytic hydrogen production from a sulfide/sulfite solution is one of the photocatalytic processes of interest recently for hydrogen production. A different situation would be where waste materials have to be eliminated or used to produce energy and useful products. Sulfide and sulfite, undesirable waste products in fossil fuel technology, should be explored as a source of hydrogen. In that case, most of the raw materials could be considered as zero or negative. For instance, a large amount of sulfide and sulfite produced in oil refineries is inconvenient, but can be used as a sacrificial material to produce hydrogen by photocatalytic reactions (Fletcher et al., 1984).

Koca and Sahin (2002) carried out different series of experiments in order to compare the amounts of hydrogen evolved by different compositions of CdS/ZnS. Table 5.7 shows the rates of hydrogen evolution, which was dependent upon the percentage of CdS and ZnS in each photocatalyst. These results show that 33% ZnS is the most active (Koca and Sahin, 2002). CdS was sensitized with a co-precipitation by using a hot mixture of $Cd(CH_3COOH)_2$, $Zn(CH_3COOH)_2$ and Na_2S; Zn^{+2} ions loaded on CdS, and this enhanced hydrogen production efficiency of the photocatalyst system.

5.5.5.2 Photocatalytic H_2 from Water over Titania Aerogels Under UV Irradiation

Aerogels are light solid materials. Their unique properties include extremely low densities combined with large porosities and surface areas. One potential application of aerogels is in photocatalytic processes like photodecomposition of organic pollutants in water or air. For such application a semiconducting material is needed as a solid matrix of aerogel. Data from literature point to titania and hybrid titania-silica aerogels as attractive photocatalysts or carriers for further doping with metals. Titania aerogels were used in photodegradation of such compounds like p-chlorophenol, 4-hydroxybenzoic acid (Malinowska et al., 2003a) and p-nitrophenol (Malinowska et al., 2003b). Dagan and Tomkiewicz (1994) proved that titania aerogels can be much more active in the photodegradation process of the simple aromatic compound, salicylic acid, in comparison with commercial P25 Degussa TiO_2.

Piotrowska and Walendziewski (2005) carried out experiments with these aerogels in the process of photocatalytic hydrogen production. The highest activity they have obtained was using titania aerogels, with platinum in the aqueous solution of methanol (10 ml CH_3OH/400 ml H_2O). After 55 min of the tests exposure they obtained 6.59 mmol of H_2 over titania aerogel. Titania aerogels impregnated with platinum show a similar efficiency in the process of photocatalytic hydrogen production by water splitting as commercial TiO_2 P25 impregnated with Pt. The obtained results indicated that titania aerogels are a promising catalytic material in photocatalytic production. The type of solvent used in aerogels preparation has no influence on catalytic activity. Platinum deposited over titania aerogel gives a higher activity water-splitting photocatalyst in comparison to palladium and nickel deposited over titania aerogel (Piotrowska and Walendziewski, 2005).

5.5.5.3 Photocatalytic H_2 from Suspension of Spinel Powders AMn_2O_4

Visible light-induced hydrogen evolution over the spinel oxide AMn_2O_4 (A=Cu, Zn) has been investigated in the presence of sulfide and/or sulfite as hole scavengers (Bessekhouad and Trari, 2002). The platinized catalysts showed a better performance, with an increase of 30% of H_2 production obtained. Over time, the oxidation of sulfide does not maintain a constant hydrogen evolution because of the formation of the yellow polysulfide S_n^{2-} ions that form an optical filter reducing the light absorption of AMn_2O_4 and competing with the reduction of H_2O molecules. SO_3^{-2} acts as a regenerative agent and keeps the solution transparent. At lower temperatures, SO_3^{-2} worked as a hole scavenger and exhibited a higher gas production. Higher specific surface area of $ZnMn_2O_4$ powder, obtained by co-precipitation, does not improve the H_2 production and suggests that both the depletion layer thickness and the carrier diffusion length are smaller than the average crystallite size.

5.5.5.4 Zirconium-Titanium Phosphates for H_2 by Photo-Induced Water Splitting

Due to the UV light absorption property of titanium phosphate containing mesoporous materials with large surface areas and ion exchange capacity, they are candidates for use as photocatalysts, especially for hydrogen generation by water decomposition (Bhaumik and Inagaki, 2001). In the case of mesoporous zirconium-titanium phosphates (ZTP), the homogeneous pore distribution including the large internal surface area makes such materials accessible to the water molecule, and a highly charged structure existing on the surface facilitates the charge separation process essential for the decomposition of water. Inagaki and co-workers examined the photocatalytic splitting of water for hydrogen production using Pt-loaded ZTP mesoporous materials under UV-vis light irradiation. The results of photocatalytic decomposition of water for hydrogen generation evidently suggest that mesoporous ZTP materials are possible candidates for photocatalysts for hydrogen production (Kapoor et al., 2005).

5.5.5.5 Light-Induced Production of H_2 from Water by Catalysis

A ruthenium-containing melanoidin (a condensation product of amino acids and carbohydrates) was found to photocatalyze hydrogen production from water under light irradiation with $\lambda < 320$ nm, in the presence of EDTA as an electron donor and methyl viologen as an electron relay. The Ru-melanoidin photocatalyst is stable under prolonged irradiation times and, due to its insolubility in water, is suitable for recycling. Ru-containing melanoidin can act as a cheap, stable and reasonably efficient photosensitizer for photocatalytic hydrogen production from water. It is also conceivable that a support mounted catalyst system would allow hydrogen production in a flow system (Serban and Nissenbaum, 2000).

Cadmium sulfide is one of the most well-studied semiconductors for solar energy utilization. It possesses a relatively narrow band-gap of approximately 2.4 eV; its conduction band-edge is more negative than the H_2/H_2O redox potential, thus it is able to evolve H_2 from water when irradiated with visible light (Buhler et al., 1984). However, the activity of CdS is generally poor in the absence of electron donors such as Na_2S and H_3PO_2, because photogenerated holes in the valence band tend to react with CdS itself (photocorrosion). In an attempt to improve its activity, $LaMnO_3$/CdS nanocomposite was developed as a new type of visible light-sensitive photocatalyst that can produce hydrogen from water containing electron donors. The nanocomposite catalyst, prepared by a reverse micelle method, was found to show high photocatalytic activity, as compared to that of CdS; on the other hand, $LaMnO_3$ showed no photocatalytic activity. The proposed energy diagrams for the $LaMnO_3$/CdS composite, based on the results obtained from photoemission yield measurements, suggest that photogenerated holes in the valence band of CdS could move to that of $LaMnO_3$, thus separating photogenerated charge carriers and improving the activity (Kida et al., 2003).

5.5.5.6 Production of Hydrogen from Photocatalytic Water Splitting by Using TiO_2

Photocatalytic water splitting using TiO_2 for hydrogen production offers a promising way for clean, low-cost and environmentally friendly production of hydrogen by solar energy. TiO_2 has been widely used in photocatalysis, because of its favorable band-gap energy (3.2 eV in anatase) and its high stability in aqueous solution under UV irradiation (Galinska and Walendziewski, 2005).

It is well-known that nano-TiO_2 is one of the suitable semiconductors for photocatalyst and has been applied in various photocatalytic reactions (Fujishima et al., 2000). However, its properties, not only the photoefficiency or activity but also the photoresponse, are not sufficient (Kawai and Sakata, 1980). Meanwhile, the high recombination ratio of photoinduced electron–hole pairs also reduces its catalytic efficiency. Therefore various modifications have been performed on nano-TiO_2 to promote its catalytic ability and develop new photocatalytic functions (Ohno et al., 1996; Litter, 1999; Navvio et al., 1999; Choi et al., 1994; Nishikawa et al., 2001; Amiridis et al., 1999).

Since the discovery of photoelectrochemical splitting of water on titanium dioxide (TiO_2) electrodes (Fujishima and Honda, 1972), semiconductor-based photocatalysis has received much attention. Although TiO_2 is superior to other semiconductors for many practical uses, two types of defects limit its photocatalytic activity. Firstly, TiO_2 has a high band-gap (E_g=3.2 eV), and it can be excited only by UV light (λ < 387 nm), which is about 4–5% of the overall solar spectrum. Thus, this restricts the use of sunlight or visible light (Kormann et al., 1988). Secondly, the high rate of electron-hole recombination at TiO_2 particles results in a low efficiency of photocatalysis (Linsebigler et al., 1995). To overcome these limitations of TiO_2, many attempts have been made, for example, depositing of noble metals, mixing metal oxides with TiO_2 and doping selective metal ions into the TiO_2 lattice (Anderson and Bard, 1995; Borgarello et al., 1982; Ranjit et al., 1999; Moon et al., 2001). The advantage of doping the metal ions into TiO_2 is the temporary trapping of the photogenerated charge carriers by the dopant and the inhibition of their recombination during migration from inside the material to the surface. The effect of metal ion doping strongly depends on many factors such as the dopant concentration, the distribution of the dopants, the configuration of doping ions and so on (Borgarello et al., 1982). There have been many studies on transition metal, rare-earth and noble metal ion dopants in TiO_2, but reports on alkaline-earth metal ion-doped TiO_2 and their photocatalytic properties have seldom been seen (Talavera et al., 1997).

Kiwi and Morrison (1984) investigated the effect of lithium doping on TiO_2, towards water photocleavage experiments. It showed that Li doping promoted conduction band electron-transfer. This effect was likely to contribute to the observed enhancement in hydrogen generation and oxygen photoadsorption with increased Li content. Displacement of the conduction band of TiO_2 samples to more positive values took place upon Li doping (Kiwi and Morrison, 1984).

Abe et al. (1999) illuminated an aqueous suspension of TiO_2/Pt with an Hg arc lamp and observed H_2 production in the absence of O_2 production. An unspecified elemental analysis excluded carbonaceous contaminants, and the source of the electron donor was concluded to be Ti^{4+} cations (Abe et al., 1999).

Sodium carbonate (Na_2CO_3) addition to a Pt/TiO_2 suspension significantly promotes stoichiometric photodecomposition of liquid water (Sayama and Arakawa, 1992a). This procedure was also applicable to many other photocatalyst systems such as M/TiO_2, M/Ta_2O_5, M/ZrO_2, M/$SrTiO_3$, etc. (M = metal or metal oxide) (Sayama and Arakawa, 1992b; Sayama and Arakawa, 1996). Water was decomposed efficiently and stoichiometrically to H_2 and O_2 using a 3 wt.% NiO_x/TiO_2 photocatalyst under real solar light irradiation by this Na_2CO_3 addition method (Arakawa and Sayama, 2000). The main role of Na_2CO_3 for water splitting is the significant acceleration of O_2 desorption from the oxide semiconductor surface via peroxycarbonate intermediates which are formed by the reaction of surface carbonate species and positive holes in the valence band area of the oxide semiconductor catalyst by photoexcited charge separation under irradiation.

Wider band-gap materials such as TiO_2, ZnO, $SrTiO_3$ and ZnS showed good photostability but lower efficiency due to limited light harvesting when using the solar light (Reber and Meier, 1984). Small and medium band-gap materials such as CdS, CdSe, MoS_2 and Cu_2O experienced severe photoanodic corrosion. Thus, one of methods to overcome these drawbacks was to mix two or more semiconducting materials with different band-gaps and band positions so that an effective charge separation could occur between the contacting particles (Serpone et al., 1984). The intercalation of nanosized Fe_2O_3, TiO_2, and CdS particles into the interlayer of layered compounds such as $HNbWO_6$ or $HTaWO_6$ has been reported (Wu et al., 1999a,

Table 5.8 Photocatalytic hydrogen production from water over Cr- and Fe-doped $La_2Ti_2O_7$ under visible light irradiation ($\lambda > 420$ nm)

Photocatalysts	H_2 production rate[a] (μmol/h)
$Pt/La_2Ti_2O_7$	0
$Pt/Cr\text{-}La_2Ti_2O_7$ (Cr/La=0.01)	15
Pt/Cr, $Sb\text{-}La_2Ti_2O_7$ (Cr/La=0.01)	4
$Pt/Cr\text{-}La_2Ti_2O_7$ (Cr/La=0.05)	8
$Pt/Fe\text{-}La_2Ti_2O_7$ (Fe/La=0.01)	10
$Pt/Fe\text{-}La_2Ti_2O_7$ (Fe/La=0.05)	5

[a]Measured in outer irradiation cell from a 500 W high-pressure Hg lamp; H_2O 100 ml + MeOH 50 ml, catalyst 0.5 g, 420 nm cutoff filter was used; 1 wt.% of Pt was loaded using H_2PtCl_6 by photoplatinization

1999b, 2001). The photoactivity improved markedly over those of unintercalated photocatalyst particles, which was ascribed to an effective separation of photogenerated electrons and holes due to their rapid diffusion (Jang et al., 2005). The intercalated nanoparticles were the production of hydrogen from water containing a sacrificial agent (CH_3OH) under visible light irradiation with quantum yields as high as 10% (Wu et al., 2001).

Wu et al. (1999b) reported that the intercalated Fe_2O_3 showed photocatalytic activity of H_2 evolution from a CH_3OH–H_2O solution by transfer of the photoelectron from Fe_2O_3 to the host layered compound, which decreased the recombination of electrons and holes under visible light irradiation. However, it was found in this work that hydrogen evolution under visible light was not due to photocatalytic decomposition of water (Wu et al., 1999b).

Photocatalytic water splitting is observed at the presence of applying EDTA and Na_2S as the sacrificial reagents. They act as effective hole scavengers; however, they are oxidized due to OH radicals, preventing oxygen formation and the recombination of the reaction of oxygen with hydrogen (Galinska and Walendziewski, 2005).

A few oxides are also active under visible light such as $HPb_2Nb_3O_{10}$ (Yoshimura et al., 1993), $MgWO_x$ (Hwang et al., 2002), and $Ni_xIn_{1-x}TaO_4$ (Zou et al., 2001), yet their activities are very low. Recently, some UV-active oxides turned into visible light photocatalysts by substitutional doping of C, N, and S, i.e., $TiO_{2-x}N_x$ (Asahi et al., 2001), $TiO_{2-x}C_x$ (Khan et al., 2002), TaON (Hitoki et al., 2002) and $Sm_2Ti_2O_5S_2$ (Ishikawa et al., 2002). Lee and co- workers studied cation-hybrid photocatalysts based on $La_2Ti_2O_7$ (Lee et al., 1984). Among various types of dopants, only Cr and Fe showed intense absorption in the visible light region ($\lambda > 420$ nm), over which H_2 was produced photocatalytically in the presence of methanol. Table 5.8 shows photocatalytic hydrogen production values from water over Cr- and Fe-doped $La_2Ti_2O_7$ under visible light irradiation ($\lambda > 420$ nm) (Hwang et al., 2004).

5.5.5.7 H_2 from Photocatalytic Water Splitting Using Photobiocatalytic Method

Hydrogen can be generated from water photocatalytically, photobiologically, and photoelectrochemically. The photobiocatalytic method is a novel method of hydrogen production by coupling an inorganic semiconductor with a bacterial enzyme (Gurunathan, 2000). Some researchers have reported on photocatalytic hydrogen production with Cu(II)/WO_3 (Maruthamuthu and Ashokkumar, 1988; Maruthamuthu et al., 1989), Cu(II)/Bi_2O_3 (Maruthamuthu et al., 1993), Rh[III]/Fe_2O_3 (Gurunathan and Maruthamuthu, 1995), Pt/SnO_2 (Gurunathan et al., 1997) using methyl viologen and reported photobiocatalytic hydrogen production using undoped Bi_2O_3 coupled with photosynthetic bacteria (Maruthamuthu et al., 1992). It is discussed that the nitrogenase enzyme of the bacterial cells is responsible for catalyzing hydrogen production. Sensitization of TiO_2 was performed in three ways:

1. Using organic dyes
2. Using Cu(II) ion doping
3. Loading with low band-gap semiconductors (CdS)

In the four components, i.e., TiO_2/MV^{2+}/electron donor/bacterial cells, each of the last three components has its own specific function and each facilitates the other's role, thereby enhancing the yield of hydrogen production. It was found that with sensitized TiO_2, there is a higher amount of hydrogen production than with the naked TiO_2. Among the sensitizers used, Rhodamine B and $Ru(bpy)_3^{2+}$ exhibited higher efficiencies compared with other sensitizers, as well as other methods of sensitization (2 and 3) (Gurunathan, 2000).

Nikandrov et al. (1988) coupled bacterial enzyme from *Thiocapsa roseopersina* to TiO_2 and observed an efficient photobiocatalytic hydrogen production. Thus a better catalyst has an origin in biology, i.e., bacterial enzyme which catalyzes the hydrogen evolution.

TiO_2 was coupled to intact bacterial cells of *Rhodopseudomonas capsulata* species as photocatalyst for hydrogen production using light of $\lambda > 400$ nm. The semiconductor-bacterial cells system is then subjected to hydrogen in aqueous solution under visible light using MV^{2+} (methyl viologen dichloride hydrate) as an electron relay with or without an electron donor (Gurunathan, 2000).

5.5.5.8 Photocatalytic H_2 from Sewage Sludge

Among various types of biomass, sewage sludge is more difficult to utilize due to its lesser quality. Nevertheless, some attempts have been made for converting sewage sludge into fuel, e.g., by thermochemical pyrolysis (Urban and Antal, 1982; Stolarek and Ledakowicz, 2001; Midilli et al., 2002) or solubilization methods (Yokoyama et al., 1987; Suzuki et al., 1988; Suzuki et al., 1990; He et al., 2000). Kida et al. (2004) investigated photocatalytic hydrogen production from a digested sewage sludge solubilized in hot-compressed water (573 K) in order to develop a low-cost sacrificial agent for CdS-based photocatalysts from biomass. H_2 evolution occurred over a $LaMnO_3$/CdS composite photocatalyst under Xe lamp irradiation from water containing the solubilized sewage sludge. The amount of evolved H_2 reached more than 30 mmol/g catalyst for a 200 h reaction; on the other hand, no H_2 was formed in the absence of the solubilized sewage sludge. The H_2 evolution rate was comparable to that when typical Na_2S–Na_2SO_3 sacrificial agents were used, suggesting the applicability of a biomass-derived sacrificial agent for photocatalysis. Organic compounds, such as methanol and formic acid, contained in the solubilized sewage sludge are responsible for the H_2 evolution observed (Kida et al., 2004).

5.5.6 Solar-Powered Hydrogen Generation

Solar radiation is the largest renewable source on earth. A comparative overview of the selected integrated systems indicated that the Sun is the primary source of energy for many of the hydrogen demonstration projects. Electrochemical water splitting powered by photovoltaic arrays produces molecular hydrogen at the cathode, while organic compound oxidation under mild conditions takes place at the anode in competition with the production of oxygen. Hydrogen can be obtained from water by directly connecting a photovoltaic module to a hydrogen generator, with a solid polymer electrolyte (SPE).

The electrolysis of water using cells with a SPE is a very efficient method of producing hydrogen. SPE water electrolysis is one of the ways for producing hydrogen with comparatively high levels of efficiency and with compact equipment. SPE is composed of a membrane, cathode and anode which produce hydrogen by providing pure water to one side of a polymer ion exchange film, which is put between the anode and cathode.

If hydrogen is produced via solar photovoltaic (PV)-powered water electrolysis, it would be possible, in principle, to provide energy on a global scale, with essentially no greenhouse gas emission and very low local pollution.

Accordingly, the operation of electrolyzers with intermittent sources of solar energy and the various possibilities for matching photovoltaic current with the characteristics of the electrolyzer was one of the recurrent design issues in such projects. Most of the electrolyzers were of the alkaline type and operated at low pressure.

Concentrated solar radiation can be used for gasification of biomass to produce hydrogen. A detailed review with many references of the technology describes solar gasification of carbonaceous materials to produce a syngas quality intermediate for production of hydrogen and other fuels (Midilli et al., 2000). Shahbazov and Usupov (1994) have shown a good yield of hydrogen from agricultural wastes using a parabolic mirror reflector.

5.5.7 Hydrogen from Hydrogen Sulfide

Sulfur is recovered from H_2S by partial oxidation by the Claus process, and low-quality steam is produced utilizing the heat of reaction. Efforts to produce both hydrogen and sulfur from hydrogen sulfide have been made in recent years through a diverse variety of technologies. These involve thermal, thermochemical, electrochemical, photochemical and plasmochemical methods (Zaman and Chakma, 1995). H_2S potentially has economic value if both sulfur and hydrogen can be recovered. Various techniques that are at different stages of development can be considered for the decomposition of H_2S and they can be basically grouped into categories of thermal, thermochemical, electrochemical, photochemical and plasmochemical methods (Baykara et al., 2005, 2007).

Thermolysis of H_2S was carried out in an open tubular reactor quartz tube with argon/H_2S feed over a wide composition spectrum (20–100% H_2S) at four temperatures (1030–1070 K). These experiments show that the reaction is essentially first order in H_2S partial pressure. Hydrogen yield also increases monotonically with feed composition at all temperatures (Adesina et al., 1995).

Petrov (1991) suggested the idea for using H_2S for production of hydrogen and sulfur in the electrolytic stage of the solar-wind-hydrogen energy systems (SWHES).

A thermodynamic model for the partial oxidation of H_2S was used to study product compositions attainable in the superadiabatic partial oxidation regime. Superadiabatic partial oxidation techniques permit attainment of operating temperatures significantly in excess of the adiabatic temperature of the incoming reactants (Slimane et al., 2004a).

Superadiabatic partial oxidation (combustion) process is the thermal decomposition of H_2S in H_2S-rich waste streams to high-purity hydrogen and elemental sulfur (Slimane et al., 2002). In the superadiabatic combustion (SAC) process, some part of the H_2S is combusted to provide the thermal energy required for the decomposition reaction, as indicated by the following two reactions:

$$2H_2S + O_2 \rightarrow 2S + H_2O \qquad \Delta H = -189\,269 \text{ kJ/mol} \tag{5.34}$$

$$H_2S \rightarrow H_2 + S \qquad \Delta H = +17\,423 \text{ kJ/mol} \tag{5.35}$$

As can be seen from Eqs. 5.34 and 5.35, each molecule of H_2S reacting with oxygen can provide enough energy to dissociate up to 10 additional molecules of H_2S in an available reactor under suitable conditions.

The SAC, also known as filtration combustion, consists of combustion of a fuel gas-oxidant mixture in a porous ceramic medium with a high thermal capacity (Kennedy et al., 1995). The intense heat exchange between burning gas mixture and the porous medium permits the accumulation of combustion energy in the solid matrix. Figure 5.5 shows hydrogen recovering from H_2S via the superadiabatic combustion (SAC) process.

5.5.7.1 Sulfur–Iodine Thermochemical Water-Splitting Process

The sulfur–iodine thermochemical water-splitting cycle (S–I cycle) developed for hydrogen production from water is fundamentally based on the following three chemical reactions (Wang, 2007):

$$H_2SO_4 \rightarrow H_2O + SO_2 + 0.5O_2 \quad (5.36)$$

$$2H_2O + SO_2 + I_2 \rightarrow H_2SO_4 + 2HI \quad (5.37)$$

$$2HI \rightarrow H_2 + I_2 \quad (5.38)$$

This process proposes to replace the H_2SO_4 decomposition with a reaction between H_2S and H_2SO_4 and the replacement gives rise to a H_2S-splitting cycle that produces H_2 and elemental S from H_2S, shown as follows:

$$H_2S + H_2SO_4 \rightarrow S + SO_2 + 2H_2O \quad (5.39)$$

$$2H_2O + I_2 + SO_2 \rightarrow H_2SO_4 + 2HI \quad (5.40)$$

$$2HI \rightarrow H_2 + I_2 \quad (5.41)$$

Combined with the reactions such as:

$$O_2 + S \rightarrow SO_2 \quad (5.42)$$

$$SO_2 + 0.5O_2 \rightarrow SO_3 \quad (5.43)$$

$$SO_3 + H_2O \rightarrow H_2SO_4 \quad (5.44)$$

This new cycle cannot only produce more H_2 and extra H_2SO_4 but also facilitate a flexible H_2 to H_2SO_4 production ratio. In gas plants, H_2 from H_2S splitting is an alternative clean fuel. Environmentally, H_2 production based on this H_2S-splitting cycle is carbon-free.

The sulfur–iodine (S–I) thermochemical water-splitting cycle is one of the most studied cycles for hydrogen (H_2) production (Huang and Raissi, 2005). The S–I cycle consists of four sections: (I) acid production and separation and oxygen purification, (II) sulfuric acid concentration and decomposition, (III) hydroiodic acid (HI) concentration, and (IV) HI decomposition and H_2 purification. Section II of the cycle is an endothermic reaction driven by the heat input from a high-temperature source. Analysis of the S–I cycle in the past thirty years has focused mostly on the utilization of nuclear power as the high-temperature heat source for the sulfuric acid decomposition step. Thermodynamic as well as kinetic considerations indicate that both the extent and rate of sulfuric acid decomposition can be improved at very high temperatures (in excess of 1275 K) available only from solar concentrators. These new sulfuric acid decomposition processes are simpler and more stable than previous processes and yield higher conversion efficiencies for the sulfuric acid decomposition and sulfur dioxide and oxygen formation. Figure 5.5 shows the sulfur–iodine cycle for hydrogen production (Huang and Raissi, 2005; Balat and Ozdemir, 2005).

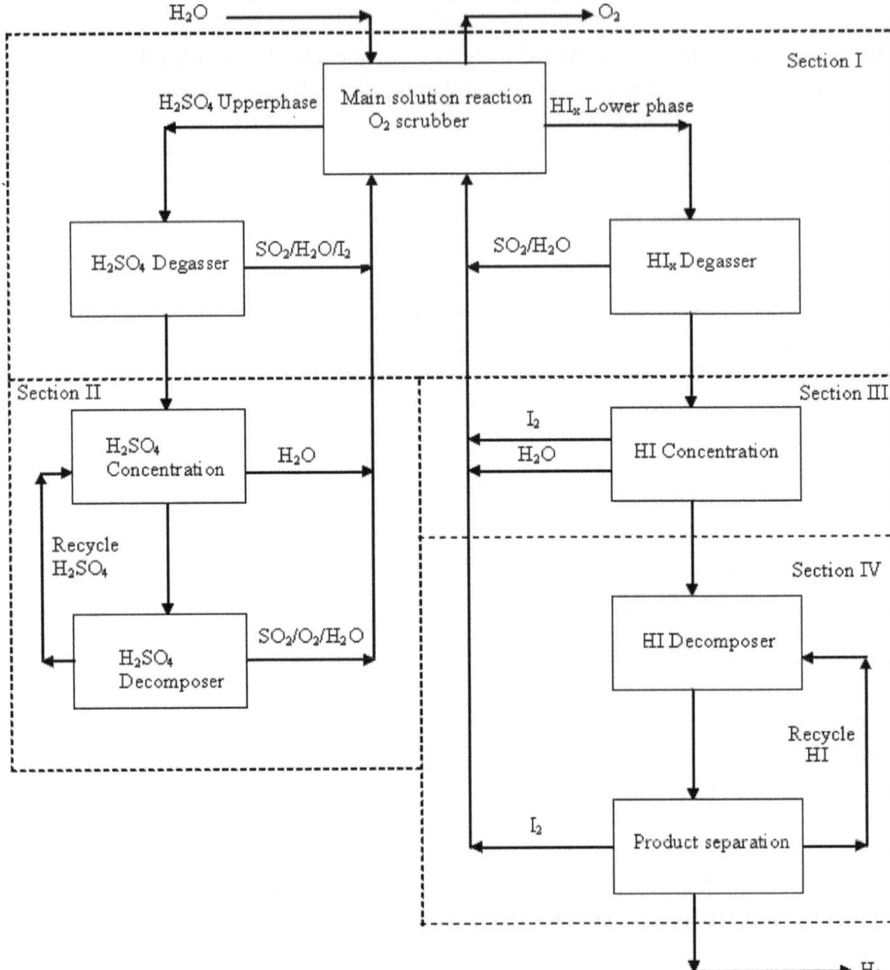

Fig. 5.5 Sulfur–iodine cycle for hydrogen production

5.5.7.2 Pyrolysis of Hydrogen Sulfide

Catalytic or non-catalytic, thermal decomposition appears to be the most direct process for obtaining both hydrogen and sulfur from H_2S. The reactions involved are highly endothermic and the equilibrium conversion corresponding to a temperature range of 1100–1400 K would be between 10–30%. The conversion level can be increased by catalysis and by removal of H_2 and S during reaction. Membrane, thermal diffusion and solar technologies need more development since they seem to be quite promising (Baykara et al., 2005).

Production of hydrogen by direct decomposition of hydrogen sulfide has been studied extensively (Clark et al., 1995; Zaman and Chakma, 1995; Luinstra, 1996). Hydrogen sulfide decomposition is a highly endothermic process and equilibrium

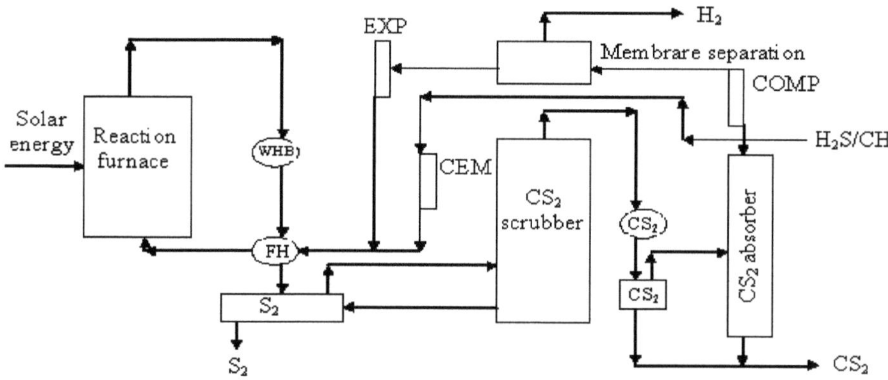

Fig. 5.6 CH_4/H_2S reforming for hydrogen production

yields are poor (Clark et al., 1995). At temperatures less than 1775 K, the thermodynamic equilibrium is unfavorable toward hydrogen formation. However, in the presence of catalysts such as platinum–cobalt (at 1275 K), disulfides of Mo or W at 1075 K (Kotera et al., 1976), or other transition metal sulfides supported on alumina (at 775–1075 K), H_2S decomposition proceeds rapidly (Kiuchi et al., 1982; Bishara et al., 1987; Al-Shamma and Naman, 1989; Megalofonos and Papayannakos, 1997).

5.5.7.3 Hydrogen Sulfide Reformation of Natural Gas

Hydrogen gas can be obtained by direct thermolysis of methane and pyrolysis of hydrogen sulfide. Direct thermal dissociation of methane and H_2S does not generate greenhouse gases. However, compared to the steam methane reforming (SMR) process, thermolysis of CH_4 and H_2S generates smaller amounts of hydrogen per mole of methane and hydrogen sulfide reacted. The reaction of H_2S with methane can be thought of as the sulfur analog of the SMR process, as indicated by the following two reactions (Hacisalihoglu et al., 2008):

$$H_2S \rightarrow H_2 + 0.5S_2 \qquad \Delta H = +79.9 \text{ kJ/mol} \tag{5.45}$$

$$CH_4 + 2S_2 \rightarrow CS_2 + 2H_2S \qquad \Delta H = -107 \text{ kJ/mol} \tag{5.46}$$

The overall reaction may be written as follows:

$$CH_4 + 2H_2S \rightarrow CS_2 + 4H_2 \qquad \Delta H = +232.4 \text{ kJ/mol} \tag{5.47}$$

The byproduct obtained from the overall reaction is carbon disulfide (CS_2). The reaction between CH_4 and H_2S given with Eq. 5.47 is the well-known methane process for production of CS_2. Most commercial CH_4–sulfur processes employ silica gel/aluminum catalyst for CS_2 production. The reaction of CH_4 with sulfur is thermodynamically favorable for CS_2 formation, and conversion is usually in the range of 90 to 95% with respect to methane (Arpe, 1989). The industrial CH_4–sulfur pro-

Table 5.9 Main hydrogen storage methods and techniques

Storage method	Storage condition
Liquid hydrogen in cryogenic tanks	At 21 K
High-pressure gas cylinders	Up to 800 bar
Adsorbed hydrogen on materials	At $T < 100$ K
Absorbed on interstitial sites in a host metal	At ambient pressure and temperature
Oxidation of reactive metals, e.g., Li, Na, with water	At ambient pressure and temperature
Chemically bonded in covalent and ionic compounds	At ambient pressure
Chemically bonded to either metal or liquid hydrides	At ambient pressure
Chemically carbon bonding	At ambient pressure

cess operates in the temperature range of 775 to 925 K and pressure range of 4 to 7 atm. Figure 5.6 shows the H_2S/CH_4 reforming process for hydrogen production (Huang and Raissi, 2005; Balat and Ozdemir, 2005).

5.6 Storage of Hydrogen

Hydrogen has the highest energy density per unit of weight of any known fuel (excluding nuclear reactions), but 1 g of hydrogen gas occupies about 11 l of space at atmospheric pressure, so its storage for vehicular use presents problems. There are a few different approaches for hydrogen transportation and storage. Hydrogen can be stored as a compressed gas (up to 800 bar), as a liquid (at 21 K) and in solid-state compounds. The first two methods are established technologies with several limitations, the most important of which is their energy-intensive character. Conventional storage systems consist of classical high-pressure tanks and insulated liquid hydrogen systems. Compressing hydrogen is similar to compressing natural gas, though as hydrogen is less dense the compressors need very high-quality seals. The main disadvantages of using hydrogen as a fuel for automobiles are huge on-board storage tanks, which are required because of hydrogen's extremely low density. Hydrogen may be stored on board a vehicle as compressed gas in ultrahigh-pressure vessels, as a liquid in cryogenic containers, or as a gas bound with certain metals in metal hydrides. Liquefaction by the Linde cycle, for example, consumes nearly 30% of the total energy contained in the hydrogen and in addition requires expensive equipment and energy to retain hydrogen in the liquid state. Up to 40% of the energy content in the hydrogen can be lost during the liquefaction operation. The liquid hydrogen storage is very expensive in comparison to other methods. The advantage of liquid hydrogen is its high energy-to-mass ratio, three times that of gasoline. However, it is difficult to store and the insulated tank required may be large and bulky (Demirbas, 2007).

Table 5.10 Characteristics of six basic hydrogen storage methods

Storage media	Pressure (bar)	Temperature (K)	Maximum capacity wt.%
Composite cylinder	800	298	13
Liquid hydrogen	1	21	100
Metal hydrides	1	298	2
Complex hydrides	1	298	18
Alkali with water	1	298	14
Physisorption	70	65	4

Hydrogen can be stored by six different methods and phenomena (Züttel et al., 2003). The main hydrogen storage methods and techniques are given in Table 5.9. The characteristics of six basic hydrogen storage methods are given in Table 5.10. The major challenge with the method of oxidation of reactive metals is the reversibility and the control of the thermal reduction process in order to produce the metal in a solar furnace. Sodium and lithium have a gravimetric hydrogen density of 3 mass% and 6.3 mass%, respectively. Pressurized hydrogen gas occupies a great deal of volume compared with, for example, gasoline with equal energy content, about 30 times larger in volume at 100 bar gas pressure. Condensed hydrogen is about 10 times denser, but is much too expensive to produce and maintain (Jensen et al., 1999). There are also obvious safety concerns with the use of pressurized or liquefied hydrogen in vehicles (David, 2005). Hence, boron has an important role here.

Storing hydrogen in a carbon structure is another form of chemical bonding. New forms of carbon structures are currently being researched and promise the best and safest approach as a reversible gas absorption technology. This research is the design and use of carbon nanostructures, either nanotubes or nanofibers.

The metal hydrides provide a safe method for fuel storage in hydrogen-powered vehicles. Charging and discharging of the hydride tanks is a process that can be repeated an indefinite number of times provided that the hydride material does not become contaminated. The hydrogen would be transported from the production plant either as a gas or in liquefied form. The liquefied hydrogen, in comparison to gaseous hydrogen, offers the advantages of low pressure and low bulk, but containment difficulties arise from the cryogenic nature of the liquefied gas (Demirbas, 2002).

Thermal compression of hydrogen using reversible metal hydride alloys offers an economical alternative to traditional mechanical hydrogen compressors. The hydrogen pressure in a metal hydride increases exponentially with increasing temperature. By employing successively higher-pressure hydride alloy stages in series, high-pressure ratios can be generated. For example, a five-stage hydride compressor will compress a 1.43 atm inlet pressure to 346.94 atm resulting in a compression ratio of 242.6 (Surmen and Demirbas, 2002).

Glass microspheres can be used to safely store high-pressure hydrogen. The glass spheres are filled by being immersed in high-pressure hydrogen gas after warming to about 675 K. The fill rate is a function of glass properties, gas permeability, temperature and pressure differential. The spheres are then cooled, locking the hydrogen inside the glass balls. The subsequent increase in temperature will release the hydrogen trapped in the spheres.

In the physisorption process a gas molecule interacts with several atoms at the surface of the solid. Once a monolayer of adsorbate molecules is formed the gaseous molecule interacts with a surface of the liquid or solid adsorbate. Therefore, the binding energy of the second layer of adsorbate molecules is similar to the latent heat of sublimation or vaporization of the adsorbate.

5.6.1 Hydrogen Storage with Metal Hydrides

Metals, intermetallic compounds, and alloys generally react with hydrogen and form mainly solid metal–hydrogen compounds (MH_n). Hydrides exist as ionic, polymeric covalent, volatile covalent and metallic hydrides. Hydrogen reacts at elevated temperatures with many transition metals and their alloys to form hydrides. Many of the MH_n show large deviations from ideal stoichiometry ($n=1, 2, 3$) and can exist as multiphase systems.

Using metal hydrides in electrochemical batteries is an old and rapidly improving method for hydrogen storage. Solid-state materials are attractive due to improved safety, a high energy density, and better energy efficiency (no compression or liquefaction). They are capable of absorbing and desorbing hydrogen with small pressure variations. Utilization of hydrides is also a promising technique for on-board hydrogen storage. A new method is the use of nanostructural materials such as carbon and boron nitride nanotubes, which are known to have the property to store gases within their structure (Fakioglu et al., 2004).

Table 5.11 Hydrogen contents of metal hydrides

Metal hydride	Wt.% hydrogen based on formula
$LiBH_4$	21.5
$Al(BH_4)_3$	16.8
$LiAlH_2(BH_4)_2$	15.2
$Mg(BH_4)_2$	14.8
$Ti(BH_4)_3$	13.0
$Fe(BH_4)_3$	11.9
$Ca(BH_4)_2$	11.5
$Zr(BH_4)_3$	8.8

When hydrogen combines with the metal alloy (in granular form or particles), an exothermic reaction occurs. The gas is thus stored in these metal particles until some heat is applied to release the hydrogen and build up the pressure in the tanks. When a metal hydride absorbs hydrogen, heat is given off. A hydride cold-start heater can be developed that instantly heats an automobile's catalytic converter when the car is started to dramatically reduce overall exhaust pollution up to 80%.

Hydrogen can be chemically bonded to either metal or liquid hydrides. Metal hydrides such as FeTi compounds are used to store hydrogen by bonding it to the surface of the material. As discussed later, the iron-titanium alloy acts as a sponge to absorb the hydrogen, thus becoming a metal hydride ($FeTiH_{1.6}$) with a physical appearance like that of a fine silvery powder. The safest method for hydrogen storage is the "metal hydride" system (Williamson and Edeskuty, 1986; Veziroglu, 1975). The low-temperature hydrides, $FeTiH_2$ and $LaNi_5H_7$, hold hydrogen loosely and evolve it at low temperatures. The high-temperature hydrides, MgH and Mg_2NiH_4, only release hydrogen at high temperatures since they have high binding enthalpies (Dinga, 1988).

Metal hydrides generate hydrogen gas via reversible pyrolysis reactions, i.e.,

$$2M + nH_2 \leftrightarrows 2MH_n + Heat \tag{5.48}$$

where M is a metal or an alloy. Such reactions are reversible and hydrogen can be stored by hydriding the metal under high pressure exothermically (Fakioglu et al., 2004). Table 5.11 shows the hydrogen contents of metal hydrides. Table 5.12 shows the characteristics of some metal hydrides. Theoretical reversible capacities of alanates such as $NaAlH_4$ and $LiAlH_4$ have been studied to some extent, where the maximum theoretical reversible capacity of $NaAlH_4$ is 5.5% by weight.

Hydrolysis is defined as the reaction of a hydride with water to liberate hydrogen gas. The reactions are as follows (Fakioglu et al., 2004):

$$MH_n + nH_2O \rightarrow M(OH)_n + nH_2 \ (5.49)$$

where M is a metal and n is its valence, or,

$$MXH_4 + 4H_2O \rightarrow 4H_2 + MOH + H_3XO_3 \tag{5.50}$$

$$NaBH_4 + 2H_2O \rightarrow NaBO_2 + 4H_2 + Heat \tag{5.51}$$

There are two classes of hydrides: metallic hydrides and complex hydrides. The main difference between these two is the transition of metals to ionic or cova-

Table 5.12 Properties of some metal hydrides

Alloy	Weight (kg)	Weight of H_2 (kg)	Energy of H_2 (MJ)	Energy (MJ/kg)
Fe Ti	563	5.4	767	1.362
Ti Mn TiF	568	6.0	848	1.492
$LaNi_5$	180	2.5	353	1.962
Fe Mn Ti	433	5.0	706	1.629

lent compounds for the complex hydrides upon absorbing hydrogen. The complex hydrides of boron represent a very interesting and challenging new hydrogen storage material.

Hydrogen storage in boron hydrides is a complex process consisting of many mechanistic steps and depends on important parameters. The boron surface has to be able to dissociate the hydrogen molecule and to allow hydrogen atoms to move easily in order to be able to store hydrogen. Metals differ in the ability to dissociate hydrogen, this ability being dependent on surface structure, morphology and purity (David, 2005).

The number of hydrogen atoms per metal atom is 2 in many cases (Zhou, 2005). This kind of complex shows the highest volumetric density and the highest gravimetric density at room temperature known today in lithiumtetrahydroboride (lithiumborohydride, $LiBH_4$). $LiBH_4$ and $NaBH_4$ can reversibly absorb/desorb hydrogen at moderate temperatures, and have therefore, received much attention (Gross et al., 2002; Bogdanovic and Schwickardi, 1997; Bogdanovic et al., 2000; Zaluska et al., 2000; Zaluska et al., 1999).

Direct synthesis from the metal, boron and hydrogen at 825–975 K and 30–150 bar H_2 has yielded the lithium salt, and such a method is generally applicable to groups IA and IIA metals (Goerrig, 1958). The reaction involving either the metal or the metal hydride, or the metal together with triethylborane in an inert hydrocarbon has formed the basis of a patent.

$$M + B + 2H_2 \rightarrow MBH_4 \tag{5.52}$$

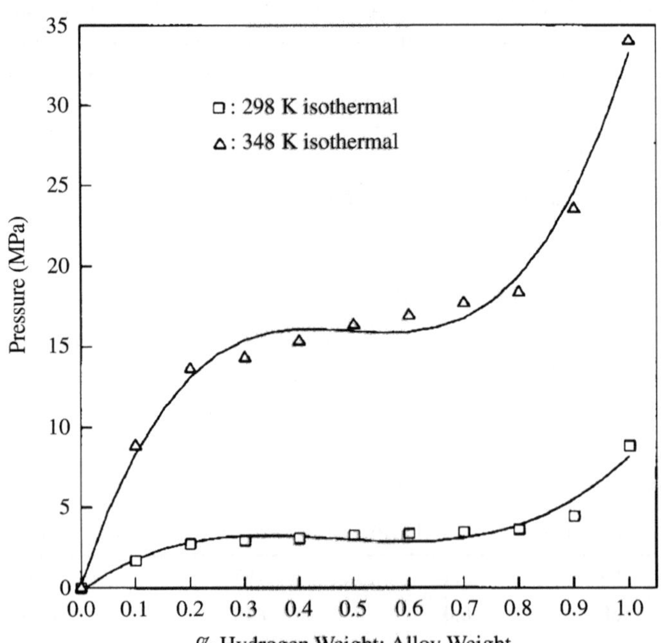

Fig. 5.7 Hydrogen absorption capacity of metal hydride alloy

M is Li, Na, K, etc., in Eq. 5.52. The decomposition of the compounds can proceed in one of the following directions:

$$MBH_4 \rightarrow M \quad + B + 2H_2 \tag{5.53}$$

$$MBH_4 \rightarrow MH \quad + B + 3/2H_2 \tag{5.54}$$

M is Li, Na, K, etc., for Eqs. 5.53 and 5.54.

The thermal decomposition, at least in the early stages, is reversible, as shown by the higher decomposition temperature in the presence of hydrogen than in vacuum or inert gas, and also by the existence of a rapid isotopic exchange between tetrahydroboride and deuterium at $T > 625$ K (Züttel et al., 2003).

The compound with the highest gravimetric hydrogen density at room temperature known today is $LiBH_4$. Therefore, this complex hydride could be the ideal hydrogen storage material for mobile applications. $LiBH_4$ desorbs 3 or 4 hydrogen in the compound upon melting at 553 K and decomposes into LiH and boron. The desorption process can be catalyzed by adding SiO_2 and significant thermal desorption was observed starting at 373 K (Züttel et al., 2003). Recently it has been shown, that the hydrogen desorption reaction is reversible and the end-products lithium hydride and boron absorb hydrogen at 980 K and 200 bar to form $LiBH_4$. The scientific understandings of the mechanism of the thermal hydrogen desorption from $LiBH_4$ and the absorption remains a challenge and more research work has to be carried out (Züttel et al., 2004).

Thermal compression of hydrogen using reversible metal hydride alloys offers an economical alternative to traditional mechanical hydrogen compressors. The hydrogen pressure in a metal hydride increases exponentially with increasing temperature. The hydrogen absorption capacity of a metal hydride alloy at 298 and 348 K is given in Fig. 5.7. The pressure rise generated in a single hydride heat exchanger is about 300%.

5.6.2 Hydrogen Absorption/Desorption with Oxygen-Contaminated Boron Film

The capability of H absorption decreased by 30–50% compared to the pure boron film without oxygen contamination. After the discharge, the depth profile of the oxygen atoms was not changed, which means that a stable oxide layer had formed. The reduction of the H absorption capability occurs probably because the formation of the boron oxide prevents H atoms from becoming trapped in the form of B–H bonding. Most of the retained H atoms can be released by heating up to 775 K with the oxygen contamination. The required temperature for H evacuation is slightly higher than that for the pure boron film. In addition, a small peak was observed at around 475 K. The reduction of H retention can be attributed to the oxide formation followed by the decrease of dangling bond, and thus, the decrease of H trapping sites (Tsuzuki et al., 1999).

5.6.3 Hydrogen Storage with Carbon Structures

In 1997 the journal *Nature* was the first publication to discuss the possibility of hydrogen storage in carbon material containing single-walled carbon nanotubes, which estimated for purified tubes a high possible storage capacity at room temperature and ambient pressure (Dillon et al., 1997). This article initiated tremendous research activity on the use of different nanocarbon structures for potential hydrogen storage.

Activated carbon has been extensively studied for its gas storage properties. This material was found to be ineffective for storing hydrogen, because only a small percentage of the surface strongly interacted with the hydrogen molecules at ambient temperatures and pressures (Dillon and Heben, 2001). The more recently discovered nanomaterials such as carbon nanotubes (CNTs) and graphitic nanofibers (GNFs) have renewed attention on carbon as an absorbent (Dillon et al., 1997; Fan et al., 1999; Chen et al., 1999; Gupta and Srivastava, 2000).

5.6.3.1 Carbon Nanotubes (CNTs)

Carbon nanotube technology is the other new direction being pursued in the search for high-capacity hydrogen storage media. Since their discovery in 1991 by the Japanese electron microscopist Sumio Iijima significant research has been undertaken to improve synthesis methods and more accurately characterize the unique mechanical and chemical properties of carbon nanotubes (Iijima, 1991). Nanotubes can exist as single-walled or as multiwalled units and can be pristine or doped usually with an alkali metal. As a result of van der Waals forces nanotubes can be bundled to form sites where hydrogen can be adsorbed inside the tubes. Carbon nanotubes are microscopic tubes of carbon, two nanometers (billionths of a meter) across, that store hydrogen in microscopic pores on the tubes and within the tube structures. There are a number of options for hydrogen storage in carbon nanotubes. Carbon nanotubes are fullerene-related structures, which consist of graphene microscopic cylinders closed at either end with caps containing pentagonal rings. The apparent advantage of carbon nanotubes is that they are theoretically capable of storing from about 4.2 to 6.5% of their own weight in hydrogen.

5.6.3.2 Carbon Nanofibers and Fullerenes

Another variant in carbon nanostructures is carbon nanofiber technology. Carbon nanofibers are grown by the decomposition of hydrocarbons or carbon monoxide over a metal catalyst, and the fibers comprise graphite sheets aligned in a set direction. Graphite nanofibers are ideal for selective gas absorption, as reported by Gupta and Srivastava (2000) who achieved hydrogen adsorption capacities of ~10 wt.% for GNFs grown by thermal cracking. Fan et al. (1999), observed vapor-grown carbon fibers exhibiting 10–13 wt.% hydrogen storage, almost twice as high as the 5.7 wt.% capacity reported by Chen et al. (1999) for the same material.

Fullerenes carbon C_{60} and C_{70} have been shown to reversibly store hydrogen. Theoretically, fullerene C_{60} can be hydrogenated up to $C_{60}H_{60}$, which corresponds to 7.7% weight of hydrogen.

5.7 Hydrogen Storage Materials

Hydrogen may be stored on board a vehicle as a gas bound with certain metals, as a liquid in cryogenic containers, or as highly compressed gas in ultrahigh-pressure (69 MPa) vessels. On-board hydrogen storage approaches presently being examined by developers include compressed hydrogen gas, cryogenic gas and liquid hydrogen, metal hydrides, high surface area adsorbents, and chemical hydrogen storage media (Ozturk and Demirbas, 2007).

The density of hydrogen in a storage material is crucial for mobile applications. Hydrogen can be stored by six different methods and phenomena (Züttel et al., 2003):

1. Liquid hydrogen in cryogenic tanks (at 21 K)
2. High-pressure gas cylinders (up to 800 bar)
3. Adsorbed hydrogen on materials with a large specific surface area (at $T < 100$ K)
4. Absorbed on interstitial sites in a host metal (at ambient pressure and temperature)
5. Oxidation of reactive metals, e.g., Li, Na, with water
6. Chemically bonded in covalent and ionic compounds (at ambient pressure)

The main challenge with storing in terms of oxidation of reactive metals is the reversibility and the control of the thermal reduction process in order to produce the metal in a solar furnace, where Na and Li have a gravimetric hydrogen density of 3 mass% and 6.3 mass%, respectively.

The density of liquid hydrogen is 70.8 kg/m^3 and that of solid hydrogen is 70.6 kg/m^3. But the condensation temperature of hydrogen at 1 bar is 21 K and the vaporization enthalpy at the boiling point is 452 kJ/kg. As the critical temperature of hydrogen is 32 K (above this temperature hydrogen is gaseous), liquid hydrogen containers are open systems to prevent strong overpressure. Therefore, heat transfer through the container leads directly to the loss of hydrogen. The continuously evaporated hydrogen may be catalytically burnt with air in the overpressure safety system of the container or collected again in a metal hydride (Schlapbach and Züttel, 2001).

The iron–titanium alloy, FeTi, acts as a sponge to absorb the hydrogen, thus becoming a metal hydride ($FeTiH_{1.6}$). Its physical appearance is like that of a fine silvery powder. The safest method for hydrogen storage is then this metal hydride system (Veziroglu, 1975). When hydrogen combines with the metal alloy (in granular form or particles), an exothermic reaction occurs. The gas is thus stored in these metal particles until some heat is applied to release the hydrogen and build up the pressure in the tanks. When a metal hydride absorbs hydrogen, heat is given off. This process could be used as a cold-start heater that instantly heats an vehicle's catalytic converter when the car is started, thereby dramatically reducing overall exhaust pollution up to 80%.

Thus, metal hydrides provide a safe method for storing fuel in hydrogen-powered vehicles. Charging and discharging of the hydride tanks is a process that can be repeated an indefinite number of times provided that the hydride material does not become contaminated.

5.7.1 Boron Hydrides as Metal Hydrides

As mentioned above, the safest method for hydrogen storage is the metal hydride system (Williamson and Edeskuty, 1986; Veziroglu, 1975). When hydrogen combines with the metal alloy, an exothermic reaction occurs. The gas is then stored in the metal particles until some applied heat releases the hydrogen and builds up pressure in the tanks. Heat is then emitted when a metal hydride absorbs hydrogen.

Boron absorbs hydrogen and forms hydrides. There are two classes of hydrides: metallic hydrides and complex hydrides, with the main difference being the transition of metals to ionic or covalent compounds for the complex hydrides upon absorbing hydrogen.

Boron chemistry is an interesting field of research, which includes plenty of boron hydrides. Some of which are stable, but some are unstable and interconvertible to each other, while producing hydrogen (Turker, 2001). Diborane (B_2H_6) can be synthesized via various reactions (Durant and Durant, 1970) and produce hydrogen by treatment with H_2O, NH_3, ROH, etc. (Durant and Durant, 1970; Schriver and Atkins, 1990). The molecular orbital bonding scheme for diborane has been discussed extensively (Lippard and Ucko, 1968). Tetraborane (B_4H_{10}) can be obtained by means of various synthetic routes and it also forms by the slow decomposition of diborane (Huhey, 1978; Cotton and Wilkinson, 1967):

$$2B_2H_6 \leftrightarrows B_4H_{10} + H_2 \tag{5.55}$$

B_4H_{10} decomposes from heat treatment (at 363–373 K) to produce diborane (Durant and Durant, 1970; Cotton and Wilkinson, 1967) or at elevated temperatures (at 453–473 K) to produce B_5H_{11} and B_5H_9. B_5H_{11} is known as unstable pentaborane because it decomposes in a few hours in the cold, mainly to H_2 and $B_{10}H_{14}$ (Durant and Durant, 1970).

The necessary condition for hydrogen storage is that the thermodynamic and kinetic conditions are fulfilled. In that case, a metal exposed to hydrogen gas absorbs hydrogen until equilibrium is established. There are a number of reaction steps that kinetically may hinder a hydrogen-storing system from reaching its thermodynamical equilibrium of hydrogen storage within a reasonable time. The reaction rate of a metal-hydrogen system is therefore a function of pressure and temperature (David, 2005).

The storage of hydrogen in boron hydrides is a complex process consisting of numerous mechanistic steps, depending on many important parameters. For instance, the boron surface must be able to dissociate the hydrogen molecule and to allow hydrogen atoms to move easily in order to be able to store hydrogen. Metals differ in the ability to dissociate hydrogen, this ability being dependent on surface structure, morphology and purity (David, 2005).

5.7.2 Hydrogen in Mechanically Milled Amorphous Boron

Hydrogen can be absorbed in amorphous boron by mechanical milling under hydrogen atmosphere, the amount of which has reached up to about 2.3 mass% after 80 h of milling. The trapped hydrogen can be desorbed only as molecular hydrogen in a temperature range of 330–1000 K. Thus the hydrogen desorption in boron is achieved at relatively low temperatures with molecular hydrogen as the sole product (Wang et al., 2003). In contrast to the case of mechanically prepared nanostructured graphite, in which hydrogen desorbs at high temperatures together with desorption of hydrocarbons, hydrogen desorption in boron can be achieved at relatively low temperatures with molecular hydrogen as the sole product. The revealed hydrogenation properties, especially the attracting dehydriding aspect, warrant further investigations of the B–H system from the viewpoint of hydrogen energy storage (Wang et al., 2003).

5.7.3 Boron Complex Hydrides

In metal hydrides, the number of hydrogen atoms per metal atom is 2 in many cases (Zhou, 2005). Groups 1–3, light metals, e.g., Li, Mg, B, Al, build a large variety of metal–hydrogen complexes. They are especially interesting because of their light weight and the number of hydrogen atoms per metal atom, which in many cases is 2. The main difference between these complex hydrides and the previously described metallic hydrides is the transition to an ionic or covalent compound of the metals upon hydrogen absorption. Table 5.13 shows the physical properties of some selected complex hydrides (Züttel et al., 2003). The highest volumetric density and the highest gravimetric density at room temperature known today belong to $LiBH_4$ (18.4 mass%). $LiBH_4$ has stability compared to other chemical hydrides, and can easily be converted to H_2. As for $NaBH_4$, along with $LiBH_4$, it can reversibly absorb/desorb hydrogen at moderate temperatures and has therefore received considerable attention (Gross et al., 2002; Bogdanovic and Schwickardi, 1997; Bogdanovic et al., 2000; Zaluska et al., 2000; Zaluski et al., 1999).

Table 5.13 Physical properties of selected complex hydrides

Formula	M (g/mol)	ρ (g/cm³)	T_m (K)	T_{dec} (K)	X (mass%)	Reference
$LiBH_4$	21.784	0.66	541	653	18.4	Schlesinger and Brown, 1940
$NaBH_4$	37.83	1.074	778	673	10.6	Knacke et al., 1991
$LiAlH_4$	37.95	0.917	> 398	398	9.5	Knacke et al., 1991
KBH_4	53.94	1.178	858	773	7.4	Knacke et al., 1991
$NaAlH_4$	54.0	1.27	451	483	7.4	Bogdanovic and Schwickardi, 1997
Mg_2NiH_4	111.3	2.72		553	3.6	Sandrock and Thomas, 2001
Mg_2FeH_6	110.5	2.72		593	5.4	Yvon, 1998
Mg_3MnH_7	134.9	2.30		553	5.2	Yvon, 1998
$BaReH_9$	332.5	4.86		< 373	2.7	Yvon, 1998

M molecular mass, ρ gravimetric density, T_m melting point, T_{dec} decomposition temperature, X gravimetric hydrogen density

Boron builds a large variety of metal-hydrogen complexes. It is especially interesting because of its light weight and the number of hydrogen atoms per metal atom. The hydrogen in the complex hydrides is often located at the corners of a tetrahedron with boron in the center. The negative charge of the anion, $[BH_4^-]$, is compensated by a cation, e.g., Li or Na. The hydride complexes of borane and the tetrahydroborates $M(BH_4)$ are interesting storage materials (M = Li, Na, K, etc.); however, they are known to be stable and decompose only at elevated temperatures and often above the melting point of the complex (Züttel et al., 2004).

Schlesinger and Brown synthesized the lithiumtetrahydroboride (lithiumborohydride) ($LiBH_4$) by the reaction of ethyllithium with diborane (B_2H_6) in 1940 (Schlesinger and Brown, 1940). It was the first report of a pure alkali metal tetrahydroboride. The direct reaction of the corresponding metal with diborane in ethereal solvents under suitable conditions produces high yields of the tetrahydroborides (Huhey, 1978):

$$2MH + B_2H_6 \leftrightarrows 2MBH_4 \tag{5.56}$$

where M is a metal such as Li, Na, K, etc.

Stasinevich and Egorenko (1968) investigated the alkali metal tetrahydroborides by means of thermal analysis in hydrogen at pressures up to 10 bar. The thermal decomposition, at least in the early stages, is reversible, as is shown by the decomposition temperature being higher in the presence of hydrogen than in vacuum or inert gas (Ostroff and Sanderson, 1957; Mesmer and Jolly, 1962) and also by the existence (Mesmer and Jolly, 1962; Brown et al., 1952) of a rapid isotopic exchange between tetrahydroboride and deuterium at $T > 625$ K (Züttel et al., 2003).

The compound with the highest gravimetric hydrogen density at room temperature known today is $LiBH_4$ (~18 mass%). Therefore, this complex hydride could be the ideal hydrogen storage material for mobile applications. $LiBH_4$ desorbs 3 or 4 hydrogen in the compound upon melting at 553 K and decomposes into LiH and B. The process of desorption can be catalyzed by adding SiO_2; significant thermal desorption was observed starting at 373 K (Züttel et al., 2003). Recently it has been shown that the hydrogen desorption reaction is reversible and the end-products lithium hydride and boron absorb hydrogen at 963 K and 200 bar to form $LiBH_4$. The scientific understanding of the mechanism of the thermal hydrogen desorption from $LiBH_4$ and the absorption remains a challenge; thus, more work must be carried out (Züttel et al., 2004). $LiBH_4$ desorbs the hydrogen only at temperatures from 353 K up to 873 K (Brown et al., 1952). The decomposition temperature of $NaAlH_4$ can be lowered by doping the hydride with TiO_2, and the reversibility of the reaction for several desorption/absorption cycles is shown. This is a good example demonstrating the potential of boron hydrides.

$Al(BH_4)_3$ is a complex hydride with a very high gravimetric hydrogen density of 17 mass%, and the highest known volumetric hydrogen density of 150 kg/m³. $Al(BH_4)_3$ has a melting point of 208 K and is the only liquid hydride at room temperature. The complex hydrides of boron represent a very interesting and challenging new hydrogen storage material.

5.8 Hydrogen Fuel for Internal Combustion Engine

Water is a readily available resource for generating of hydrogen. With 70% of the world's surface being covered by water, it is one of the most abundant resources on this planet. For more than a decade, hydrogen as an alternative to traditional energy sources such as oil and natural gas has been the focus of research and development efforts in all technologically advanced countries of the world. In the pursuit for a cleaner world, hydrogen power is perhaps the cleanest fuel alternative.

Hydrogen exhibits the highest heating value of all chemical fuels (Demirbas, 1998). The energy from 1 kg hydrogen is equivalent to 2.1 kg natural gas and 2.8 kg petroleum. A hydrogen engine has both the advantages of the high thermal efficiency of a diesel engine (Das, 1996) and the excellent operating performance of a petrol engine (Gren and Glasson, 1992). In the automotive world, hydrogen can be utilized in two ways: hydrogen can be internally combusted to generate power or it can be combined with oxygen in a fuel cell to generate electricity.

Internal combustion engines that run on hydrogen are virtually identical to their gasoline predecessors. The difference is that water is the byproduct from combustion with oxygen rather than harmful pollutants. Because the engine runs on hydrogen gas, there are some slight changes in the engine design. Hydrogen has a very wide explosive range (from 4 to 75% in air) in which it can be combined with oxygen to explode. This explosive range is significantly larger than that of gasoline. Therefore, on hydrogen internal combustion engines (HICEs) there is an electronic

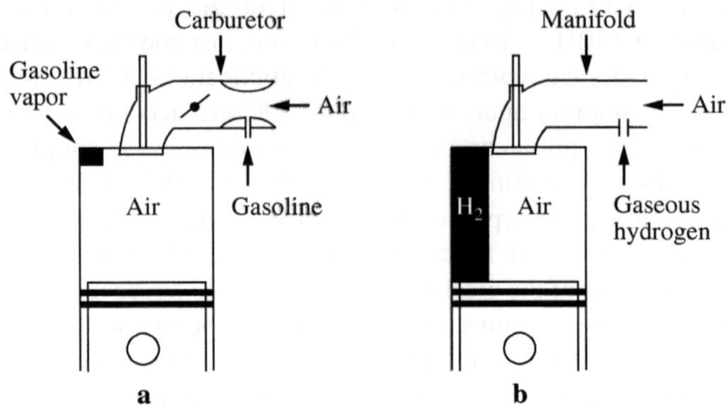

Fig. 5.8 **a** Carburetor in a liquid gasoline-fueled Otto engine. **b** Manifold in gaseous hydrogen-fueled modified engine systems

throttle system which controls the air coming into the engine so that the engine can always run at a lean mixture of air and fuel. Another characteristic of HICEs is that they use high-compression pistons such that more power can be achieved from the combustion process.

The performance of hydrogen fuel cell vehicles must be comparable or superior to today's gasoline vehicles in order to achieve widespread commercial success. The most critical issues facing the automotive fuel cell developer are overall power plant efficiency, interfacing the power plant with the fuel's infrastructure, achieving ultralow or zero emissions, and power plant costs. Fuel cells are very useful as power sources in remote locations, such as spacecraft, remote weather stations, large parks, rural locations, and in certain military applications. There are numerous prototype or production cars and buses based on fuel cell technology being researched or manufactured. A practical commercial fuel cell automobile is not expected until at least 2015 according to the automobile industry.

The hydrogen would be transported from the production plant either as a gas or in liquefied form. The liquefied hydrogen, in comparison to gaseous hydrogen, offers the advantages of low pressure and low bulk, but containment difficulties arise from the cryogenic nature of the liquefied gas (Demirbas, 2002). Figure 5.8a,b shows a carburetor in a liquid gasoline-fueled Otto engine and the manifold in a gaseous hydrogen-fueled modified engine system.

5.8.1 Advantages of Hydrogen as an Engine Fuel

Hydrogen can be used as a motor fuel, whereas neither nuclear nor solar energy can be used directly. Nuclear power requires heavy shielding to keep the neutrons away from people, but is too heavy for cars. It can be used in ships especially submarines and aircraft carriers. Hydrogen fuel is hydrogen gas with small amounts of oxygen and other materials added. Benefits include cleaner air, cleaner water, and better health. Hydrogen is also a renewable resource.

Hydrogen has many properties that make it ideal as a fuel for internal combustion engines in automobiles. When hydrogen is burned in air the main product is water. Should greenhouse warming turn out to be an important problem, the key advantage of hydrogen is that carbon dioxide (CO_2) is not produced when hydrogen is burned. Hydrogen combustion does not produce toxic products such as hydrocarbons (C_xH_y), carbon monoxide (CO), oxides of sulfur (SO_x), organic acids, and CO_2. Acid rain and the CO_2 greenhouse effect could effectively be eliminated.

Hydrogen contains 2.75 times as much energy as the same weight of gasoline. Hydrogen has a high flame speed, wide flammability limits, and a high detonation temperature with lean burning, which give improved engine efficiency.

Hydrogen is a gas at 20.13 K; therefore, there is no problem in starting an engine at the coldest winter temperatures, i.e., instant start-up.

As with any gasoline engine, efficiency depends on driving conditions. At freeway speeds the hydrogen engine has demonstrated at least a 20% increase in efficiency. Hydrogen engines demonstrate the efficient operations of a diesel and the high rpm characteristics of a gasoline motor.

The thermodynamic cycle for hydrogen is much closer to the ideal Otto cycle than for either a gasoline or a diesel engine. In addition, the compression ratio can be higher.

5.8.2 Disadvantages of Hydrogen as an Engine Fuel

The main disadvantages of using hydrogen as a fuel for automobiles are huge onboard storage tanks, which are required because of hydrogen's extremely low density. Hydrogen may be stored on board a vehicle as compressed gas in ultrahigh-pressure vessels, as a liquid in cryogenic containers, or as a gas bound with certain metals in metal hydrides. It is expensive to bring hydrogen to mass production. The easiest and cheapest way to refine hydrogen is by electrolysis, which uses electricity to process hydrogen from water. However, most of our cheap electricity comes from fossil fuels, which pollute the environment.

Hydrogen can be used as a fuel directly in an internal combustion engine not much different from the engines used with gasoline. The problem is that while hydrogen supplies three times the energy per pound of gasoline, it has only one-tenth the density when the hydrogen is in a liquid form and much less when it is stored as a compressed gas. This means that hydrogen fuel tanks must be large.

Demonstrations of hydrogen-powered vehicles have usually used compressed hydrogen gas. However, because of the low density, compressed hydrogen will not give a car as useful a range as gasoline. It may be even worse than using lead-acid batteries. Hydrogen can achieve a reasonable density adsorbed in metal hydrides, but then the weight of the metals makes the system very heavy.

Hydrogen does not occur freely in nature in useful quantities. It has to be made, usually by splitting water to get the hydrogen. This requires all the energy you are going to get from burning the hydrogen and a bit more on account of inefficiencies. Therefore, hydrogen is an energy transfer medium rather than a primary source of energy.

Hydrogen is the lightest of the elements with an atomic weight of 1.0. Liquid hydrogen has a density of 0.07 grams per cubic centimeter, whereas water has a density of 1.0 g/ml and gasoline about 0.75 g/ml. A hydrogen fuel tank will have three times the size of a gasoline tank. Also it must be insulated, and this will add to its bulk. This seems entirely bearable. These facts give hydrogen both advantages and disadvantages. The advantage is that it stores approximately 2.6 times the energy per unit mass as gasoline, and the disadvantage is that it needs about four times the volume for a given amount of energy. A 15-gallon automobile gasoline tank contains 90 pounds of gasoline. The corresponding hydrogen tank would be 60 gallons, but the hydrogen would weigh only 34 pounds.

Liquid hydrogen is cold enough to freeze air, and accidents have occurred from pressure build-up following plugged valves. Limited accident experience suggests that the danger is somewhat less with hydrogen than with gasoline, because the hydrogen dissipates rapidly. Although, the release of hydrogen into a confined space like a garage risks an explosion.

5.9 Liquefaction and Compression of Hydrogen

The liquefaction of hydrogen above 20 K is not possible with known non-catalytic methods. The advantage of liquid hydrogen is its high energy mass ratio, three times that of gasoline, but up to 40% of the energy content in the hydrogen can be lost during the liquefaction operation.

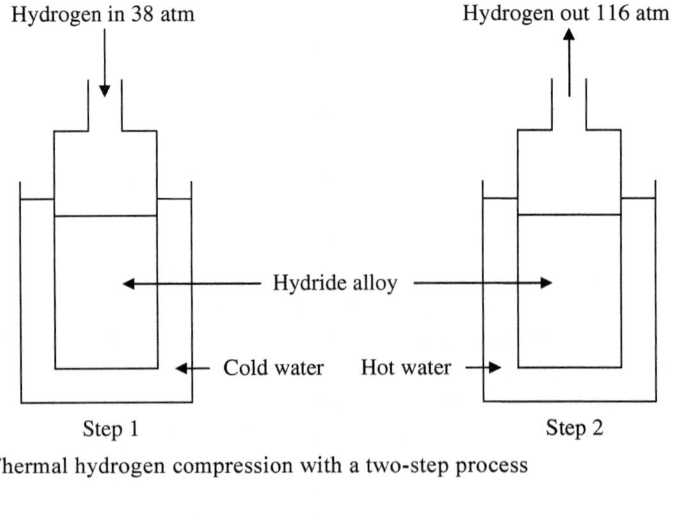

Fig. 5.9 Thermal hydrogen compression with a two-step process

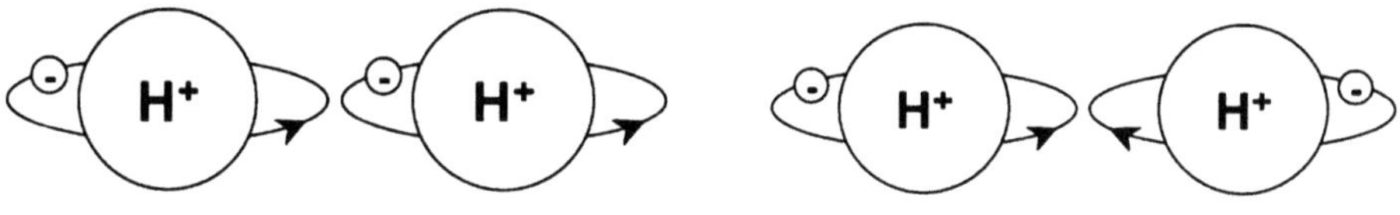

Ortho-hydrogen *Para*-hydrogen

Fig. 5.10 *Ortho*- and *para*-hydrogen structures

Energy is needed to compress gases. The compression work depends on the thermodynamic compression process. The ideal isothermal compression cannot be realized. Even more energy is needed to compact hydrogen by liquefaction. Low density and extremely low boiling point of hydrogen increases the energy cost of compression or liquefaction.

The activated hydrogen molecules can be attracted to each other more than normal hydrogen molecules in the reaction medium and these attractions may result easily in liquefaction of hydrogen in reaction conditions (Fig. 5.12).

Thermal hydrogen compression with a two-step process is given in Fig. 5.9. By employing successively higher-pressure hydride alloy stages in series, high-pressure ratios can be generated. For example, a five-stage hydride compressor will compress a 0.14 MPa inlet pressure to 35 MPa, resulting in a compression ratio of 25 MPa (Surmen and Demirbas, 2002).

A major concern in liquid hydrogen storage is minimizing hydrogen losses from liquid boil-off. Because liquid hydrogen is stored as a cryogenic liquid that is at its boiling point, any heat transfer to the liquid causes some hydrogen to evaporate. Liquefaction by the Linde cycle, for example, consumes nearly 30% of the total energy contained in the hydrogen and in addition requires expensive equipment and energy to retain hydrogen in the liquid state. Up to 40% of the energy content in the hydrogen can be lost during liquefaction operation. Figure 5.10 shows the *ortho*- and *para*-structures of hydrogen. Normal hydrogen includes 75% *ortho*- and 25% *para*-hydrogen, while liquid hydrogen is composed of 0.2% *ortho*- and 99.8% *para*-structures. The source of this heat can be *ortho*-to-*para* conversion, mixing or pumping energy, radiant heating, convection heating or conduction heating. The liquid hydrogen storage is very expensive in comparison to other methods.

5.9.1 Nanocatalytic Liquefaction of Hydrogen

The experiments of nanocatalytic liquefaction of hydrogen were carried out using a platinum-supported carbon nanolayer. In the experiments, H_2PtCl_6, $6H_2O$ or H_2PtCl_6 was used as a Pt precursor. To prepare the Pt catalytic reaction nanolayer, the required amounts of H_2PtCl_6 were mixed with 5 wt.% carbon prepared by burning naphthalene in air (Demirbas, 2008).

All the runs were performed in a 100-ml cylindrical autoclave. The catalyzed carbon and hydrogen gas was loaded into an autoclave. The autoclave was cooled

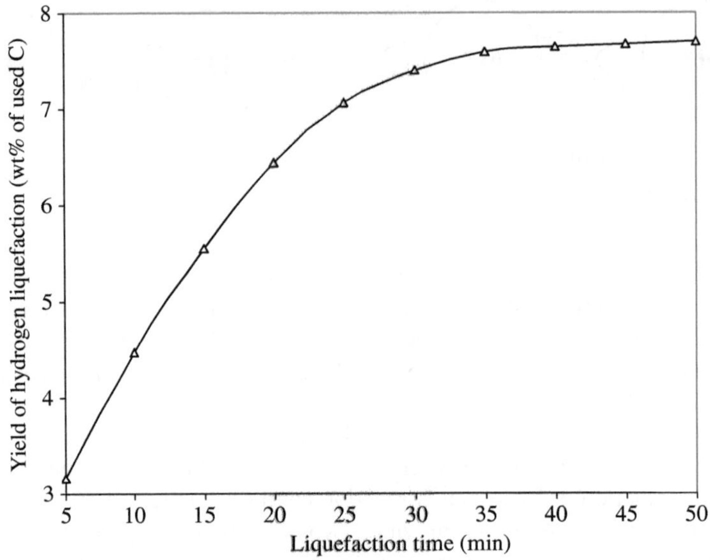

Fig. 5.11 Plot for yield of hydrogen liquefaction on Pt-supported carbon versus time

using an external liquid nitrogen bath. The nominal reaction time is 30 min. After each run, non-liquefied hydrogen gas was vented and then the autoclave was heated with an external heater for determining the liquefaction hydrogen in the reaction conditions (Demirbas, 2008).

Figure 5.11 shows the plot for yield of hydrogen liquefaction on Pt-supported carbon versus liquefaction time. The yield of hydrogen liquefaction increased with increasing liquefaction time. The nominal liquefaction time was 30 min in reaction conditions. The yield of liquefied hydrogen was 7.4% weight of Pt-catalyzed carbon.

Hydrogen consists of a mixture of two types of molecule, *ortho*-hydrogen, in which the two proton spins are parallel, and *para*-hydrogen, in which they are antiparallel. There are a number of possible methods for measuring the proportion of *ortho*- to *para*-molecules in a sample of hydrogen gas (Farkas, 1935; Stewart and Squires, 1955; Silvera and Pravica, 1998).

At hydrogen's boiling point of 20 K, the equilibrium concentration is almost all *para*-hydrogen, but at room temperature or higher, the equilibrium concentration is 25% *para*-hydrogen and 75% *ortho*-hydrogen (Noganow, 1992). The uncatalyzed conversion from *ortho*- to *para*-hydrogen proceeds very slowly, so without a catalyzed conversion step, the hydrogen may be liquefied, but may still contain significant quantities of *ortho*-hydrogen. This *ortho*-hydrogen will eventually be converted into the *para*-form in an exothermic reaction (Timmerhaus and Flynn, 1989).

Figure 5.12 shows the structure of the activated hydrogen molecule. The Pt catalyst may cause the increase of activation of the hydrogen molecule. As mentioned before, since the activated hydrogen molecules are more likely to be attracted each other than normal hydrogen molecules in the reaction medium, these attractions may result in liquefaction of hydrogen.

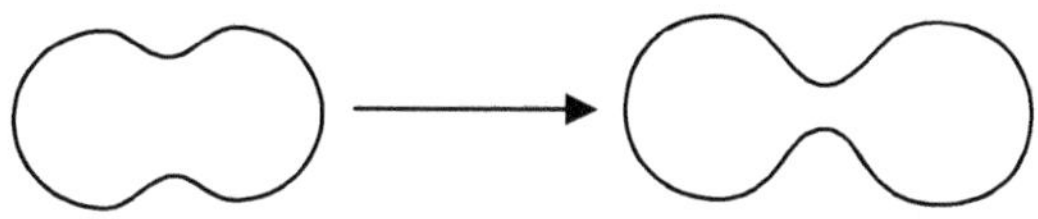

Fig. 5.12 Structure of activated hydrogen molecule

Table 5.14 Catalysts prepared from different precursors

Precursor (wt.%)	Catalyst	Composition	Loading ratio
$H_2PtCl_6 . 6H_2O$	Pt	Single	5.0
$RuCl_3 . 3H_2O + H_2PtCl_6 . 6H_2O$	Ru	Composite	2.5
	Pt		2.5
K_2PtCl_4	Pt	Single	5.0
$RuCl_3 . 3H_2O + K_2PtCl_4$	Ru	Composite	2.5
	Pt		2.5

Carbon-supported platinum (Pt) and platinum–ruthenium (Pt–Ru) alloy are one of the most popular electrocatalysts in polymer electrolyte fuel cells (PEFC). Pt supported on electrically conducting carbons, preferably carbon black, is being increasingly used as an electrocatalyst in fuel cell applications (Parker et al., 2004). Carbon-supported Pt could be prepared at loadings as high as 70 wt.% without a noticeable increase of particle size. Unsupported and carbon-supported nanoparticle Pt–Ru_{adatom} (Pt–Ru_{ad}) catalysts prepared using the surface reductive deposition technique were evaluated as anode catalysts in liquid feed PEM-DMFCs. It was found that the surface composition of unsupported Pt–Ru_{ad} nanoparticles has a significant influence on their activities as anode catalysts in direct methanol fuel cells (Cao and Bergens, 2004). Carbon-supported Pt–Ru_{ad} catalysts display higher mass activities than unsupported Pt–Ru_{ad}. The electrochemical deposition of Pt–Ru nanoparticles on carbon nanotube electrodes and their electrocatalytic properties have been investigated by He et al. (2004). Table 5.14 shows the catalysts Pt and Ru, prepared from various precursors serving as electrolytic electrodes in the fuel cells (Chaurasia et al., 2003).

Activated carbon has been extensively studied for its gas storage properties. This material was found to be ineffective for storing hydrogen, because only a small percentage of the surface strongly interacted with the hydrogen molecules at ambient temperatures and pressures (Dillon and Heben, 2001). The more recently discovered nanomaterials such as carbon nanotubes (CNTs) and graphitic nanofibers (GNFs) have renewed attention on carbon as an absorbent (Dillon et al., 1997; Fan et al., 1999; Chen et al., 1999; Gupta and Srivastava, 2000).

Another technology being pursued in the search for high-capacity hydrogen storage media is that of carbon nanotubes. Since their discovery in 1991 by Sumio

Iijima, significant research has been undertaken to improve synthesis methods and more accurately characterize the unique mechanical and chemical properties of carbon nanotubes (Iijima, 1991). Nanotubes can exist as single-walled or as multi-walled units and can be pristine or doped usually with an alkali metal. As a result of van der Waals forces nanotubes can be bundled to form sites where hydrogen can be adsorbed inside the tubes. Carbon nanotubes are microscopic tubes of carbon, two nanometers (billionths of a meter) across, that store hydrogen in microscopic pores on the tubes and within the tube structures. There are a number of options for hydrogen storage in carbon nanotubes. Carbon nanotubes are fullerene-related structures, which consist of graphene microscopic cylinders closed at either end with caps containing pentagonal rings. The apparent advantage of carbon nanotubes is that they are theoretically capable of storing from about 4.2 to 6.5% of their own weight in hydrogen.

Carbon nanofiber technology is another variant in carbon nanostructures. Carbon nanofibers are grown by the decomposition of hydrocarbons or carbon monoxide over a metal catalyst, and the fibers comprise graphite sheets aligned in a set direction. Graphite nanofibers are ideal for selective gas absorption, with results reported by Gupta and Srivastava (2000) who achieved hydrogen adsorption capacities of ~10 wt.% for GNFs grown by thermal cracking. Fan et al. (1999) observed vapor-grown carbon fibers exhibiting 10–13 wt.% hydrogen storage, almost twice as high as the 5.7 wt.% capacity reported by Chen et al. (1999) for the same material.

Summary

Hydrogen energy is a promising alternative solution because it is clean and environmentally safe. It also produces negligible levels of greenhouse gases and other pollutants when compared with the fossil fuel energy sources they replace.

Hydrogen can be obtained by direct electrolysis, by direct thermal conversion, thermochemically, photochemically, photoelectrochemically, and biochemically from water.

There are a few different approaches for hydrogen transportation and storage. Hydrogen can be stored as a compressed gas (up to 800 bar), as a liquid (at 21 K) and in solid-state compounds. The first two methods are well-established technologies with several limitations, the most important of which is their energy-intensive character. Conventional storage systems consist of classical high-pressure tanks and insulated liquid hydrogen systems. Compressing hydrogen is similar to compressing natural gas, though as hydrogen is less dense the compressors need very high-quality seals. The main disadvantages of using hydrogen as a fuel for automobiles are huge on-board storage tanks, which are required because of hydrogen's extremely low density. Hydrogen may be stored on board a vehicle as compressed gas in ultrahigh-pressure vessels, as a liquid in cryogenic containers, or as a gas bound with certain metals in metal hydrides.

References

Abe, T., Suzuki, E., Nagoshi, K., Miyashita, K., Kaneko, M. 1999. Electron source in photoinduced hydrogen production on Pt-supported TiO_2 particles. J Phys Chem B 103:1119–1123.

Adesina, A.A., Meeyoo, V., Foulds, G. 1995. Thermolysis of hydrogen sulphide in an open tubular reactor. Int J Hydrogen Energy 20:777–783.

Al-Shamma, L., Naman, S.A. 1989. Kinetic study for thermal production of hydrogen from H_2S by heterogeneous catalysis of vanadium sulfide in a flow system. Int J Hydrogen Energy 14:173–179.

Amiridis, M.D., Duevel, R.V., Wachs, I.E. 1999. The effect of metal oxide additives on the activity of V_2O_5/TiO_2 catalysts for the selective catalytic reduction of nitric oxide by ammonia. Appl Catal B Environ 20:111–122.

Anderson, C., Bard, A.J. 1995. Improved photocatalyst of TiO_2/SiO_2 prepared by a sol-gel synthesis. J Phys Chem 99:9882–9885.

Arakawa, H., Sayama, K. 2000. Solar hydrogen production: significant effect of Na_2CO_3 addition on water splitting using simple oxide semiconductor photocatalysts. Catal Surv Jpn 4:75–80.

Arni, S. 2004. Hydrogen-rich gas production from biomass via thermochemical pathways. Energy Edu Sci Technol 13:47–54.

Arpe, H. (Ed.) 1989 Ullmann's Encyclopedia of industrial chemistry. Hydrogen sulfide. 5th edn. Vol. A13, VCH, Weinheim, pp. 467–485.

Asahi, R., Morikawa, T., Ohwaki, T., Aoki, K., Tao, Y. 2001. Visible-light photocatalysis in nitrogen-doped titanium oxides. Science 293:269–271.

Ashokkumar, M., Kudo, A., Saito, N., Sakata, T. 1994. Semiconductor sensitization by RuS_2 colloids on TiO_2 electrodes. Chem Phys Lett 229:383–388.

Ashokkumar, M., Maruthamuthu, P. 1989. Factors influencing the photocatalytic efficiency of WO_3 particles. J Photochem Photobiol A 49:249–258.

Bak, T., Nowotny, J., Rekas, M., Sorrell, C.C. 2002. Photoelectrochemical hydrogen generation from water using solar energy: materials-related aspects. Int J Hydrogen Energy 27:991–1022.

Balat, M., Ozdemir, N. 2005. New and renewable hydrogen production processes. Energy Sources 27:1285–1298.

Bamwenda, G.R., Arakawa, H. 2000. Cerium dioxide as a photocatalyst for water decomposition to O-2 in the presence of Ce-aq(4+) and Fe-aq(3+) species. J Mol Catal A 161:105–113.

Baykara, S.Z., Ayfer Kale, A., Veziroglu, T.N. 2005. Possibilities for hydrogen production from H_2S in Black Sea. Proceedings International Hydrogen Energy Congress and Exhibition IHEC 2005, Istanbul, Turkey, 13–15 July 2005.

Baykara, S.Z., Figen, E.H., Kale, A., Veziroglu, T.N. 2007. Hydrogen from hydrogen sulphide in Black Sea. Int J Hydrogen Energy 32:1246–1250.

Berry, G.D., Pasternak, A.D., Rambach, G.D., Smith, J.R., Schock, R.N. 1996. Hydrogen as a future transportation fuel. Energy 21:289–303.

Bessekhouad, Y., Trari, M. 2002. Photocatalytic hydrogen production from suspension of spinel powders AMn_2O_4 (A=Cu and Zn). Int J Hydrogen Energy 27:357–362.

Bhaumik, A., Inagaki, S. 2001. Mesoporous titanium phosphate molecular sieves with ion-exchange capacity J Am Chem Soc 123:691–696.

Bishara, A., Salman, O.A., Khraishi, N., Marafi, A. 1987. Thermochemical decomposition of hydrogen sulfide by solar energy. Int J Hydrogen Energy 12:679–685.

Bogdanovic, B., Brand, R.A., Marjanovic, A., Schwikardi, M., Tolle, J. 2000. Metal doped sodium aluminum hydrides as potential new hydrogen storage materials. J Alloys Comp 302:36–58.

Bogdanovic, B., Schwickardi, M. 1997. Ti-doped alkali metal aluminum hydrides as potential novel reversible hydrogen storage materials. J Alloys Comp 253–254:1–9.

Borgarello, E., Kiwi, J., Graetzel, M., Pelizzetti, E., Visca, M. 1982. Visible-light induced water cleavage in colloidal solutions of chromium-doped titanium-dioxide particles. J Am Chem Soc 104:2996–3002.

Brown, W.G., Kaplan, L., Wilzbach, K.E. 1952. The exchange of hydrogen gas with lithium and sodium borohydrides. J Am Chem Soc 74:1348.

Buhler, N., Meier, K., Reber, J.F. 1984. Photochemical hydrogen-production with cadmium sulfide suspensions. J Phys Chem 88:3261–3268.

Buntkowsky, G., Walaszek, B., Adamczyk, A., Xu, Y., Limbach, H.-H., Chaudret, B. 2006. Mechanism of nuclear spin initiated para-H_2 to ortho-H_2 Conversion. Phys Chem Chem Phys 16:1929–1935.

Caglar, A., Ozmen, H. 2000. Hydrogen: as an attractive fuel in future. Energy Edu Sci Technol 6:1–18

Cao, D., Bergens, S.H. 2004. Pt–Ru_{adatom} nanoparticles as anode catalysts for direct methanol fuel cells. J Power Sources 134:170–180.

Cerveramarch, S.C., Borrell, L., Gimenez, J., Simarro, R., Andujar, J.M. 1992. Solar hydrogen photoproduction from sulfide sulfite substrate. Int J Hydrogen Energy 17:683–688.

Chaurasia, P.B.L., Ando, Y., Tanaka, T. 2003. Regenerative fuel cell with chemical reactions. Energy Convers Manage 44:611–628.

Chen, P., Wu, X., Lin, J., Tan, K.L. 1999. High H_2 uptake by alkali-doped carbon nanotubes under ambient pressure and moderate temperatures. Science 285:91–93.

Choi, W., Termin, A., Hoffmann, M. 1994. The role of metal ion dopants in quantum-sized TiO_2. J Phys Chem 98:13669–13679.

Choudhary, T.V., Goodman, D.W. 1999. Stepwise methane steam reforming: a route to CO-free hydrogen. Catal Lett 59:93–94.

Choudhary, T.V., Goodman, D.W. 2000. CO-free production of hydrogen via stepwise steam reforming of methane. J Catal 192:316–321.

Clark, P.D., Dowling, N.I., Hyne, J.B., Moon, D.L. 1995. Production of hydrogen and sulfur from hydrogen sulfide in refineries and gas processing plants. Alberta Sulphur Res Quart Bull 32:11–28.

Cotton, F.A., Wilkinson, G. 1967. Advanced Inorganic Chemistry. Interscience, New York.

Dagan, G., Tomkiewicz, M. 1994. Preparation and characterization of TiO_2 aerogels for use as photocatalysts. J NonCryst Sol 175:294–302.

Das, L.M. 1996. Hydrogen–oxygen reaction mechanism and its implication to hydrogen engine combustion. Int J Hydrogen Energy 21:703–715.

David, E. 2005. An overview of advanced materials for hydrogen storage. J Materials Proc Technol 162–163:169–177.

De, G.C., Roy, A.M., Bhattacharya, S.S. 1996. Effect of n-Si on the photocatalytic production of hydrogen by Pt-loaded CdS and CdS/ZnS catalyst. Int J Hydrogen Energy 21:19–23.

Demirbas, A. 1998. Determination of combustion heat of fuels by using non-calorimetric experimental data. Energy Edu Sci Technol 1:7–12.

Demirbas, A. 2002. Fuel properties of hydrogen, liquefied petroleum gas (LPG), and compressed natural gas (CNG) for transportation. Energy Sources 24:601–610.

Demirbas, A. 2007. Storage and transportation opportunities of hydrogen. Energy Sources Part B 2:287–295.

Demirbas, A. 2008. Nano-catalytic liquefaction of hydrogen. Energy Sources Part A 30:1540–1547.

Desilvestro, J., Neumannspallart, M. 1985. Photoredox reactions on semiconductors at open circuit reduction of Fe^{3+} on WO_3 electrodes and particle suspensions. J Phys Chem 89:3684–3689.

Dillon, A.C., Heben, M.J. 2001. Hydrogen storage using carbon adsorbents: past, present and future. Appl Phys A 72:133–142.

Dillon, A.C., Jones, K.M., Bekkedahl, T.A., Kiang, C.H., Bethune, D.S., Heben, M.J. 1997. Storage of hydrogen in single-walled carbon nanotubes. Nature 386:377–378.

Dinga, G.P. 1988. Hydrogen: The ultimate fuel and energy carrier. J Chem Educ 65:688–691.

Domen, K., Kudo, A., Onishi, T. 1986. Mechanism of photocatalytic decomposition of water into H-2 and O-2 over NiO – $SrTiO_3$. J Catal 102:92–98.

Domen, K., Naito, S., Onishi, T., Tamaru, K. 1982. Photocatalytic decomposition of liquid water on a NiO–$SrTiO_3$ catalyst. Chem Phys Lett 92:433–434.

Durant, P.J., Durant, B. 1970. Introduction to advanced inorganic chemistry. Longman, London.

Fakioglu, E., Yurum, Y., Veziroglu, T.N. 2004. A review of hydrogen storage systems based on boron and its compounds. Int J Hydrogen Energy 29:1371–1376.

Fan, Y.-Y., Liao, B., Liu, M., Wei, Y.-L. Liu, M.-Q., Cheng, H.-M. 1999. Hydrogen uptake in vapor grown carbon nanofibers. Carbon 37:1649–1652.

Faherburch, A.L., Bube, R.H. 1983. Fundamentals of Solar Cells. Academic, New York.

Farkas, A. 1935. Orthohydrogen, parahydrogen and heavy hydrogen. Cambridge University Press, Cambridge.

Fletcher, E.A., Noring, J.E., Murray, J.P. 1984. Hydrogen-sulfide as a source of hydrogen. Int Hydrogen Energy 9:587–593.

Fonash, S.J. 1981. Solar cells device physics. Academic, New York.

Fujishima, A., Honda, K. 1972. Electrochemical photolysis of water at a semiconductor electrode. Nature 238:37–38.

Fujishima, A., Rao, T.N., Tryk, D.A. 2000. Titanium dioxide photocatalysis. J Photochem Photobiol C Photochem Rev 1:1–21.

Galinska, A., Walendziewski, J. 2005. Photocatalytic water splitting over Pt TiO_2 in the presence of sacrificial reagents. Energy Fuels 19:1143–1147.

Goerrig, D. 1958. DBP German Patent 1,077,644.

Green, M.A. 1982. Solar cells. Prentice-Hall, Englewood Cliffs, NJ.

Gringue, D., Horowitz, G., Garnier, E. 1987. Stabilization of CdS and CdSe photoelectrodes modified by a catalyst-containing polythiophene coating. Ber Buns Phys Chem 91:402–405.

Gross, K.J., Thomass, G.J., Jensen, C.M. 2002. Catalyzed alanates for hydrogen storage. J Alloys Comp 330–332:683–690.

Gupta, B.K., Srivastava, O.N. 2000. Synthesis and hydrogenation behaviour of graphitic nanofibres. Int J Hydrogen Energy 25:825–830.

Gurunathan, K. 2000. Photobiocatalytic production of hydrogen using sensitized TiO–MV^{2+} system coupled *Rhodopseudomonas* capsulate. J Mol Catalysis A Chem 156:59–67.

Gurunathan, K., Maruthamuthu, P. 1995. Photogeneration of hydrogen using visible-light with undoped doped alpha-Fe_2O_3 in the presence of methyl viologen. Int J Hydrogen Energy 20:287–295.

Gurunathan, K., Maruthamuth, P., Sastri, M.V.C. 1997. Photocatalytic hydrogen production by dye-sensitized Pt/SnO_2 and Pt/SnO_2/RuO_2 in aqueous methyl viologen solution. Int J Hydrogen Energy 22:57–62.

Hacisalihoglu, B., Demirbas, A.H., Hacisalihoglu, S. 2008. Hydrogen from gas hydrate and hydrogen sulfide in the Black Sea. Energy Edu Sci Technol 21:109–115.

He, P., Gu, G., Shao, L., Zhang, Y. 2000. Research on low temperature thermo-chemical conversion to oil process from sewage sludge. Water Sci Tech 42:301–308.

He, Z., Chen, J., Liu, D., Zhou, H., Kuang, Y. 2004. Electrodeposition of Pt–Ru nanoparticles on carbon nanotubes and their electrocatalytic properties for methanol electrooxidation. Diamond Related Mater 13:1764–1770.

Herrmann, J.M., Disdier, J., Pichat, P. 1986. Photoassisted platinum deposition on TiO_2 powder using various platinum complexes. J Phys Chem 90:6028–6034.

Hitoki, G., Takata, T., Kondo, J., Hara, M., Kobayashi, H., Domen, K. 2002. An oxynitride, TaON, as an efficient water oxidation photocatalyst under visible light irradiation (lambda $\lambda \leq 500$ nm). Chem Commun 16:1698–1699.

Hohhmann, P. 2002. Tomorrow's energy: hydroen, fuel cells, and the prospects for a cleaner planet. MIT Press, Cambridge, MA, London, UK, 2002.

Höhlein, B., von Andrian, S., Grube, Th., Menzer, R. 2000. Critical assessment of power trains with fuel-cell systems and different fuels. J Power Sources 86:243–249.

Huang, C., Raissi, T.-A. 2005. Analysis of sulfur–iodine thermochemical cycle for solar hydrogen production. Part I: decomposition of sulfuric acid. Solar Energy 78:632–646.

Huhey, E.J. 1978. Inorganic chemistry. Harper and Row, New York.

Hwang, D.W., Kim, H.G., Jang, J.S., Bae, S.W., Ji, S.M., Lee, J.S. 2004. Photocatalytic decomposition of water–methanol solution over metal-doped layered perovskites under visible light irradiation. Catal Today 93–5:845–850.

Hwang, D.W., Kim, J., Park, T.J., Lee, J.S. 2002. Mg-doped WO_3 as a novel photocatalyst for visible light-induced water splitting. Catal Lett 80:53–57.

Iijima, S. 1991. Helical microtubules of graphitic carbon. Nature 354:56–57.

Inoue, Y., Kubokawa, T., Sato, K. 1991. Photocatalytic activity of alkali-metal titanates combined with Ru in the decomposition of water. J Phys Chem 95:4059–4063.

Ishihara, T., Nishiguchi, H., Fukamachi, K., Takita, Y. 1999. Effects of acceptor doping to $KTaO_3$ on photocatalytic decomposition of pure H_2O. J Phys Chem B 103:1–3.

Ishikawa, A., Takata, T., Kondo, J.N., Hara, M., Kobayashi, H., Domen, K. 2002. Oxysulfide $Sm_2Ti_2S_2O_5$ as a stable photocatalyst for water oxidation and reduction under visible light irradiation (lambda $\lambda \leq 650$ nm). J Am Chem Soc 124:13547–13553.

Jang, J.S., Kim, H.G., Reddy, V.R., Bae, S.W., Ji, S.M., Lee, J.S. 2005. Photocatalytic water splitting over iron oxide nanoparticles intercalated in $HTiNb(Ta)O_5$ layered compounds. J Catal 231:213–222.

Jensen, C.M., Zidan, R., Mariels, N., Hee, A., Hagen, C. 1999. Advanced titanium doping of sodium aluminum hydride: segue to a practical hydrogen storage material? Int J Hydrogen Energy 24:461–465.

Kamat, P.V. 1989. Photoelectrochemistry in particulate systems 9. Photosensitized reduction in a colloidal TiO_2 system using anthracene-9-carboxylic acid as the sensitizer. J Phys Chem 93: 859–864.

Kamat, P.V., Fox, M.A. 1983. Photo-sensitization of TiO_2 colloids by Erythrosin-B in acetonitrile. Chem Phys Lett 102:379–384.

Kapoor, M.P., Inagaki, S., Yoshida, H. 2005. Novel zirconium-titanium phosphates mesoporous materials for hydrogen production by photoinduced water splitting. J Phys Chem B 109:9231–9238.

Kato, H., Kudo, A. 1998. New tantalite photocatalysts for water decomposition into H_2 and O_2. Chem Phys Lett 295:487–492.

Kato, H., Kudo, A. 1999a. Photocatalytic decomposition of pure water into H_2 and O_2 over$SrTa_2O_6$ prepared by a flux method. Chem Lett 11:1207–1208.

Kato, H., Kudo, A. 1999b. Highly efficient decomposition of pure water into H_2 and O_2 over $NaTaO_3$ photocatalysts. Catal Lett 58:153–155.

Kawai, T., Sakata, T. 1980. Photocatalytic hydrogen production from liquid methanol and water. J Chem Soc Chem Commun 15:694–695.

Kennedy, I.H., Dunnwald, D. 1983. Photooxidation of organic-compounds at doped alpha Fe_2O_3 electrodes. J Electrochem Soc 130:2013–2016.

Kennedy, L.A., Fridman, A.A., Saveliev, A.V. 1995. Superadiabatic combustion in porous media: wave propagation, instabilities, new type of chemical reactor. Fluid Mechanics Res 22:1–25.

Khan, S.U.M., Al-Shahry, M., Ingler, W.B. 2002. Efficient photochemical water splitting by a chemically modified n-TiO_2. Science 297:2243–2245.

Kida, T., Guan, G., Yoshida, A. 2003. $LaMnO_3$/CdS nanocomposite: a new photocatalyst for hydrogen production from water under visible light irradiation. Chem Phys Lett 371:563–567.

Kida, T., Guan, G., Yamada, N., Ma, T., Kimura, K., Yoshida, A. 2004. Hydrogen production from sewage sludge solubilized in hot-compressed water using photocatalyst under light irradiation. Int J Hydrogen Energy 29:269–274.

Kiuchi, H., Nakamura, T., Funaki, K., Tanaka, T. 1982. Recovery of hydrogen from hydrogen sulfide with metals and metal sulfides. Int J Hydrogen Energy 7:477–482.

Kiwi, J., Morrison, C. 1984. Heterogeneous photocatalysis-dynamics of charge-transfer in lithium-doped anatase-based catalyst powders with enhanced water photocleavage under ultraviolet-irradiation. J Phys Chem 88:6146–6152.

Knacke, O., Kubaschewski, O., Hesselmann, K. 1991. Thermochemical Properties of Inorganic Substances, 2nd edn. Springer, Berlin Heidelberg New York.

Koca, A., Sahin, M. 2002. Photocatalytic hydrogen production by direct sun light from sulfide/sulfite solution. Int J Hydrogen Energy 27:363–367.

Koca, A., Sahin, M. 2003. Photocatalytic hydrogen production by direct sunlight: a laboratory experiment. J Chem Edu 80:1314–1315.

Kohno, M., Kaneko, T., Ogura, S., Sato, K., Inoue, Y. 1998. Dispersion of ruthenium oxide on barium titanates ($Ba_6Ti_{17}O_{40}$, $Ba_4Ti_{13}O_{30}$, $BaTi_4O_9$ and $Ba_2Ti_9O_{20}$) and photocatalytic activity for water decomposition. J Chem Soc Faraday Trans 94:89–94.

Kormann, C., Bahnemann, D.W., Hoffmann, M.R. 1988. Preparation and characterization of quantum-size titanium-dioxide. J Phys Chem 92:5196–5201.

Koroneos, C., Dompros, A., Roumbas, G., Moussiopoulos, N. 2005. Advantages of the use of hydrogen fuel as compared to kerosene. Res Conser Recyc 44:99–113.

Kotera, Y., Todo, N., Fukuda, K. 1976. Process for production of hydrogen and sulfur from hydrogen sulfide as raw material. US Patent No. 3,962,409.

Kudo, A. 2001. Development of photocatalyst materials for water splitting with the aim at photon energy conversion. J Ceram Soc Jpn 109:81–88.

Kudo, A., Tanaka, A., Domen, K., Maruya, K., Aika, K., Onishi, T. 1988. Photocatalytic decomposition of water over NiO–$K_4Nb_6O_{17}$ catalyst. J Catal 111:67–76.

Kudo, A., Kato, H., Nakagawa, S. 2000. Water splitting into H_2 and O_2 on new $Sr_2M_2O_7$ (M=Nb and Ta) photocatalysts with layered perovskite structures: factors affecting the photocatalytic activity. J Phys Chem B 104:571–575.

Lee, J., Kato, T., Fujishima, A., Honda, K. 1984. Photoelectrochemical oxidation of alcohols on polycrystalline zinc-oxide. Bull Chem Soc Jpn 57:1179–1183.

Linsebigler, A., Lu, G., Yates, J.T. 1995. Co-chemisorption on TiO_2-oxygen vacancy site influence on co-adsorption. J Chem Phys 103:9438–9443.

Lippard, S.J., Ucko, D.A. 1968. Transition metal borohydride complexes. II. Th reaction of copper(I) compounds with boron hydride anions. Inorg Chem 7:1051–1058.

Litter, M.I. 1999. Heterogeneous photocatalysis: transition metal ions in photocatalytic systems. Appl Catal B Environ 23:89–114.

Luinstra, E. 1996. Hydrogen from H_2S: a review of the leading processes. Proceedings of 7th GRI Sulfur Recovery Conference, Gas Research Institute, Chicago, pp. 149–165.

Mackor, A., Blasse, G. 1981. Visible-light induced photocurrents in $SrTiO_3$-$LaCrO_3$ single crystalline electrodes. Chem Phys Lett 77:6–8.

Malinowska, B., Walendziewski, J., Robert, D., Weber, J.V., Stolarski, M. 2003a. Titania aerogels: preparation and photocatalytic tests. Int J Photoenergy 5:147–152.

Malinowska, B., Walendziewski, J., Robert, D., Weber, J.V., Stolarski, M. 2003b. The study of photocatalytic activities of titania and titania-silica aerogels. Appl Catal B Environ 46:441–451.

Maruthamuthu, P., Ashokkumar, M. 1988. Hydrogen generation using Cu(II)/WO_3 and oxalic acid by visible-light. Int J Hydrogen Energy 13:677–680.

Maruthamuthu, P., Ashokkumar, M., Gurunathan, K., Subramanian, E., Sastri, M.V.C. 1989. Hydrogen evolution from water with visible radiation in presence of Cu(II)/WO_3 and electron relay. Int J Hydrogen Energy 14:525–528.

Maruthamuthu, P., Gurunathan, K., Subramanian, E., Sastri, M.V.C. 1993. Visible-light induced hydrogen-production with Cu(II)/Bi_2O_3 and Pt/ Bi_2O_3/RuO_2 from aqueous methyl viologen solution. Int J Hydrogen Energy 18:9–13.

Maruthamuthu, P., Muthu, S., Gurunathan, K.,Ashokkumar, M., Sastri, M.V.C. 1992. Photobiocatalysis: hydrogen evolution using a semiconductor coupled with photosynthetic bacteria. Int J Hydrogen Energy 17:863–866.

Megalofonos, S.K., Papayannakos, N.G. 1997. Kinetics of catalytic reaction of methane and hydrogen sulphide over MoS2. J Appl Catal A General 165:249–258.

Mesmer, R.E., Jolly, W.L. 1962. The exchange of deuterium with solid potassium hydroborate. J Am Chem Soc 84:2039–2042.

Midilli, A., Olgun, H., Ayhan, T. 2000. Solar hydrogen production from hazelnut shells. Int J Hydrogen Energy 25:723–732.

Midilli, A., Dogru, M., Akay, G., Howarth, C.R. 2002. Hydrogen production from sewage sludge via a fixed bed gasifier product gas. Int J Hydrogen Energy 27:1035–1041.

Midilli, A., Ay, M., Dincer, I., Rosen, M.A. 2005. On hydrogen and hydrogen energy strategies I: current status and needs. Renew Sust Energy Rev 9:255–271.

Miller, E., Rocheleau, R., Khan, S. 2004. A hybrid multijunction photoelectrode for hydrogen production fabricated with amorphous silicon/germanium and iron oxide thin films. Int J Hydrogen Energy 29:907–914.

Miranda, R. 2004. Hydrogen from lignocellulosic biomass via thermochemical processes. Energy Edu Sci Technol 13:21–30.

Moon, J., Takagi, H., Fujishiro, Y., Awano, M. 2001. Preparation and characterization of the Sb doped TiO_2 photocatalysts. J Mater Sci 36:949–955.

Moon, S.C., Mametsuka, H., Tabata, S., Suziki, E. 2000. Photocatalytic production of hydrogen from water using TiO_2 and B/TiO_2. Catal Today 58:125–132.

Namon, S.A., Aliwi, S.M., Alemara, K. 1986. Hydrogen-production from the splitting of H_2S by visible-light irradiation of vanadium sulfides dispersion loaded with RuO_2. Int J Hydrogen Energy 11:33–38.

Navvio, J.A., Testa, J.J., Djedjeian, P., Padron, J.R., Rodriguez, D., Litter, M.I. 1999. Iron-doped titania powders prepared by a sol-gel method. Appl Catal A General 178:191–203.

Ni, M., Leung, K.H.M., Leung, D.Y.C., Sumathy, K. 2007. A review and recent developments in photocatalytic water-splitting using TiO_2 for hydrogen production. Renew Sustain Energy Rev 11:401–425.

Ni, M., Leung, K.H.M., Sumathy, K., Leung, D.Y.C. 2004. Water electrolysis: a bridge between renewable resources and hydrogen. Proceedings of the International Hydrogen Energy Forum, Beijing, PRC, pp. 475–480.

Nikandrov, V.V., Shlyk, M.A., Zorin, N.A., Gogotov, I.N., Krosnovsky, A.A. 1988. Efficient photoinduced electron-transfer from inorganic semiconductor TiO_2 to bacterial hydrogenase. Febs Lett 234:111–114.

Nishikawa, T., Nakajima, T., Shinohara, Y. 2001. An exploratory study on effect of the isomorphic replacement of TiO_2. J Mol Struct Theochem 545:67–74.

Noganow, L.S. 1992. Hydrogen. McGraw-Hill Encyclopedia of Science & Technology, 7th edn. Vol. 8. McGraw-Hill, New York, pp. 581–588.

Ogura, S., Kohno, M., Sato, K., Inoue, Y. 1997. Photocatalytic activity for water decomposition of RuO_2-combined $M_2Ti_6O_{13}$ (M=Na, K, Rb, Cs). Appl Surf Sci 121:521–524.

Ohno, T., Saito, S., Fujihara, K., Matsumura, M. 1996. Photocatalyzed production of hydrogen and iodine from aqueous solutions of iodide using platinum-loaded TiO_2 powder. Bull Chem Soc Jpn 69:3059–3064.

Okamoto, K., Yamamoto, Y., Tanaka, H., Itaya, A. 1985. Kinetics of heterogeneous photocatalytic decomposition of phenol over anatase TiO_2 powder. Bull Chem Soc Jpn 58:2023–2028.

Oosawa, Y. 1984. Photocatalytic decomposition of aqueous hydroxylamine solution over anatase and precious metal anatase. J Phys Chem 88:3069–3074.

Ostroff, A., Sanderson, R. 1957. Oxidation and thermal decomposition of sodium and potassium borohydrides. J Inorg Nuclear Chem 4:230–231.

Ozturk, T., Demirbas, A. 2007. Boron compounds as hydrogen storage materials. Energy Sources Part A 29:1415–1423.

Parker, S.F., Taylor, J.W., Albers, P., Lopez, M., Sextl, G., Lennon, D., McInroy, A.R., Sutherland, I.W. 2004. Inelastic neutron scattering studies of hydrogen on fuel cell catalysts. Vibration Spectrosc 35:179–182.

Petrov, K. 1995. The Black Sea and hydrogen energy. Int J Hydrogen Energy 16:805–808.

Plass, J.H., Barbir, F., Miller, H.P., Veziroglu, T.N. 1990. Economics of hydrogen as a fuel for surface transportation. Int J Hydrogen Energy 15:663–668.

Piotrowska, A., Walendziewski, J. 2005. Photocatalytic hydrogen production from the water over titania aerogels under UV irradiation. Proceedings International Hydrogen Energy Congress and Exhibition IHEC 2005, Istanbul, Turkey.

Ranjit, K.T., Cohen, H., Willner, I., Bossmann, S., Braun, A. 1999. Lanthanide oxide-doped titanium dioxide: effective photocatalysts for the degradation of organic pollutants. J Mater Sci 34:5273–5280.

Reber, J.F., Meier, K. 1984. Photochemical production of hydrogen with zinc-sulfide suspensions. J Phys Chem 88:5903–5913.

Rifkin, J. 2002. The hydrogen economy. Jeremy P. Tarcher/Penguin, New York.

Rocheleau, R.E., Vierthaler, M. 1994. Optimization of multijunction a-Si:H solar cells using an integrated optical/electrical model, *in* Proceedings of the 21st World Conference on Photovoltaic Energy Conversion, pp. 567–570. Institute for Electrical and Electronics Engineers, Honolulu, HI.

Rostrup-Nielsen, J.R. 1984. Catalytic steam reforming, *in* Catalysis: science and technology, Vol. 5, J.R. Anderson, M. Boudart. Springer, Berlin Heidelberg New York, pp. 1–117.

Sandrock, G., Thomas, G. 2001.The IEA/DOE/SNL on-line hydride databases. Appl Phys A Mater Sci Proc 72A:153–155.

Sato, J., Saito, N., Nishiyama, H., Inoue, Y. 2001. New photocatalyst group for water decomposition of RuO_2-loaded p-block metal (In, Sn, and Sb) oxides with d^{10}configuration. J Phys Chem B 105:6061–6063.

Sayama, K. Arakawa, H. 1992a. Significant effect of carbonate addition on stoichiometric photodecomposition of liquid water into hydrogen and oxygen from platinum titanium (IV) oxide suspension. J Chem Soc Chem Commun 2:150–152.

Sayama, K., Arakawa, H. 1992b. Remarkable effect of Na_2CO_3 addition on photodecomposition of liquid water into H_2 and O_2 from suspension of semiconductor powder loaded with various metals. Chem Lett 2:253–256.

Sayama, K., Arakawa, H. 1996. Effect of carbonate addition on the photocatalytic decomposition of liquid water over a ZrO_2 catalyst. J Photochem Photobiol A Chem 94:67–68.

Sayama, K., Arakawa, H. 1993. Photocatalytic decomposition of water and photocatalytic reduction of carbon-dioxide over ZrO_2 catalyst. J Phys Chem 97:531–533.

Sayama, K., Arakawa, H. 1994. Effect of Na_2CO_3 addition on photocatalytic decomposition of liquid water over various semiconductor catalysts. J Photochem Photobiol A Chem 77:243–247.

Schlapbach, L., Züttel, A. 2001. Hydrogen-storage materials for mobile applications. Nature 414:353–358.

Schlesinger, H.J., Brown, H.C., 1940. Metallo borohydrides. III. Lithium borohydride. J Am Chem Soc 62:3429–3435.

Schriver, D.F., Atkins, P.W. 1990. Langford CH. Inorganic Chemistry. ELBS, Oxford.

Serban, A., Nissenbaum, A. 2000. Light induced production of hydrogen from water by catalysis with ruthenium melanoidins. Int J Hydrogen Energy 25:733–737.

Serpone, N., Borgarello, E., Gratzel, M. 1984. Visible-light induced generation of hydrogen from H_2S in mixed semiconductor dispersions: improved efficiency through inter-particle electron transfer. Chem Commun 6:342–344.

Shahbazov, J., Usupov, I. 1994. Non-trading sources of energy for hydrogen. Int J Hydrogen Energy 19:863–864.

Silvera, I.F., Pravica, M.G. 1998. Hydrogen at megabar pressures and the importance of ortho–para concentration. J Phys Condens Matter 10:11169–11177.

Slimane, R.B., Lau, F.S., Dihu, R., Bingue, J.P., Saveliev, A.V., Fridman, A.A., Kennedy, L.A. 2002. Production of hydrogen by superadiabatic decomposition of hydrogen sulfide, 14th World Hydrogen Energy Conference, Montreal, Quebec, Canada, 9–14 June 2002.

Slimane, R.B., Lau, F.S., Khinkis, M., Bingue, J.P., Saveliev, A.V., Kennedy, L.A. 2004a. Conversion of hydrogen sulfide to hydrogen by superadiabatic partial oxidation: thermodynamic consideration. Int J Hydrogen Energy 29:1471–1477.

So, W.W., Kim, K.J., Moon, S.J. 2004. Photo-production of hydrogen over the CdS–TiO_2 nanocomposite particulate films treated with $TiCl_4$. Int J Hydrogen Energy 29:229–234.

Stasinevich, D.S., Egorenko, G.A. 1968. A thermographic investigation of sodium hydroborate-sodium hydride system. Russ J Inorg Chem 133:341–343.

Stewart, A.T., Squires, G.L. 1955. Analysis of ortho- and para-hydrogen mixtures by the thermal conductivity method. J Sci Instrum 32:26–29.

Stolarek, P., Ledakowicz, S. 2001. Thermal processing of sewage sludge by drying, pyrolysis, gasification and combustion. Water Sci Tech 44:333–339.

Surmen, Y., Demirbas, A. 2002. Thermochemical conversion of residual biomass to hydrogen for Turkey. Energy Sources 24:403–411.

Suzuki, A., Nakamura, T., Yokoyama, S. 1990. Effect of operating parameters on thermochemical liquefaction of sewage sludge. J Chem Eng Jpn 23:6–11.

Suzuki, A., Nakamura, T., Yokoyama, S., Ogi, T., Koguchi, K. 1988. Conversion of sewage sludge to heavy oil by direct thermochemical liquefaction. J Chem Eng Jpn 21:288–293.

Takata, T., Shinohara, K., Tanaka, A., Hara, M., Kondo, J.N., Domen, K. 1997a. A highly active photocatalyst for overall water splitting with a hydrated layered perovskite structure. J Photochem Photobiol A: Chem 106:45–49.

Takata, T., Furumi, Y., Shinohara, K., Tanaka, A., Hara, M., Kondo, J.N., Domen, K. 1997b. Photocatalytic decomposition of water on spontaneously hydrated layered perovskites. Chem Mater 9:1063–1064.

Talavera, R.R., Vargas, S., Arroyo-Murillo, R., Montiel-Campos, R., Haro-Poniatowski, E. 1997. Modification of the phase transition temperatures in titania doped with various cations. J Mater Res 12:439–443.

Timmerhaus, C., Flynn, T.M. 1989. Cryogenic engineering. Plenum, New York.

Tryk, D.A., Fujishima, A., Honda, K. 2000. Recent topics in photoelectrochemistry: achievements and future prospects. Electrochim Acta 45:2363–2376.

Tsuzuki, K., Eiki, H., Inoue, N., Sagara, A., Noda, N., Hirohata, Y., Hino; T. 1999. Hydrogen absorption/desorption behavior with oxygen contaminated boron film. J Nuclear Mat 266–269:247–250.

Turker, L. 2001. Diborane–tetraborane conversion in C60 vesicles-a theoretical study. Int J Hydrogen Energy 26:837–842.

Urban, D.L., Antal, M.J. 1982. Study of the kinetics of sewage pyrolysis using DSC and DTA. Fuel 61:799–806.

Veziroglu, T. 1975. Hydrogen Energy, Part B. Plenum, New York.

Viswanathan, B. 2006. An introduction to energy sources. Indian Institute of Technology, Madras, India.

Wang, H. 2007. Hydrogen production from a chemical cycle of H_2S splitting. Int J Hydrogen Energy 32:3907–3914.

Wang, P., Orimo, S., Tanabe, K., Fujii, H. 2003. Hydrogen in mechanically milled amorphous boron. J Alloys Comp 350:218–221.

Williamson, K., Edeskuty, F. 1986. Recent developments in hydrogen technology, Vol. I–II. CRC, Boca Raton, FL.

Wu, J., Uchida, S., Fujishiro, Y., Yin, S., Sato, T. 1999a. Synthesis and photocatalytic properties of $HTaWO_6$/(Pt, TiO_2) and $HTaWO_6$/(Pt, Fe_2O_3) nanocomposites. Int J Inorg Mater 1:253–258.

Wu, J., Uchida, S., Fujishiro, Y., Yin, S., Sato, T. 1999b. Synthesis and photocatalytic properties of $HNbWO_6/TiO_2$ and $HNbWO_6/Fe_2O_3$ nanocomposites. J Photochem Photobiol A Chem 128:129–133.

Wu, J., Yin, S., Lin, Y., Lin, J.M., Huang, M.L., Sato, T. 2001. Hydrothermal synthesis of $HNbWO_6$/MO series nanocomposites and their photocatalytic properties. J Mat Sci 36:3055–3059.

Yokoyama, S., Suzuki, A., Murakami, M., Ogi, T., Koguchi, K., Nakamura, E. 1987. Liquid fuel production from sewage sludge by catalytic conversion using sodiumcarbonate. Fuel 66:1150–1155.

Yoshimura, J., Ebina, Y., Kondo, J., Domen, K., Tanaka, A. 1993. Visible-light induced photocatalytic behavior of a layered perovskite type niobate. J Phys Chem B 97:1970–1973.

Yvon, K. 1998. Complex transition-metal hydrides. Chimia 52:613–619.

Zaluska, A., Zaluski, L., Strom-Oslen, J.O. 2000. Sodium alanates for reversible hydrogen storage. J Alloys Comp 298:125–134.

Zaluski, L., Zaluska, A., Strom-Olsen, J.O. 1999. Hydrogenation properties of complex alkali metal hydrides fabricated by mechano-chemical synthesis. J Alloys Comp 290:71–78.

Zaman, J., Chakma, A. 1995. Production of hydrogen and sulphur from hydrogen sulphide. Fuel Proc Technol 41:159–198.

Zhou, L. 2005. Progress and problems in hydrogen storage methods. Renew Sustain Energy Rev 9:395–408.

Zou, Z.G., Ye, J.H., Sayama, K., Arakawa, H. 2001. Direct splitting of water under visible light irradiation with an oxide semiconductor photocatalyst. Nature 424:624–627.

Züttel, A., Wenger, P., Rentsch, S., Sudan, P., Maurona, P., Emmenegger, C. 2003. $LiBH_4$ a new hydrogen storage material. J Power Sources 118:1–7.

Züttel, A., Wenger, P., Sudan, P., Maurona, P., Orimo, S. 2004. Hydrogen density in nanostructured carbon, metals and complex materials. Mater Sci Eng B 108:9–18.

Chapter 6
Biohydrogen

6.1 Introduction

Hydrogen is the element of greatest abundance in the universe; however, its production from renewable resources remains a major challenge. Biohydrogen produced from biorenewables is a promising alternative for a sustainable energy source. Biohydrogen is a renewable biofuel produced from biorenewable feedstocks by chemical, thermochemical, biological, biochemical, and biophotolytical methods. As a sustainable energy supply with minimal or zero use of hydrocarbons, hydrogen is a promising alternative to fossil fuel. It is a clean and environmentally friendly fuel, which produces water instead of greenhouse gases when combusted.

The topic of renewable resources, biohydrogen in particular, includes energy security, environmental concerns, foreign exchange savings, and socioeconomic issues related to the rural sectors of all countries in the world. Biohydrogen is an environmentally friendly alternative automotive fuel that can be used in an internal combustion engine. Due to its environmental merits, the share of biohydrogen in the automotive fuel market will grow fast in the next decade. There are several reasons for biohydrogen to be considered as a relevant technology by both industrialized and developing countries.

The most known industrial methods for producing hydrogen include steam reformation of natural gas, coal gasification, and splitting water with electricity typically generated from carbonaceous fuels. These energy-intensive industrial processes release carbon dioxide and other greenhouse gases and pollutants as byproducts. In addition to thermochemical hydrogen production by gasification, hydrogen can be produced biologically. There are currently two proposed pathways to use organisms for hydrogen production. Direct production through "photobiohydrogen" utilizes microorganisms capable of using solar photons to separate oxygen from water. Some microorganisms produce hydrogen naturally, and biotechnologies based on these microbial systems could lead to the development of clean, renewable sources of hydrogen.

The biorenewable feedstocks can be converted to petroleum-substitute fuels. The majority of the bioengineering strategies for biochemically derived fuels involve options for the disposition of organic matter produced via photosynthate. The bulk of the presently exploited photosynthate is directed toward the production of wood, food, and feed. During processing and consumption, biomass materials are generated which can be used for energy production via combustion, pyrolysis or biochemical conversions to ethanol, hydrogen, methane, and isopropanol. A second option is to engineer the photosynthetic apparatus to provide hydrogen. The third strategy is the cultivation of crops as energy sources via the foregoing processes. Biomass has the potential to accelerate the realization of hydrogen as a major fuel of the future. However, obtaining hydrogen from biomass has major challenges. There are no completed technology demonstrations. The yield of hydrogen is low from biomass since the hydrogen content in biomass is low to begin with (approximately 6% versus 25% for methane) and the energy content is low due to the 40% oxygen content of biomass.

6.2 Definition

The term biohydrogen (from the Greek words *bio* meaning *life* + *hydro* meaning *water* + *genes* meaning *born of* or *forming*) refers to non-fossilized and biodegradable organic material originating from plants, animals and microorganisms derived from biological sources. Biohydrogen is defined as hydrogen produced biologically (mostly by bacteria) as a biofuel from waste organic materials. Fermentative biohydrogen production plants are proposed industrial plants for the production of hydrogen. They would typically involve processes such as thermophillic fermentation, dark fermentation and/or photofermentation and gas cleaning. Biohydrogen production can also involve an element of anaerobic digestion where the methane from biogas is converted through steam reforming into hydrogen.

Hydrogen produced by microalga and bacteria is biohydrogen (Benemann, 1998). There are currently no practical biohydrogen production processes. However, several concepts have promise for near to long-term process development (Benemann, 1998).

Biohydrogen can comprehensively be described as hydrogen, which is produced chemically, thermochemically, biologically, biochemically, and biophotolytically from all biomass materials.

6.3 History

Sir E. Ray Lankester (1847–1929) was a prominent biologist who did research on a wide variety of organisms, from protozoa to mammals (Gest and Blankenship, 2004). In 1939 Hans Gaffron (German researcher), while working at the University of Chicago, obtained fermentative and photochemical hydrogen from algae (Gaffron, 1939). Gaffron and Rubin first reported that *Scenedesmus,* a green microalga, evolved molecular hydrogen under light conditions after being kept in anaerobic and dark conditions (Gaffron and Rubin, 1942).

Gest and co-workers studied the metabolism of photosynthetic bacteria: the photochemical production of molecular hydrogen by growing cultures of photosynthetic bacteria (Gest and Kamen, 1949a; Ormerod and Gest, 1962). Gest and Kamen discovered the light-dependent production of H_2 and N_2 fixation by *Rsp. rubrum*. Subsequent studies revealed that many anoxygenic phototrophs have N_2 fixation capacity (Gest and Kamen, 1949b; Kamen and Gest, 1949), and that *Rsp. rubrum* can use H_2 and CO_2 (as the sole carbon source) for photoautotrophic growth in a synthetic medium (Ormerod et al., 1961). Hydrogen gas is a product of the mixed acid fermentation of *Escherichia coli*, the butylene glycol fermentation of *Aerobacter*, and the butyric acid fermentations of *Clostridium* spp. (Alba et al., 1973).

The production of biohydrogen using bacteria and microalga started some 35 years ago with a meeting on "Biological Energy Conservation." Benemann and co-workers studied the hydrogen evolution from water using *Clostridium kluyveri* hydrogenase in 1973. Spinach chloroplast preparations were mixed with *Clostridium kluyveri* hydrogenase and ferredoxin. These findings indicate that hydrogen evolution from water and sunlight by photosynthetic processes could be a method for solar energy conversion. In the summer of 2001, researchers manipulated the photosynthetic process of spinach plants to produce hydrogen. But these biological means of hydrogen production are known only as laboratory experiments. Intense research persists to better understand ways to improve these hydrogen production methods. Biological hydrogen production has been known for over a century and research directed at applying this process to a practical means of hydrogen fuel production has been carried out for over a quarter century. Dark fermentation of biomass or wastes presents an alternative route to biological hydrogen production that has been little studied (Hallenbeck and Benemann, 2002).

Researchers have investigated hydrogen production with anaerobic bacteria since the 1980s (Nandi and Sengupta, 1998). Biohydrogen seems particularly suitable for relatively small-scale, decentralized systems, integrated with agricultural and industrial activities or waste processing facilities for the future renewable energy demands (Kotay and Das, 2008).

Direct hydrogen conversion from sunlight may offer higher theoretical efficiencies (Akkerman et al., 2002) compared to the 0.2 to 2.6% photosynthetic efficiencies in energy crops today (Klas, 2004). However, the scientific and technical feasibility of the process requires more development, and the costs of a photobioreactor will also be a challenge. Indirect biohydrogen involves oxygen inhibition of the electron transfer step and hydrogenase production of hydrogen from carbohydrates (Masukawa et al., 2002).

Several reviews have examined the potential of biological hydrogen production (Madamwar et al., 2000; Melis and Happe, 2001; Levin et al., 2004; Nath and Das, 2004; Prince and Kheshgi, 2005). Although microorganisms produce hydrogen by different mechanisms, the step can be represented by the simple chemical reaction

$2H^+ + 2e \rightarrow H_2$. This reaction is known to be catalyzed by either nitrogenase or hydrogenase enzymes. Although alternative biological hydrogen production pathways exist, each with its own set of advantages and disadvantages, the following discussion on biohydrogen production will focus on challenges that must be overcome to improve one type of biological hydrogen production known as biophotolysis. Biophotolysis holds potential for the scale of hydrogen production necessary to meet future energy demand. One advantage of biophotolysis is that it is a more efficient conversion of solar energy to hydrogen. Theoretically, the maximal energetic efficiency for direct biophotolysis is about 40% (Prince and Kheshgi, 2005). Appropriate enzymes then perform an electrochemical process that unites two electrons from the oxygen-producing step with two H^+ ions to make hydrogen as the other product. Indirect hydrogen production through photosynthesis in microalgae uses a dark fermentation step where the carbohydrate produced by photosynthesis is converted to hydrogen gas.

Hydrogen-intensive research work has already been carried out on the advancement of these processes, such as the development of genetically modified microorganisms, metabolic engineering, improvement of the reactor designs, use of different solid matrices for the immobilization of whole cells, biochemical assisted bioreactor, development of two-stage processes, etc., for higher H_2 production rates. Pilot plant trials of the photofermentation processes require more attention. Use of cheaper raw materials and efficient biological hydrogen production processes will surely make them more competitive with the conventional H_2 generation processes in the near future (Das and Veziroglu, 2008).

6.4 Hydrogen from Biorenewables via Biological Processes

Biological generation of biohydrogen technologies provide a wide range of approaches to generate hydrogen, including direct biophotolysis, indirect biophotolysis, photofermentations, and dark fermentation by a combination of these processes. It is important to develop hydrogen generation methods that are renewable like biohydrogen. Biohydrogen production processes are found to be more environmentally friendly and less energy intensive as compared to thermochemical and electrochemical processes.

There are three types of microorganisms that generate hydrogen: cyanobacteria, anaerobic bacteria, and fermentative bacteria. The cyanobacteria directly decompose water to hydrogen and oxygen in the presence of light energy by photosynthesis. Photosynthetic bacteria use organic substrates like organic acids. Anaerobic bacteria use organic substances as the sole source of electrons and energy, converting them into hydrogen. Biohydrogen can be generated using bacteria such as *Clostridia* by controlling temperature, pH, reactor hydraulic retention time (HRT) and other factors of the treatment system.

Biological hydrogen can be generated from plants by biophotolysis of water using microalgae (green algae and cyanobacteria), fermentation of organic compounds, and photodecomposition of organic compounds by photosynthetic bacteria. To produce hydrogen by fermentation of biomass, a continuous process using a non-sterile substrate with a readily available mixed microflora is desirable (Hussy et al., 2005). A successful biological conversion of biomass to hydrogen depends strongly on the processing of raw materials to produce feedstock, which can be fermented by the microorganisms.

Biochemical and biological processes have many advantages:

1. No unwanted and hazardous byproducts are formed.
2. They occur at ambient temperatures and pressures.
3. No special equipment is needed.
4. All are renewable energy sources.
5. They are ecofriendly processes.

Green chemistry offers cleaner processes for energy abatement. Some of these energy conversion processes are:

1. Photosynthesis
2. Glycolysis
3. Nitrogen fixation
4. Fermentation processes

Although some of the steps in photosynthesis are still not fully understood, the overall photosynthetic reaction has been known since the 1800s. Jan van Helmont began the research of the process in the mid-1600s when he carefully measured the mass of the soil used by a plant and the mass of the plant as it grew. After noticing that the soil mass changed very little, he hypothesized that the mass of the growing plant must come from the water, the only substance he added to the potted plant. Photosynthesis is the most important biochemical process in which plants, algae, and some bacteria harness the energy of sunlight to produce food. Organisms that produce energy through photosynthesis are called photoautotrophs. Photosynthesis is a process in which green plants utilize the energy of sunlight to manufacture carbohydrates from carbon dioxide and water in the presence of chlorophyll (Viswanathan, 2006).

The initial process in photosynthesis is the decomposition of water into oxygen and hydrogen and oxygen will be released. The simplified photosynthesis pathways are given in Fig. 6.1 and Eqs. 6.1 and 6.2. The hydrogen and the carbon and oxygen of carbon dioxide are then converted into formaldehyde (Fig. 6.1). Then, a series of increasingly complex compounds results finally in a stable organic compound, hexose (typically glucose) (Eq. 6.1). This phase of photosynthesis utilizes stored energy and therefore can proceed in the dark. Hexose is polymerized into hexosan (typically glucosan) or carbohydrates (Eq. 6.2).

Glucose formation:

$$\underset{\text{Formaldehyde}}{6CH_2O} \rightarrow \underset{\text{Glucose}}{C_6H_{12}O_6} \tag{6.1}$$

Glucosan formation via polymerization:

$$\underset{\text{Glucose}}{nC_6H_{12}O_6} \rightarrow \underset{\text{Glucosan}}{(C_6H_{12}O_6)n} \tag{6.2}$$

In general, the results of this process are the reverse of those in respiration, in which carbohydrates are oxidized to release energy, with the production of carbon dioxide and water. The intermediary reactions before glucose is formed involve several enzymes, which react with the coenzyme ATP (adenosine triphosphate) to produce various molecules. The main product, glucose, is the fundamental building block of carbohydrates (e.g., sugars, starches, and cellulose). Cellulose is used to build the rigid cell walls that are the principal supporting structure of plants. Plants are photoautotrophs, which mean they are able to synthesize food directly from inorganic compounds using light energy, instead of eating other organisms or relying on material derived from them. The energy for photosynthesis ultimately comes from absorbed photons and involves a reducing agent, which is water in the case of plants, releasing oxygen as a waste product. The light energy is converted to chemical energy, in the form of ATP and NADPH, using the light-dependent reactions and is then available for carbon fixation (Viswanathan, 2006).

Energy and carbon are obtained by organisms either directly or indirectly via the photosynthetic conversion of solar energy. The majority of the bioengineering strategies for biochemically derived fuels involve options for the disposition of organic matter produced via photosynthate. During processing and consumption, waste organic materials are generated which can be used for energy production via combustion, pyrolysis or biochemical conversions to ethanol, hydrogen, methane, and isopropanol.

Nitrogen fixation is the process by which nitrogen is taken from its relatively inert molecular form in the atmosphere and converted into nitrogen compounds useful for other chemical processes. Hydrogen gas is also released during the biological nitrogen fixation process.

A promising method is the biological production of hydrogen by fermentation. Fermentative hydrogen production is the fermentative conversion of organic substrate to biohydrogen manifested by a diverse group of bacteria using multienzyme systems involving three steps similar to anaerobic conversion. Fermentative/hydrolytic microorganisms hydrolyze complex organic polymers to monomers, which further convert to a mixture of lower molecular weight organic acids and alcohols by obligatory H_2-producing acidogenic bacteria. The production of biohydrogen occurs in the second and third step of anaerobic degradation. Thereby, methanogenic bacteria must be inhibited to avoid the consumption of H_2 to produce CH_4. The fermentative production of biohydrogen is characterized by a lower technical complexity compared to photofermentation. At that light-independent hydrogen production

$$CO_2 + H_2O + \text{Solar energy} \xrightarrow{T\,>285\text{ K, Chlorophyll}} CH_2O + O_2$$

Fig. 6.1 Formaldehyde formation by the photosynthesis process

anaerobic microorganisms are always involved, converting the organic compounds to organic acids, hydrogen and carbon dioxide. The advantages of the fermentative hydrogen production are the broad spectrum of applicable substrates as well as high hydrogen production yields (Nandi and Sengupta, 1998; Hawkes et al., 2002).

Hydrogen production from the bacterial fermentation of sugars has been examined in a variety of reactor systems. Hexose concentration has a greater effect on H_2 yields than the hydraulic retention time (HRT). Flocculation also was an important factor in the performance of the reactor (Van Kinkel and Logan, 2005). Anaerobic fermentation was conducted to improve hydrogen fermentation of food waste in a leaching-bed reactor by heat-shocked anaerobic sludge, and also to investigate the effect of dilution rate on the production of hydrogen and metabolites in hydrogen fermentation (Han and Shin, 2004).

Direct photobiological hydrogen production by photosynthetic microorganisms is an active, developing field nowadays. Realization of technical processes for large-scale photobiological hydrogen production from water, using solar energy, would result in a major, novel source of sustainable, environmentally friendly and renewable energy. The unique biological process of photosynthesis in which solar energy is used to split water is combined with the natural capacity to combine obtained products into hydrogen, catalyzed by enzymes called hydrogenases. In nature, only cyanobacteria and green algae possess water oxidizing photosynthesis and hydrogen production, providing the option to form hydrogen from the Sun and water (Prince and Kheshgi, 2005; Viswanathan, 2006).

An increase in the production of hydrogen from biomass-derived glucose and attainment of the maximum molar yield of hydrogen, can be achieved through the enzymes of the pentose phosphate cycle in conjunction with a hyperthermophilic hydrogenase. This process includes centers on three $NADP^+$ ($NADP^+$ to form NADPH, nicotinamide adenine dinucleotide phosphate) dependent enzymes, glucose-6 phosphate dehydrogenase (G6PDH), 6-phosphogluconate dehydrogenase (6PGDH) and hydrogenase from *Pyrococcus furiosus*. The dehydrogenases are currently obtained from mesophilic sources (Viswanathan, 2006).

The enzymatic conversion of cellulosic waste to hydrogen via an in vitro enzymatic pathway involves the conversion of potential glucose sources such as cellulose by cellulases and plant sap (i.e., sucrose) by invertase and glucose isomerase to glucose. Glucose, the sugar produced by photosynthesis, is also renewable, unlike fossil fuels such as oil. The glucose substrate is then oxidized and the cofactor, $NADP^+$ is simultaneously reduced. The presence of a pyridine-dependent hydrogenase in this system causes the regeneration and recycling of $NAD(P)^+$ with the concomitant production of molecular hydrogen. The overall aim is to increase the production of hydrogen from biomass-derived glucose and achieve the maximum molar yield of hydrogen by employing the enzymes of the pentose phosphate pathway in conjunction with the hydrogenase from *Pyrococcus furiosus*. The main advantage over hydrogen production by fermentation is that close-to-theoretical yields of hydrogen from sugar would be possible (Das and Veziroglu, 2008; Viswanathan, 2006).

Hydrogen is produced in microorganisms by enzymes capable of reducing free protons to molecular hydrogen. Examples of these enzymes include hydrogenases and the nitrogenases. The production of hydrogen by these enzymes is usually coupled to some other biochemical processes. The energy used by these enzymes is usually in multiple steps from an organism's central energy input and is provided in the form of electron carriers such as ferredoxin or NADPH and energy yielding molecules like ATP. Obtaining useful amounts of hydrogen from microorganisms will require increasing the efficiency of hydrogenases and overcoming other obstacles. One problem is that some hydrogenases and nitrogenases are inhibited by oxygen. Oxygen is produced by photosystem II (PSII) during oxygenic photosynthesis. The driving force for reaction energy can be obtained by the breaking of bonds (the multistep breakdown of glucose to generate ATP and CO_2) or from electronic excitation. For example, plants, algae, cyanobacteria and photosynthetic bacteria can use light energy to raise electrons into higher energy states. In the case of plants, algae and cyanobacteria, the source of excitable electrons is water. The excited electrons are stripped from water, which then splits into oxygen and protons (Das and Veziroglu, 2008; Viswanathan, 2006).

6.4.1 *Hydrogen Production via Microbial Fermentation of Biomass*

In microbial fermentation of biomass, different waste materials can be used as substrates. A new and unique process has been developed when substrates such as carbohydrates are fermented by a consortium of bacteria; they produce hydrogen and carbon dioxide. Highly concentrated organic wastewater is one of the most abundantly available biomasses, which can be exploited for microbial conversion into hydrogen (Nath and Das, 2003). Municipal solid wastes and digested sewage sludge have the potential to produce large amounts of hydrogen by suppressing the production of methane by introducing low voltage electricity into the sewage sludge. The substrate from the acidogenesis of fruit and vegetable market wastes gives higher hydrogen evolution rates (about threefold) compared to synthetic medium. A mixed culture of photosynthetic anaerobic bacteria provides a method of utilization of a variety of resources for biohydrogen production (Miyaka et al., 1990).

6.4.2 *Anaerobic Hydrogen Production*

Anaerobic hydrogen production proceeds from photofermentation as well as without the presence of light. Anaerobic bacteria use organic substances as the sole source of electrons and energy, converting them into hydrogen.

$$\text{Glucose} + 2H_2O \rightarrow 2\text{Acetate} + 2CO_2 + 4H_2 \qquad (6.3)$$

$$\text{Glucose} \rightarrow \text{Butyrate} + 2CO_2 + 2H_2 \qquad (6.4)$$

The reactions involved in hydrogen production (Eqs. 6.3 and 6.4) are rapid and these processes do not require solar radiation, making them useful for treating large quantities of wastewater by using a large fermenter.

Since they cannot utilize light energy, the decomposition of organic substrates is incomplete. Further decomposition of the acetic acid is not possible under anaerobic conditions. Nevertheless, these reactions are still suitable for the initial steps of wastewater treatment and hydrogen production, followed by further waste treatment stages.

A new fermentation process that converts valueless organic waste streams into hydrogen-rich gas has been developed by Van Ginkel et al. (2001). The process employs mixed microbial cultures readily available in nature, such as compost, anaerobic digester sludge, soil, etc., to convert organic wastes into hydrogen-rich gas. The biodegradation efficiencies of the pollutants were examined by changing hydraulic retention time (HRT) as a main operating variable. An enriched culture of hydrogen-producing bacteria such as *Clostridia* was obtained by heat treatment, pH control and HRT control of the treatment system. The biohydrogen fermentation technology could enhance the economic viability of many processes utilizing hydrogen as a fuel source or as raw materials. Figure 6.2 shows the basic components of an anaerobic digestion system.

Anaerobic fermentative microorganisms, cyanobacteria and algae are suitable in biological production of hydrogen via hydrogenase due to reversible hydrogenases (Adams, 1990). Cyanobacteria and algae can carry out photoevolution of hydrogen catalyzed by hydrogenases. The reactions are similar to electrolysis involving splitting of water into oxygen and hydrogen (Gaffron, 1940).

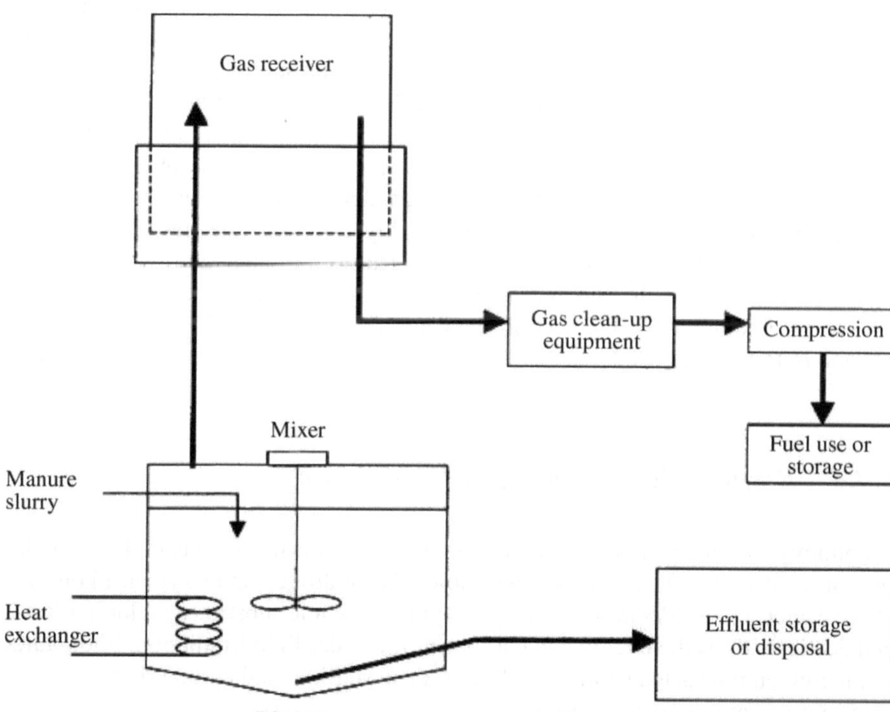

Fig. 6.2 Basic components of an anaerobic digestion system

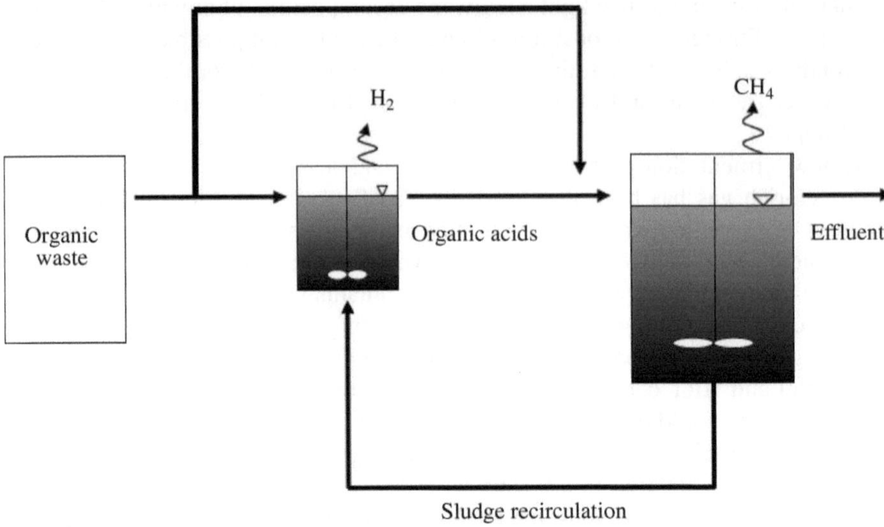

Fig. 6.3 Schematic diagram of the two-stage reactor system

Hydrogen-producing bacteria (*Clostridia*) were found to have growth rates about five to ten times higher than that of methane-producing bacteria (Van Ginkel et al., 2001; Sung, 2004). In a continuous flow bioreactor system, hydrogen production showed a declining trend at the later stage of reactor operation. Based on these findings, it is hypothesized that *Clostridia* may have gone through a phenomenon known as "degeneration" in which they lose their ability to produce hydrogen. Therefore, inoculating fresh mixed cultures may be a feasible way to maintain a sustainable hydrogen production. Based on this hypothesis, a two-stage anaerobic reactor has been proposed. The first-stage reactor is designed as a hydrogen-producing reactor whereas the second-stage reactor will be employed to cultivate fresh seed culture to perpetually supply the first one. Figure 6.3 shows the schematic diagram of the two-stage reactor system. The hydrogen-producing reactor has a total volume of 5 l with active volume of 3 l. The second stage has a total volume of 22 l, and the active volume is 18 l (Sung, 2004).

6.4.3 *Biophotolytic Hydrogen Production*

Biophotolysis is the action of light on biological systems that results in the dissociation of a substrate, usually water into molecular hydrogen and oxygen. Photosynthetic bacteria (e.g., *Rhodobactor*) can use broad organic substrates including lactic and acetic acids as the energy and carbon source under light irradiation. Photoautotrophic green microalgae and cyanobacteria use sunlight and carbon dioxide as the sole sources for energy and carbon.

Cyanobacteria are potential microbial species for hydrogen production via direct biophotolysis. By using nitrogenase and/or bi-directional hydrogenase, both heterocystous nitrogen-fixing strains and unicellular non-nitrogen-fixing strains are able to evolve hydrogen under special conditions. Hydrogen production increased significantly when replacing the air with argon (Howarth and Codd, 1985; Lindblad et al., 2002).

Under special conditions, the reducing equivalents (ferredoxin) can also be used by hydrogenase or nitrogenase to reduce protons for evolution of molecular hydrogen. Using Fe-hydrogenase and reduced ferredoxin is the substrate for hydrogen evolution. Anaerobic adaptation is needed for the cells to induce reversible Fe-hydrogenase. In order to overcome the inhibitive effect of oxygen on hydrogenase, various methods have been investigated with limited success, such as spatial separation of oxygen from hydrogen, immobilization of chloroplasts, electron carrier and hydrogenase, oxygen scavenging, and gas purging (Miura, 1995; Miura et al., 1997).

A major cost factor in large-scale hydrogen production is the photobioreactor that is usually characterized by surface area for light irradiation (Benemann, 1997). A hydrogen productivity based on reactor surface area can be found from the volumetric productivity and a critical optical length. Photobiological production of hydrogen from water (biophotolysis) requires efficient biological converters, microalgae, and low-cost photobioreactors. Microalgal strains must be developed that exhibit high hydrogen production rates and photosynthetic efficiencies in dense cultures at full solar intensities. The photobioreactor must expose the hydrogen-producing cultures to sunlight, while allowing recovery of the gas. A preliminary analysis is presented of a two-stage process in which microalgae are cultivated in large open ponds to produce a high carbohydrate biomass that then produces hydrogen in tubular photobioreactors. Photobioreactors constructed of inexpensive, commercially available glass tubes are proposed for such applications. Photobiological hydrogen production requires long-term research and development (Benemann, 1997).

Biological hydrogen production is often conducted in two stages under different atmospheric conditions, the first stage for cell growth followed by the second stage for hydrogen evolution. Nitrogen starvation is often applied at the end of the growth stage as an efficient metabolic stress to induce the activity of nitrogenase.

The atmosphere plays an important role in hydrogen evolution by cyanobacteria and could be a cost factor in large-scale hydrogen production. A nitrogen-free gas phase such as argon plus carbon dioxide gives a high hydrogen evolution rate (Weissman and Benemann, 1977).

Chlamydomonas reinhardtii, a facultative photoautotrophic and photoheterotrophic microalga, is the representative of green microalgae for biohydrogen research (Tsygankov, 2006; White and Melis, 2006). Other algal species such as *Chlorococcum littorale* and *Platymonas subcordiformis* are also investigated for hydrogen evolution (Guan et al., 2004). Gaffron and Robins first observed that unicellular microalga *Scenedesmus obliquus* could either use hydrogen as an electron donor in carbon dioxide fixation or evolve hydrogen under anaerobic conditions in dark or light environments (Gaffron and Robins, 1942). *Chlamydomonas reinhardtii* evolves large quantities of hydrogen gas when deprived of sulfur (Vijayaraghavan and Soom, 2006).

6.4.4 Dark Fermentative Hydrogen Production

Fermentative bacteria producing hydrogen may be cultivated in pure or mixed cultures selected from natural sources such as anaerobically digested sewage sludge or soil. The major pathways for microbial hydrogen production are driven by the anaerobic metabolism of pyruvate, formed during the catabolism of various substrates. Production of hydrogen by fermentation has been studied for a large group of pure fermentative bacteria, such as *Clostridia.* The biohydrogen production from bagasse based on dark fermentation was investigated using microorganisms like *Bacillus licheniformis*, *Clostridium pasteurianum* and *Enterobacter cloacae* (Vijayaraghavan and Soom, 2006).

Dark fermentation is the fermentative conversion of organic substrate to biohydrogen, it is a complex process manifested by a diverse group of bacteria by a series of biochemical reactions involving three steps similar to anaerobic conversion. Dark fermentation differs from photofermentation because it proceeds without the presence of light.

Dark fermentation for hydrogen production from wastes could be competitive with fossil fuel-derived hydrogen, providing a plausible approach to practical biohydrogen production. However, most of the research on biohydrogen production from organic wastes over the past two decades has focused on the use of photosynthetic bacteria (Sasikala et al., 1993). Benemann (1998) concluded that the dark fermentation of wastes is a promising process for biohydrogen production. The biohydrogen from carbohydrates was investigated using hyper thermophilic bacteria based on dark anaerobic fermentation (Claassen et al., 1999) and a high-rate anaerobic trickling filter (Groenestijn et al., 2002).

Scenedesmus, a green microalga, evolved molecular hydrogen under light conditions after being kept in anaerobic and dark conditions (Gaffron and Rubin, 1942). The responsible enzyme for hydrogen evolution is a reversible hydrogenase because it catalyzes the reaction in both directions (Mertens and Liese, 2004). Benemann and Weare (1974) reported that a nitrogen-fixing cyanobacterium, *Anabaena cylindrica,* evolved hydrogen and oxygen gas simultaneously in an argon atmosphere for several hours. The enzyme involved in hydrogen evolution is nitrogenase, which is responsible for the reduction of nitrogen into ammonia (Masukawa et al., 2002).

In photosynthesis, the reduced carbon is stored as endogenous carbohydrates, such as starch in microalgae and glycogen in cyanobacteria (Dauvillee et al., 2006). Following Gaffron and Rubin's discovery (Gaffron and Robins, 1942), studies on the mechanisms involved in hydrogen evolution have found that the electrons or reducing equivalents of hydrogenase and nitrogenase do not always come from water, but may sometimes originate from the intracellular energy reserve including carbohydrates (Miura et al., 1995; Antal and Lindblad, 2005). The algal fermentation conditions for the hydrogen production were investigated to improve the photobiological hydrogen production by a combination of a marine green alga *Chlamydomonas* sp. MGA161 and a marine photosynthetic bacterium *Rhodopseudomonas* sp. W-1S. The scaling-up of our biophotolysis system with algal-bacterial combination was carried out. The stably sustained starch accumulation and degradation in the algal cells and hydrogen photoproduction from algal fermentation products by photosynthetic bacteria occurred in an alternating light-dark cycle (Miura et al., 1995).

The stored energy is released through fermentation of the endogenous carbohydrates in dark conditions, and the excess reducing power can be deposited by hydrogenase on protons forming molecular hydrogen (Gfeller and Gibbs, 1984). Hydrogen evolution from endogenous carbon reserve under dark anaerobic conditions looks very similar to the conventional anaerobic hydrogen fermentation, but the endogenous carbon reserve is made in vivo during photosynthesis. In this sense, the electrons or reducing equivalents in indirect biophotolysis are derived from water by photoautotrophic cells. This indirect biophotolysis, therefore, consists of two stages in series: photosynthesis for carbohydrate accumulation, and dark fermentation of the carbon reserve for hydrogen production. In this way, the oxygen and hydrogen evolutions are temporally and/or spatially separated (Benemann, 1996). Miura and co-workers proposed a process of hydrogen production via indirect biophotolysis by using natural light/dark cycles (Miura et al., 1997). According to this proposal, carbon dioxide is reduced to starch by photosynthesis in daytime, and the starch thus formed, is fermented to hydrogen gas and organic acids under anaerobic conditions during nighttime. The organic acids and other fermentative products can be further used for hydrogen evolution by photosynthetic bacteria under light irradiation (Melis and Melnicki, 2006).

Few studies have been conducted on fermentative hydrogen production by green microalgae. Compared with green microalgae, unicellular non-nitrogen-fixing cyanobacteria have attracted more research interest for hydrogen production via indirect biophotolysis. *Spirulina platensis* is a filamentous cyanobacterium cultivated at large commercial scales as a food supplement with high proteins, antioxidants and other nutrients. *S. platensis* NIES-46 accumulates high glycogen content (50% dry mass) in nitrogen-limited conditions and evolves molecular hydrogen at a moderate rate of 0.11 mmol/g dry wt/h after induction in dark and nitrogen atmosphere (Aoyama et al., 1987). For 1 mol of hydrogen evolved, the cells also release 1.4 mol of acetate, 0.65 mol of ethanol, 0.4 mol of formate and 0.1 mol of lactate as electron acceptors.

A unicellular non-nitrogen-fixing cyanobacterium *Gloeocapsa alpicola* CALU743 showed a high activity of reversible hydrogenase induced in dark anaerobic conditions (Serebryakova et al., 1998). In dark fermentation, 1 mol glucose and the strain produces 3.92 mol of hydrogen, 1.83 mol of CO_2, 1.97 mol of acetate, and small amounts of lactate and ethanol. The hydrogen yield is very close to the theoretical value when acetate is the sole fermentative product. The energy obtained in 4 mol of hydrogen is about 40% of the energy in 1 mol of glucose. The remainder of glucose energy is mainly left in 2 mol of acetic acid. The overall energy efficiency would therefore be significantly improved if acetate could be reused to form endogenous carbon reserve by facultative photoautotrophic cells in photosynthesis.

Carbohydrates are the preferred organic carbon source for hydrogen-producing fermentations. Glucose in the biomass yielded 2 and 4 mol H_2 per mol glucose with butyrate and acetic acid as their respective fermentation product. Hydrogen can be produced sustainably by anaerobic bacterial growth on carbohydrate-rich substrates, giving organic fermentation end-products such as hydrogen and carbon dioxide.

Microbial community composition dynamics was studied during hydrogen fermentation from glucose in a fluidized-bed bioreactor (FBR) aimed at obtaining insight into the hydrogen fermentation microbiology and factors resulting in the instability of biofilm processes. FBR hydrogen production performance was characterized by an instable pattern of the prompt onset of hydrogen production followed by a rapid decrease. Gradual enrichment of organisms increased the diversity of the FBR attached and suspended growth phase bacterial communities during the operation. A high glucose loading rate favored the hydrogen production by an *Escherichia coli* affiliated strain. Decrease in H_2 production, associated with a shift from acetate–butyrate to acetate–propionate production, was due to changes in the FBR attached and suspended growth phase bacterial community compositions. During the shift, organisms, including potential propionate producers, were enriched in the communities while the proportion trend of *C. butyricum* decreased. Suggestions of reactions taking place during anaerobic, dark fermentation of glucose in the fluidized-bed bioreactor include (Koskinen et al., 2007):

Ethanol formation:

$$C_6H_{12}O_6 \rightarrow 2CH_3CH_2OH + 2CO_2 \tag{6.5}$$

Acetate formation:

$$C_6H_{12}O_6 + 4H_2O \rightarrow 2CH_3COO^- + 4H_2 + 2HCO^{3-} + 4H^+ \tag{6.6}$$

Propionate formation:

$$C_6H_{12}O_6 + 2H_2 \rightarrow 2CH_3CH_2COO^- + 2H^+ + 2H_2O \tag{6.7}$$

Butyrate formation:

$$C_6H_{12}O_6 + 2H_2O \rightarrow CH_3CH_2CH_2COO^- + 2H_2 + 2HCO^{3-} + 3H^+ \tag{6.8}$$

6.5 Hydrogen from Biorenewables via Thermochemical Processes

Hydrogen can be produced from biorenewable feedstocks via thermochemical conversion processes such as pyrolysis, gasification, steam gasification, steam reforming of bio-oils, and supercritical water gasification (SWG) of biomass.

The strategy is based on producing hydrogen from biomass pyrolysis using a co-product strategy to reduce the cost of hydrogen. It has been suggested that only this strategy could compete with the cost of the commercial hydrocarbon-based technologies (Wang et al., 1998). This strategy will demonstrate how hydrogen and biofuel are economically feasible and can foster the development of rural areas when practiced on a larger scale. The process of biomass to activated carbon is an alternative route to hydrogen with a valuable co-product that is practiced commercially. The yield of hydrogen that can be produced from biomass is relatively low,

12–14% based on the biomass weight (Demirbas, 2005a). The proposed second process is fast pyrolysis of biomass to generate bio-oil and catalytic steam reforming of the bio-oil to hydrogen and carbon dioxide.

In the pyrolysis and gasification processes, the water–gas shift is used to convert the reformed gas into hydrogen, and pressure swing adsorption is used to purify the product. The cost of hydrogen production from supercritical water gasification of wet biomass was several times higher than the current price of hydrogen from steam methane reforming (Demirbas, 2005a). The pyrolysis was carried out at moderate temperatures and steam gasification at the highest temperatures. The pyrolysis-based technology, in particular, because it has co-product opportunities, has the most favorable economics.

The yield from steam gasification increases with increasing water-to-sample ratio. The yields of hydrogen from the pyrolysis and the steam gasification increase with increase of temperature. In general, the gasification temperature is higher than that of pyrolysis and the yield of hydrogen from the gasification is higher than that of the pyrolysis. The highest yields (% dry and ash-free basis) were obtained from the pyrolysis (46%) and steam gasification (55%) of wheat straw, while the lowest yields from olive waste. The yield of hydrogen from supercritical water extraction was considerably high (49% by volume) at lower temperatures.

The catalytic aqueous phase reforming might prove useful for the generation of hydrogen-rich gas from carbohydrates extracted from renewable biomass and biomass waste streams. The biomass-derived hydrocarbons are suitable to hydrogen generation from biomass, as well as for the reforming.

It is believed that in the future biomass can become an important sustainable source of hydrogen. Biomass has the advantage of low environmental impact compared with that for fossil fuels. The price of hydrogen obtained by direct gasification of lignocellulosic biomass, however, is about three times higher than that for hydrogen produced by steam reforming of natural gas (Spath et al., 2000).

Hydrogen from organic wastes has generally been based on the following thermochemical reactions:

1. Production of hydrogen by pyrolysis of biomass

$$\text{Biomass} \rightarrow H_2 + CO + CO_2 + \text{Hydrocarbon gases} + \text{Tar} + \text{Char} \tag{6.9}$$

2. Production of hydrogen by catalytic steam gasification of biomass

$$\text{Biomass} + H_2O \rightarrow H_2 + CO_2 + CO \tag{6.10}$$

3. Production of hydrogen by air gasification of biomass

$$\text{Biomass} + H_2O + \text{Air} \rightarrow H_2 + CO_2 + CO + N_2 \tag{6.11}$$

Steam reforming is the major process for the production of hydrogen presently. This process has several disadvantages. For example, it is a thermally inefficient process (about 90% including the convection zone) and there are mechanical and maintenance issues. The process is difficult to control and reforming plants require a large capital investment. Hence to meet the increasing demand for this future fuel, alternatives to reforming processes are essential.

Biomass-derived synthesis gas can provide a renewable route to hydrogen. A novel bacterial process has been proposed as an alternative to the conventional high-temperature catalytic process for the production of hydrogen from synthesis gas via the water–gas shift (WGS) reaction. Hydrogen can be produced via pyrolysis or gasification of biomass resources such as agricultural residues like peanut shells, consumer wastes including plastics and waste grease, or biomass specifically grown for energy uses. Biomass pyrolysis produces a liquid product (bio-oil) that contains a wide spectrum of components that can be separated into valuable chemicals and fuels, including hydrogen.

6.5.1 Potential of Renewable Hydrogen Production

Hydrogen is produced today dominantly from carbonaceous raw materials. Milne et al. (2002) reported approximately 95% of the hydrogen produced comes from fossil fuels conversion, such as natural gas reforming. With the growing concern about global climate changes, the currently used way of producing hydrogen will have to change, as CO_2 carbon emission is always associated with such processes. Therefore alternative methodologies should be identified to overcome the shortage of this process. Using biomass as a hydrogen carrier in the near and mid-term appears viable in terms of technology, economy and environment as well as availability of sources. Even if without the subsequent sequestration of carbon, the whole process is still environment friendly as biomass is a CO_2-neutral resource in the life cycle (Chen et al., 2003).

Biomass has been considered as potential feedstock for gasification to produce syngas (H_2 + CO). An alternative approach to the production of H_2 from biomass begins with fast pyrolysis of biomass (Wang et al., 1997). Production of hydrogen gas by gasification of carbon with superheated steam was reported (Lower, 1963). The different methods of hydrogen production are: production from solid biomass (e.g., pellets of dedicated energy crops, waste biomass), fermentation of liquid manure and biological hydrogen production. The advantage of direct hydrogen production from biomass is that renewable energy sources can be utilized without the need of electrolysis thus leading to higher system efficiency and a more favorable overall result. Hydrogen can be extracted from hydrogen-rich biomass sources like wood chips and agricultural waste.

Biomass is available and obtainable in the world through the rational collection of byproducts from agricultural and forestry industries. The potential supply of biomass is significantly large as an energy crop.

Biomass can be used to produce hydrogen or hydrogen-rich gas via different technical pathways, i.e., anaerobic digestion, fermentation, metabolic processing, high-pressure supercritical conversion, gasification and pyrolysis. Compared with other pathways, gasification and pyrolysis appear technoeconomically viable at the current stage. A certain number of efforts have already been made to test gasification or pyrolysis of biomass for the production of hydrogen (Milne et al., 2002).

Those efforts can be categorized based on the types of biomass used, types of reactors (gasifiers or pyrolyzer), the addition of catalysts, end-products (e.g., hydrogen itself, hydrogen intermediates, hydrogen-rich gases) and gasifying medium used. The biomass types tested by previous researchers are residue coming from extraction of grass and lucern for protein, agricultural wastes (rice straw, wheat stalk), wood particles, municipal solid waste, almond shells, sawdust, black liquor, lignin, hazelnut shells, cotton and tobacco wastes, and plastic/rubber wastes (Demirbas et al., 1996; Caglar and Demirbas, 2002; Chen et al., 2003). The use of different catalysts such as dolomite, zeolites, potassium minerals, nickel-based compounds, platinum-based compounds, and metal oxides have been summarized in literature (Milne et al., 2002; Chen et al., 2002). As for the gasifying medium, the tested medium is steam, air, and oxygen. In addition, the gasification or pyrolysis process needs heating sources, and the used sources are electrical energy, combustion heat of fossil fuel or biomass char, solar energy, thermonuclear, and hydrothermal energy. Indirect pyrolysis of biomass to hydrogen through intermediates (pyrolysis oils) was also interestingly investigated (Chornet, 2001; Czernik et al., 2001).

6.5.2 Production of Hydrogen from Biomass via Pyrolysis

Pyrolysis dates back to at least ancient Egyptian times, when tar for caulking boats and certain embalming agents were made by pyrolysis. In the 1980s, researchers found that the pyrolysis liquid yield could be increased using fast pyrolysis where a biomass feedstock is heated at a rapid rate and the vapors produced also condensed rapidly (Mohan et al., 2006).

Pyrolysis is the thermal decomposition of organic matter occurring in the absence of oxygen or when significantly less oxygen is present than required for complete combustion. Pyrolysis is the basic thermochemical process for converting biomass to a more useful fuel. Biomass is heated in the absence of oxygen, or partially combusted in a limited oxygen supply, to produce a hydrocarbon-rich gas mixture, an oil-like liquid and a carbon-rich solid residue. The products of pyrolysis can be gaseous, liquid, and/or solid. Flash pyrolysis describes the rapid, moderate temperature (675–875 K) pyrolysis that produces liquids. Biomass is heated at rates of 100–10 000 K/s and the vapor residence time is normally less than 2 s. The oil products are maximized at the expense of char and gas.

Pyrolysis is a process similar to gasification except generally optimized for the production of fuel liquids (pyrolysis oils) that can be used straight or refined for higher quality uses such as engine fuels, chemicals, adhesives, and other products. Pyrolysis typically occurs at temperatures in the range of 675 to 975 K. Pyrolysis and combustion of pyrolysis-derived fuel liquids and gases also produce the same categories of end-products as direct combustion of solids.

Fast pyrolysis utilizes biomass to produce a product that is used both as an energy source and a feedstock for chemical production. Considerable efforts have been made to convert wood biomass to liquid fuels and chemicals since the oil crisis in the mid-1970s. Most work has been performed on wood, because of its consistency and comparability between tests. However, nearly 100 types of biomass have been tested, ranging from agricultural wastes such as straw, olive pits, and nut shells to energy crops such as miscanthus and sorghum. The following are affected by the pyrolysis reaction rate and the yield of the volatiles: the biomass composition and structure, heating rate, residence time, catalyst, and particle size. Although very fast and very slow pyrolyses of biomass produce markedly different products, the variety of heating rates, temperatures, residence times, and feedstock varieties found in the literature make generalizations difficult to define, in regards to trying to critically analyze the literature (Mohan et al., 2006).

Rapid heating and rapid quenching produced the intermediate pyrolysis liquid products, which condense before further reactions break down higher molecular weight species into gaseous products. High reaction rates minimize char formation. Under some conditions, no char is formed. At higher fast pyrolysis temperatures, the major product is gas. Many researchers have attempted to exploit the complex degradation mechanisms by conducting pyrolysis in unusual environments (Mohan et al., 2006). The main pyrolysis variants are listed in Table 6.1.

Pyrolysis is the simplest and almost certainly the oldest method of processing one fuel in order to produce a better one. Pyrolysis can also be carried out in the presence of a small quantity of oxygen (gasification), water (steam gasification) and hydrogen (hydrogenation). Hydrogen from organic materials has generally been based on the following pyrolysis reactions (Demirbas and Arin, 2004; Demirbas, 2008a):

$$\text{Cellulose} + H_2O + \text{Air} \rightarrow H_2 + CO + CH_4 \quad (6.12)$$

$$\text{Solid waste} \rightarrow CO + H_2 \quad (6.13)$$

$$\text{Biomass} + H_2O + \text{Air} \rightarrow II_2 + CO_2 \quad (6.14)$$

If the purpose is to maximize the yield of liquid products resulting from biomass pyrolysis, a low-temperature, high heating rate, short gas residence time process would be required. For high char production, a low-temperature, low heating rate process would be chosen. If the purpose is to maximize the yield of fuel gas resulting from pyrolysis, a high-temperature, low heating rate, long gas residence time process would be preferred. It is believed that as the pyrolysis reaction progresses the carbon residue (semichar) becomes less reactive and forms stable chemical structures, and consequently the activation energy increases as the conversion level of biomass increases (Tran and Charanjit, 1978).

Five biomass samples (hazelnut shell, cotton cocoon shell, tea factory waste, olive husk and spruce wood) were pyrolyzed in a laboratory-scale apparatus designed for the purpose of pyrolysis (Demirbas, 2001, 2002a). Figure 6.4 shows the simple experimental setup of pyrolysis. The main element of the experimental device is a vertical cylindrical reactor of stainless steel, 127.0 mm in height, 17.0 mm inner diameter and 25.0 mm outer diameter inserted vertically into an electrically heated tubular furnace and provided with an electrical heating system power source, with a heating rate of about 5 K/s. The biomass samples ground

Table 6.1 Pyrolysis methods and their variants

Method	Residence time	Temperature, K	Heating rate	Products
Carbonation	Days	675	Very low	Charcoal
Conventional	5–30 min	875	Low	Oil, gas, char
Fast	0.5–5 s	925	Very high	Bio-oil
Flash-liquid[a]	< 1 s	< 925	High	Bio-oil
Flash-gas[b]	< 1 s	< 925	High	Chemicals, gas
Hydropyrolysis[c]	< 10 s	< 775	High	Bio-oil
Methanopyrolysis[d]	< 10 s	> 975	High	Chemicals
Ultrapyrolysis[e]	< 0.5 s	1275	Very high	Chemicals, gas
Vacuum pyrolysis	2–30 s	675	Medium	Bio-oil

[a]Flash-liquid: Liquid obtained from flash pyrolysis accomplished in a time of < 1 s
[b]Flash-gas: Gaseous material obtained from flash pyrolysis within a time of < 1 s
[c]Hydropyrolysis: Pyrolysis with water
[d]Methanopyrolysis: Pyrolysis with methanol
[e]Ultrapyrolysis: Pyrolysis with very high degradation rate

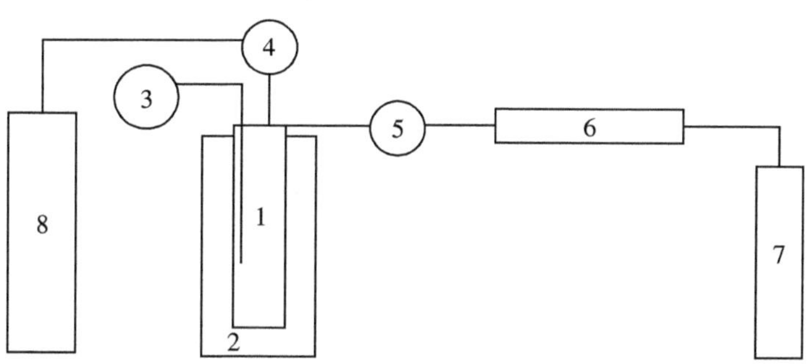

Fig. 6.4 Simple diagram of pyrolysis. *1* Stainless steel tube, *2* electrical heater, *3* temperature control monitor, *4* nitrogen pressure control monitor, *5* product exit valve, *6* condenser, *7* oily products collecting vessel, *8* nitrogen tube

to 0–0.25 mm were filled in plastic sacks and nitrogen gas was blown over them to prevent possible oxidations. In each run, the sample (5 g of biomass) is introduced into the center of the tubular reactor.

The structural and elemental analyses (contents of C, H and O) of biomass samples are given in Table 6.2. Table 6.3 shows the yields of products from three biomass samples by pyrolysis at different temperatures. As seen in Table 6.3, the pyrolysis process yields gaseous products, tarry liquid, and tar. In general, the yields of gaseous products from each run increase with increasing pyrolysis temperature. The yields of gaseous products from the pyrolysis increase from 34.4 to 43.9% for hazelnut shell, from 35.1 to 47.4% for tea waste and from 32.1 to 40.7% for spruce wood with increasing of temperature from 700 to 950 K.

Table 6.2 Structural and elemental analyses of biomass samples (wt.% of dry, ash and nitrogen-free basis)

Sample	Hemicelluloses	Cellulose	Lignin	Extractives	C	H	O
Hazelnut shell	28.9	25.5	42.6	3.3	52.1	5.7	42.2
Tea waste	22.4	31.8	41.2	4.6	50.7	5.2	44.0
Olive husk	21.8	23.3	46.9	8.1	53.8	6.5	39.7
Cotton cocoon shell	12.0	34.7	49.7	3.6	53.0	6.2	40.8
Spruce wood	25.9	43.8	28.5	1.8	52.4	6.4	41.4

Table 6.3 Yields of products from biomass samples by pyrolysis at different temperatures (wt.% dry and ash-free)

Sample	Product	700 K	750 K	800 K	850 K	900 K	950 K
Hazelnut shell	Gas	34.4	36.1	38.0	39.4	41.3	43.9
	Liquid	32.9	31.7	30.4	29.6	28.4	26.9
	Char	32.7	32.2	31.6	31.0	30.3	29.2
Tea waste	Gas	35.1	37.8	40.3	42.9	45.3	47.4
	Liquid	35.5	34.4	33.3	32.1	30.4	29.0
	Char	29.4	27.8	26.4	25.0	24.3	23.6
Spruce wood	Gas	32.1	33.6	35.3	36.9	38.8	40.7
	Liquid	28.5	27.7	26.8	25.9	24.7	23.5
	Char	29.4	28.7	27.9	27.2	26.5	25.8

The compositions of gaseous products from three biomass samples by pyrolysis at different temperatures are given in Table 6.4. In general, the yields of hydrogen from pyrolysis tests of the biomass samples increase with increasing pyrolysis temperature. The total hydrogen-rich gaseous products increased with increasing pyrolysis temperature for the biomass sample (Demirbas, 2001). As can be seen in Table 6.4, the percentage of hydrogen in gaseous products from the samples of hazelnut shell, tea waste and spruce wood increased from 30.3 to 37.6%, 27.2 to 35.3% and 26.4 to 33.3% by volume, respectively, while the final pyrolysis temperature was increased from 700 to 950 K. The largest hydrogen yield (37.6%) was obtained from hazelnut shell at a temperature of 950 K (Table 6.4).

The pyrolysis yields of H_2 + paraffins obtained from biomass samples using different percentages of additives at different temperatures are presented in Table 6.5 (Caglar and Demirbas, 2002a, 2002b). The chemicals ($ZnCl_2$, Na_2CO_3 and K_2CO_3) were used as additives in the experiments. The yields of hydrogen + paraffins

Table 6.4 Compositions of gaseous products from biomass samples by pyrolysis at different temperatures (% by volume)

Sample	Gaseous product	700 K	750 K	800 K	850 K	900 K	950 K
Hazelnut shell	H_2	30.3	31.2	31.8	32.8	34.4	37.6
	CO	22.6	24.7	26.4	27.1	27.4	28.3
	CO_2	33.1	30.2	29.4	25.0	23.3	21.0
	CH_4	6.5	6.0	6.3	6.5	7.1	6.1
	Olefins	1.0	1.1	0.8	0.8	0.7	0.9
Tea waste	H_2	27.2	28.2	29.7	30.8	32.1	35.3
	CO	23.6	25.5	26.8	27.6	28.4	29.3
	CO_2	34.0	31.2	29.8	25.5	24.3	22.0
	CH_4	6.8	7.0	6.4	6.6	7.2	6.2
	Olefins	1.1	1.2	0.9	0.9	0.8	1.0
Spruce wood	H_2	26.4	27.2	28.6	29.6	31.3	33.3
	CO	24.1	26.0	27.4	29.1	29.2	30.3
	CO_2	34.5	31.7	30.3	26.9	25.1	23.0
	CH_4	6.9	7.0	6.4	6.7	7.3	6.3
	Olefins	1.2	1.2	0.9	1.0	0.9	1.1

(mainly CH_4) from the biomass samples generally increased with increasing the pyrolysis temperature and percentage of additive. The highest (70.3% by volume) yield of hydrogen + paraffins was obtained from $ZnCl_2$ run with 13% additive of biomass sample.

Figures 6.5 to 6.7 show the effect of temperature on yields of H_2 + paraffins obtained from biomass samples by pyrolysis. As can be seen in Figs. 6.5 to 6.7, the percentage of H_2 + paraffins in gaseous products obtained from the samples of hazelnut shell, tea waste and spruce wood increased, while the final pyrolysis temperature was increased from 700 to 950 K.

Black liquor, a byproduct of the papermaking alkaline Kraft process (or sulfate method), is an important liquid fuel in the pulp and paper industry. Chemically, black liquor is a mixture of several basic elements where the largest fractions are carbon, oxygen, sodium and sulfur. The black liquor composition was assumed to be $C_{10}H_{12.5}O_7Na_{2.4}S_{0.36}$ (Backman et al., 1993; Salmenoja, 1993). The black liquors can be converted to gaseous, liquid and char products by pyrolysis. Table 6.6 shows the average chemical composition of gaseous products from the precipitate from black liquor in the pyrolysis reactor at different temperatures (Demirbas and Caglar, 1998).

Table 6.5 Pyrolysis yields of H_2 + paraffins obtained from cotton cocoon shell and olive husk by using different percents of additives at different temperatures (% by volume)

Sample	Additive	Amount of additive (wt.% of sample)	Temperature (K)			
			775	925	975	1025
Cotton cocoon shell	None	–	30.4	32.0	36.3	44.4
Olive husk	None	–	48.2	54.1	52.4	54.5
Cotton cocoon shell	$ZnCl_2$	7	46.4	48.1	50.0	44.9
Olive husk	$ZnCl_2$	7	57.7	58.6	61.2	61.7
Cotton cocoon shell	Na_2CO_3	7	33.3	33.8	41.5	44.5
Olive husk	Na_2CO_3	7	47.0	55.3	58.2	57.0
Cotton cocoon shell	K_2CO_3	7	40.8	43.4	45.7	45.6
Olive husk	K_2CO_3	7	48.5	51.3	53.2	61.7
Cotton cocoon shell	$ZnCl_2$	10	52.1	53.4	55.2	57.5
Olive husk	$ZnCl_2$	10	57.3	61.1	61.9	69.4
Cotton cocoon shell	Na_2CO_3	20	34.9	42.2	50.2	49.3
Olive husk	Na_2CO_3	20	46.6	49.0	53.5	58.6
Cotton cocoon shell	K_2CO_3	20	46.1	42.0	47.7	52.5
Olive husk	K_2CO_3	20	49.6	53.3	58.8	59.9
Cotton cocoon shell	$ZnCl_2$	13	56.1	59.3	58.4	59.9
Olive husk	$ZnCl_2$	13	58.2	61.0	67.2	70.3
Cotton cocoon shell	Na_2CO_3	33	41.2	44.5	47.8	50.7
Olive husk	Na_2CO_3	33	48.0	51.9	60.8	62.9
Cotton cocoon shell	K_2CO_3	33	43.0	47.2	52.6	51.9
Olive husk	K_2CO_3	33	46.0	51.4	59.1	62.6
Cotton cocoon shell	Na_2CO_3	47	50.9	45.6	45.7	45.6
Olive husk	Na_2CO_3	47	48.2	49.3	50.5	48.2
Cotton cocoon shell	K_2CO_3	47	48.2	45.1	45.0	47.3
Olive husk	K_2CO_3	47	47.2	47.5	55.8	57.4

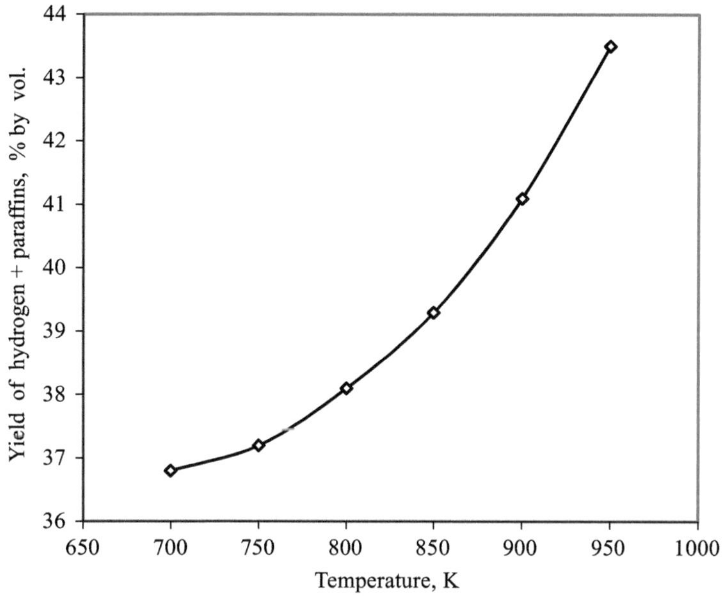

Fig. 6.5 Pyrolysis yields of H_2 + paraffins from hazelnut shells at different temperatures

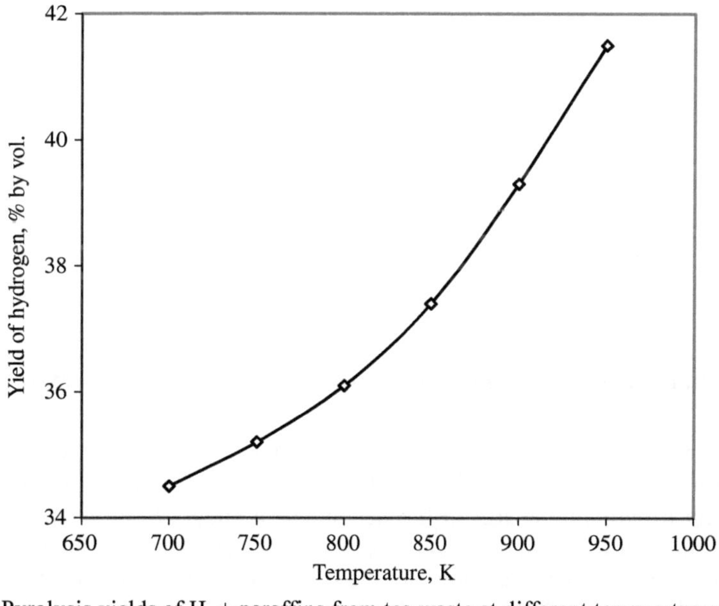

Fig. 6.6 Pyrolysis yields of H_2 + paraffins from tea waste at different temperatures

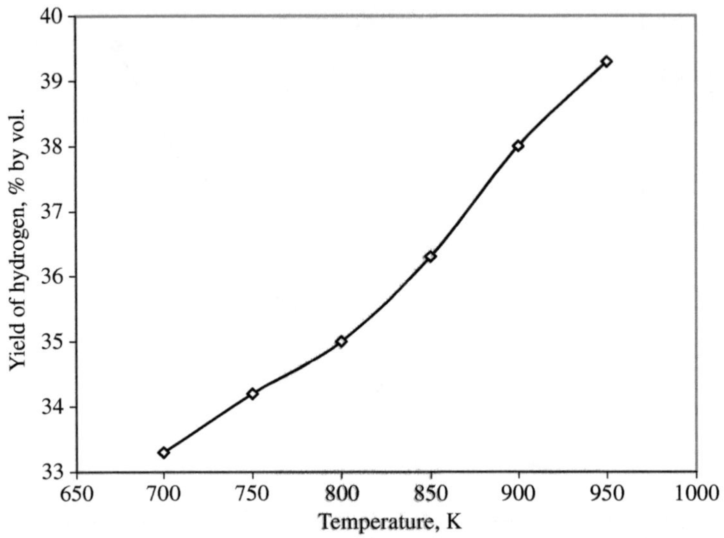

Fig. 6.7 Pyrolysis yields of H_2 + paraffins from spruce wood at different temperatures

Table 6.6 Average chemical composition of gaseous products from the precipitate from black liquor at different temperatures

Temperature (K)	Gas composition (% by volume)			
	H	CO	CO	CH_4
295–675	11.3 ± 0.9	29.0 ± 0.8	26.5 ± 0.4	13.1 ± 0.7
295–825	14.1 ± 0.8	40.8 ± 0.5	19.9 ± 0.4	12.7 ± 0.6
295–825[a]	31.7 ± 1.9	18.7 ± 0.7	30.1 ± 0.6	8.5 ± 0.3
295–925	18.4 ± 0.5	36.0 ± 0.4	15.9 ± 0.5	14.7 ± 1.1
295–925[a]	36.4 ± 0.7	14.6 ± 0.3	32.3 ± 0.6	9.9 ± 0.4

[a]3% K_2CO_3 of the samples used as catalyst

6.5.3 *Production of Hydrogen from Biomass via Gasification*

Biomass gasification has attracted the highest interest among the thermochemical conversion technologies as it offers higher efficiencies in relation to combustion, whereas flash pyrolysis is still in the development stage. When heated in a controlled atmosphere, biomass converts to synthesis gas, which primarily consists of carbon monoxide (CO), carbon dioxide (CO_2), and hydrogen (H_2). Gasification technology has been under intensive development over the last two decades. Large-scale demonstration facilities have been tested and commercial units are in operation worldwide.

Power generation from gaseous products from biomass gasification is found to be the most promising biomass gasification technology. Gasification technologies provide the opportunity to convert renewable biomass feedstocks into clean fuel gases or synthesis gases. The synthesis gas includes mainly hydrogen and carbon monoxide, which is also called syngas (H_2 + CO). Biosyngas is a gas rich in CO and H_2 obtained by gasification of biomass (Maschio et al., 1994; Demirbas, 2002b).

Gasification, one of thermochemical conversion routes, is widely recognized at present because its end-product gas can find flexible application by industries or by home users, particularly in decentralized energy production coupled with microturbine/gas, turbine/engine, boilers, even fuel cells (Chen et al., 2004). Gasification of biomass is a well-known technology that can be classified depending on the gasifying agent: air, steam, steam–oxygen, air–steam, O_2-enriched air, etc.

Gasification describes the process in which oxygen-deficient thermal decomposition of organic matter primarily produces synthesis gas. Gasification is a combination of pyrolysis and combustion. Gasification typically refers to conversion of solid or liquid carbon-based materials by direct internal heating provided by partial oxidation using substoichiometric air or oxygen to produce fuel gases (synthesis gas, producer gas), principally CO, H_2, methane, and lighter hydrocarbons in association with CO_2 and N_2 depending on the process used. Alternative configurations using either indirect heating methods such as externally fired burners or autothermal methods using exothermic reducing reactions have been demonstrated.

Gasification has more potential for near-term commercial application than other thermochemical processes. Benefits of gasification over combustion include more flexibility in terms of energy applications, more economical and thermodynamic efficiency at smaller scales, and potentially lower environmental impact when combined with gas cleaning and refining technologies. An efficient gasifier will decompose high molecular weight organic compounds released during pyrolysis into low molecular weight, non-condensable compounds in a process referred to as tar cracking. Undesirable char that is produced during gasification will participate in a series of endothermic reactions at temperatures above 800°C, which converts carbon into a gaseous fuel. Typically gaseous products include CO, H_2, and CH_4. Fischer–Tropsch processes may be used to upgrade gaseous products to liquid fuels through the use of catalysts (Sect. 3.8). Gasification requires feedstocks that contain less than 10% moisture.

Gasification is a form of pyrolysis, carried out in the presence of a small quantity of oxygen at high temperatures in order to optimize the gas production. The resulting gas, known as producer gas, is a mixture of carbon monoxide, hydrogen and methane, together with carbon dioxide and nitrogen. The gas is more versatile than the original solid biomass (usually wood or charcoal): it can be burnt to produce process heat and steam, or used in gas turbines to produce electricity.

Biomass gasification technologies are expected to be an important part of the effort to meet these goals of expanding the use of biomass. Gasification technologies provide the opportunity to convert renewable biomass feedstocks into clean fuel gases or synthesis gases. Biomass gasification is the latest generation of biomass energy conversion processes, and is being used to improve the efficiency, and to reduce the investment costs of biomass electricity generation through the use of gas turbine technology. High efficiencies (up to about 50%) are achievable using combined-cycle gas turbine systems, where waste gases from the gas turbine are recovered to produce steam for use in a steam turbine. Economic studies show that biomass suffocation plants can be as economical as conventional coal-fired plants.

Biomass gasification technologies have historically been based upon partial oxidation or partial combustion principles, resulting in the production of a hot, dirty, low heating value gas that must be directly ducted into boilers or dryers. In addition to limiting applications and often compounding environmental problems, these technologies are an inefficient source of usable energy.

The comparison of the environmental impact of biomass use in gasifiers and incinerators is very important when considering the effective use of biomass. However, high alkali content in biomass can form compounds with low melting temperature during combustion. The low melting ash constituents can induce in-bed agglomeration, in addition to fouling and corrosion problems.

The energy crisis of the 1970s brought a renewed interest to the topic of biomass. The technology was perceived as a relatively cheap indigenous alternative for small-scale industrial and utility power generation in those developing countries that suffered from high world market petroleum prices and had sufficient sustainable biomass resources. In the beginning of the 1980s at least 10 (mainly European) manufacturers were offering small-scale wood and charcoal-fired power plants (up to approximately 250 kW_{el}). At least four developing countries (Philippines, Brazil, Indonesia, India) started gasifier implementation programs based on locally developed technologies. Figure 6.8 shows the system for power production by means of biomass gasification. The gasification system of biomass in fixed-bed reactors provides the possibility of combined heat and power production in the power range of 100 kWe up to 5 MWe. A system for power production by means of fixed-bed gasification of biomass consists of the main unit gasifier, gas cleaning system and engine.

Commercial gasifiers are available in a range of sizes and types, and run on a variety of fuels, including wood, charcoal, coconut shells and rice husks. Power output is determined by the economic supply of biomass, which is limited to 80 MW in most regions. The producer gas is affected by various gasification processes from various biomass feedstocks. Table 6.7 shows composition of gaseous products from various biomass fuels by different gasification methods.

The relative simplicity of the gasification system enables its operation to be within the technical expertise of most operators who are experienced with conventional boilers and furnaces, and results in favorable project economics. Its modular design allows a wide range of scale-up or scale-down possibilities, so the systems can vary in size from about one ton per hour of residue to 20 tons per hour or larger, with the size being limited only by biomass availability.

The system can gasify a wide variety of biomass wastes and other organic materials generated by many industries. It has gasified, and in most cases has data on expected fuel composition, char analyses, and emissions analyses from the syngas produced for the following feedstocks: hardwood and pine saw dust, bark/hogged fuel, sander/grinder dust from panel board mills, pulp and paper mill sludge, whole

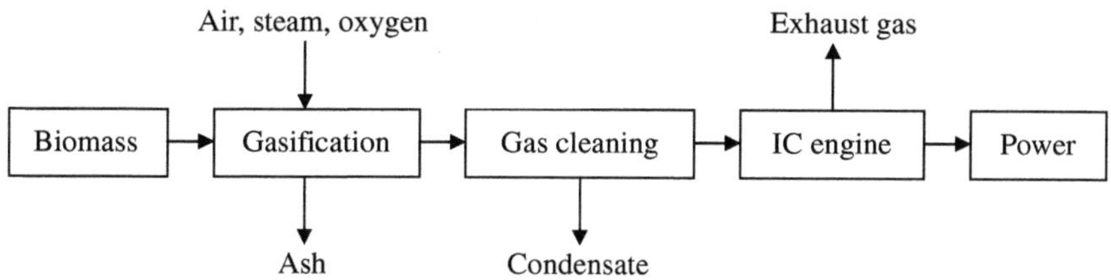

Fig. 6.8 System for power production by means of biomass gasification

Table 6.7 Composition of gaseous products from various biomass fuels by different gasification methods (% by volume)

H_2	CO_2	O_2	CH_4	CO	N_2
10–19	10–15	0.4–1.5	1–7	15–30	43–60

and ground rice hulls, sugar cane bagasse, sewage sludge, the cellulosic fraction of municipal solid waste, and several grades of lignite and subbituminous coal.

The main steps in the gasification process are:

- Step 1. Biomass is delivered to a metering bin from which it is conveyed with recycled syngas or steam, without air or oxygen into the gasifier.
- Step 2. The material is reformed into a hot syngas that contains the inorganic (ash) fraction of the biomass and a small amount of unreformed carbon.
- Step 3. The sensible heat in the hot syngas is recovered to produce heat for the reforming process.
- Step 4. The cool syngas passes through a filter and the particulate in the syngas is removed as a dry, innocuous waste. The clean syngas is then available for combustion in engines, turbines, or standard natural gas burners with minor modifications.

Fixed-bed gasifiers are the most suitable for biomass gasification. Fixed-bed gasifiers are usually fed from the top of the reactor and can be designed in either updraft or downdraft configurations. The product gases from these two gasifier configurations vary significantly. At a larger scale, fixed-bed gasifiers can encounter problems with bridging of the biomass feedstock. This leads to uneven gas flow. Achieving uniform temperatures throughout the gasifier at large scales can also be difficult due to the absence of mixing in the reaction zone. Most fixed-bed gasifiers are air-blown and produce low-energy gases (Stevens, 2001).

With fixed-bed updraft gasifiers, the air or oxygen passes upward through a hot reactive zone near the bottom of the gasifier in a direction counter-current to the flow of solid material. Exothermic reactions between air/oxygen and the charcoal in the bed drive the gasification process. Heat in the raw gas is transferred to the biomass feedstock as the hot gases pass upward, and biomass descending through the gasifier sequentially undergoes drying, pyrolysis, and finally gasification. Fixed-bed updraft gasifiers can be scaled up; however, they produce a product gas with very high tar concentrations. This tar should be removed for the major part from the gas, creating a gas-cleaning problem.

Figure 6.9 shows a schematic diagram of an updraft fixed-bed gasifier. In the drying zone, feedstock descends into the gasifier and moisture is removed using the heat generated in the zones below by evaporation. In the distillation zone, pyrolysis and partial oxidation takes place using the thermal energy released by the partial oxidation of the pyrolysis products. Tar yields can range from 0.1% (downdraft) to 20% (updraft) or greater in the product gases. The oxidation reactions of the volatiles are very rapid and the oxygen is consumed before it can diffuse to the surface of the char. In the reduction zone (often referred to as gasification zone) the char is converted into product gas by reaction with the hot gases from the upper zones. Depending on the end-use, it is necessary to cool and clean the gas in order to remove as much water vapor, dust and pyrolytic products as possible from the gas, especially if it is to be used in an internal combustion engine.

Figure 6.10 shows a schematic diagram of a downdraft fixed-bed gasifier. Fixed-bed downdraft gasifiers were widely used in World War II for operating vehicles and trucks. During operation, air is drawn downward through a fuel bed; the gas in this case contains relatively less tar compared with the other gasifier types. Fixed-bed downdraft gasifiers are limited in scale and require a well-defined fuel, making them not fuel-flexible.

Fluidized-bed (FB) gasifiers are a more recent development that takes advantage of the excellent mixing characteristics and high reaction rates of this method of gas-solid contacting. Examples of the FB gasifier systems are the bubbling fluidized-bed (BFB) gasifiers, the entrained-bed (EB) gasifiers, and the circulating fluidized-bed (CFB) gasifiers.

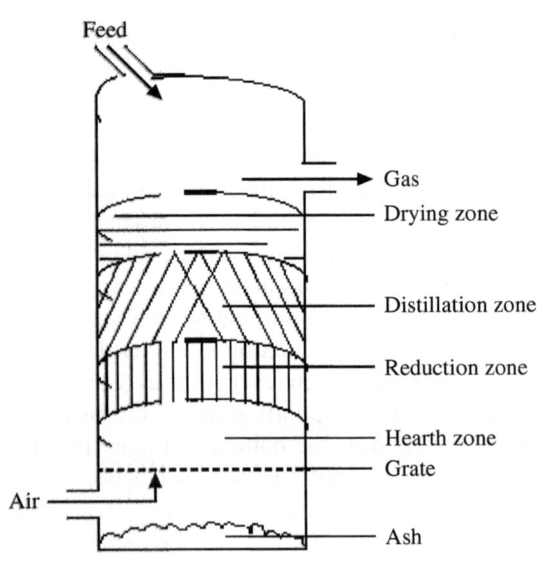

Fig. 6.9 Schematic diagram of an updraft fixed-bed gasifier

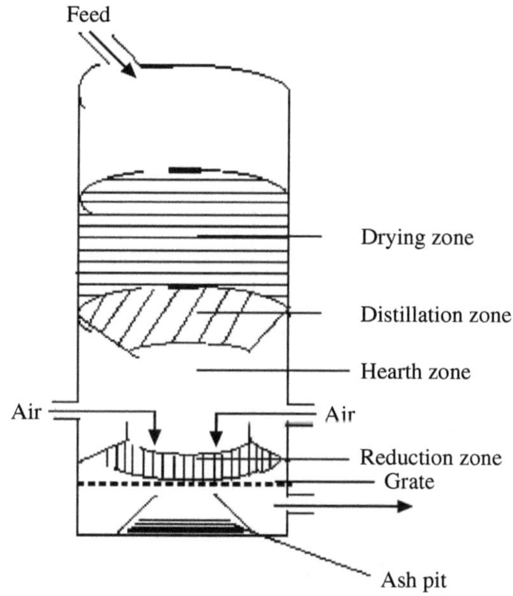

Fig. 6.10 Schematic diagram of a downdraft fixed-bed gasifier

The FB gasifiers are typically operated at 1050–1250 K (limited by the melting properties of the bed material) and are therefore not generally suitable for coal gasification, as due to the lower reactivity of coal compared to biomass, and higher required temperature (> 1550 K). Heat to drive the gasification reaction can be provided in a variety of ways in FB gasifiers. Direct heating occurs when air or oxygen in fluidizing gas partially oxidizes the biomass and heat is released by the exothermic reactions that occur. Indirect heating methods such as internal heat exchangers, using pre-heated bed material, or other means can also be used to drive the gasification reactions (Stevens, 2001). The BFB gasifier tends to produce a gas with tar content between that of the updraft and downdraft gasifiers. Some pyrolysis products are swept out of the fluid bed by gasification products, but are then further converted by thermal cracking in the freeboard region (Warnecke, 2000). The CFB gasifiers employ a system where the bed material circulates between the gasifier and a secondary vessel. The CFB gasifiers are suitable for fuel capacity higher than 10 MWth (Susta et al., 2003). The FB gasifier, and the CFB gasifier systems are given in Fig. 6.11.

6.5.3.1 Biomass Gasification Chemistry

Gasification of solid fuels to yield a mixture of H_2 and CO (syngas), followed by water–gas shift conversion to produce H_2 and CO_2, are well-established processes (Patil, 1987). Gasification is a complex thermochemical process that consists of a number of elementary chemical reactions, beginning with the partial oxidation of a biomass fuel with a gasifying agent, usually air, oxygen, or steam. The chemical reactions involved in gasification include many reactants and many possible reac-

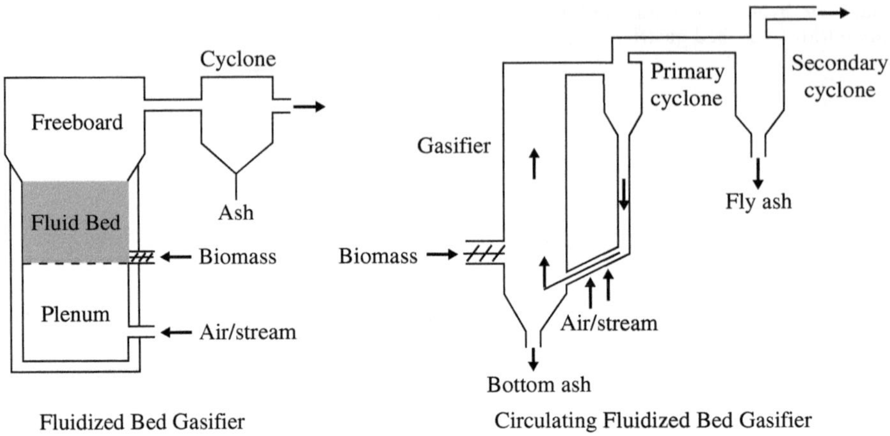

Fig. 6.11 Fluidized-bed and circulating fluidized-bed gasifier systems

tion paths. While the reactions that take place in a gasifier are complex, they can be categorized as follows: flash evaporation of inherent moisture, devolatization of higher organics, heavy hydrocarbon cracking, pyrolysis, and steam reforming.

Most biomass gasification systems utilize air or oxygen in partial oxidation or combustion processes. Volatile matter partially oxidizes to yield the combustion products H_2O and CO_2, plus heat to continue the endothermic gasification process. A product gas from thermal decomposition is composed of CO, CO_2, H_2O, H_2, CH_4, other gaseous hydrocarbons (CHs), tars, char, inorganic constituents, and ash. Gas composition of product from the biomass gasification depends heavily on the gasification process, the gasifying agent, and the feedstock composition. A generalized reaction describing biomass gasification is as follows:

$$\text{Biomass} + O_2 \rightarrow CO, CO_2, H_2O, H_2, CH_4 + \text{other (CHs)} + \text{tar} + \text{char} + \text{ash} \quad (6.15)$$

The relative amount of CO, CO_2, H_2O, H_2, and (CHs) depends on the stoichiometry of the gasification process. If air is used as the gasifying agent, then roughly half of the product gas is N_2. The air/fuel ratio in a gasification process generally ranges from 0.2 to 0.35 and if steam is the gasifying agent, the steam/biomass ratio is around 1. The actual amount of CO, CO_2, H_2O, H_2, tars, and (CHs) depends on the partial oxidation of the volatile products, as shown in Eq. 6.15.

$$C_nH_m + (n/2+m/4)\, O_2 \leftrightarrows nCO + (m/2)\, H_2O \quad (6.16)$$

Most biomass gasification systems utilize air or oxygen in partial oxidation or combustion processes. These processes suffer from low thermal efficiencies and low Btu gas because of the energy required to evaporate the moisture typically inherent in the biomass and the oxidation of a portion of the feedstock to produce this energy.

The processing of synfuels from biomass will lower the energy cost, improve the waste management and reduce harmful emissions. This triple assault on plant operating challenges is a proprietary technology that gasifies biomass by reacting it with steam at high temperatures to form a clean-burning syngas. The molecules in the biomass (primarily carbon, hydrogen and oxygen) and the molecules in the steam (hydrogen and oxygen) re-organize to form this syngas.

In essence, the system embodies a fast, continuous process for pyrolizing or thermally decomposing biomass and steam reforming the resulting constituents. The entire process occurs in a reducing environment, not an oxidizing environment like other biomass gasifiers. While the reactions that take place in the gasifier are complex, they can be categorized as follows: flash evaporation of inherent moisture, devolatization of higher organics, heavy hydrocarbon cracking, pyrolysis, and steam reforming. The major thermochemical reactions include the following:

Steam and methane:

$$CH_4 + H_2O \leftrightarrows CO + 3H_2 \tag{6.17}$$

Water–gas shift:

$$CO + H_2O \leftrightarrows CO_2 + H_2 \tag{6.18}$$

Carbon char to methane:

$$C + 2H_2 \leftrightarrows CH_4 \tag{6.19}$$

Carbon char oxides (Boudouard reaction):

$$C + CO_2 \leftrightarrows 2CO \tag{6.20}$$

The process is extremely efficient, achieving over 97% conversion of biomass carbon to useful syngas without producing troublesome tars, oils, or contaminated effluents. The process is also quite robust in its ability to handle feedstocks with varying degrees of inherent moisture, ranging from bone-dry wood (in which case moisture is actually added to the feedstock) to organic sludge with moisture contents of over 60%. Unlike traditional partial oxidation systems where any moisture in the feedstock results in an energy efficiency penalty, the process utilizes a portion of the moisture to produce hydrogen and other combustible gases. At 40% moisture in the feedstock, the system achieves a cold gas efficiency of about 80%.

Char gasification is the rate-limiting step in the production of gaseous fuels from biomass. Arrhenius kinetic parameters were determined for the reaction of chars prepared by pyrolysis of cottonwood at 1275 K with steam and carbon dioxide. Results indicate that both reactions are approximately zero order with respect to char; the overall reaction rate is fairly constant throughout and declines only when the char is nearly depleted. This suggests that the reaction rate depends on such factors as total available active surface area or interfacial area between the char and catalyst particles. These parameters would remain relatively constant during the gasification process. Sodium and potassium catalysts were equally effective for the gasification of wood char. The iron and nickel transition metals provided the highest initial catalytic activity, but lost their activity well before the char completely reacted. Softwood and hardwood chars exhibited similar gasification behavior. Results indicate that the mineral (ash) content and composition of the original biomass material, and pyrolysis conditions under which char is formed significantly influence the char gasification reactivity (Demirbas, 2000).

The char yield in a gasification process can be optimized to maximize carbon conversion or the char can be thermally oxidized to provide heat for the process. Char is partially oxidized or gasified according to the following reactions:

$$C + ½O_2 \leftrightarrows CO \quad (6.21)$$

$$C + H_2O \leftrightarrows CO + H_2 \quad (6.22)$$

$$C + CO_2 \leftrightarrows 2CO \text{ (Boudouard reaction)}$$

The gasification product gas composition, particularly the H_2:CO ratio, can be further adjusted by reforming and shift chemistry. Additional hydrogen is formed when CO reacts with excess water vapor according to the water–gas shift reaction:

$$CO + H_2O \leftrightarrows CO_2 + H_2$$

Carbon chars to methane reaction:

$$C + 2H_2 \leftrightarrows CH_4 \quad (6.23)$$

6.5.3.2 Electricity from Cogenerative Biomass Firing Power Plant

The future of modern biomass electricity generation lies in biomass integrated gasification/gas turbine technology, which offers high energy conversion efficiencies. The electricity is produced by direct combustion of biomass, advanced gasification and pyrolysis technologies, which are almost ready for commercial-scale use. Biomass power plants (BPPs) use technology that is very similar to that used in coal-fired power plants. For example, biomass plants use similar steam-turbine generators and fuel delivery systems. BPP efficiencies are about 25%. Electricity costs are in the 6–8 ¢/kWh range. The average BPP is about 20 MW in size, with a few dedicated wood-fired plants in the 40–50 MW range. As the biomass-to-electricity industry grows, it will be characterized by larger facilities of 50–150 MW capacity, with a gas turbine/steam combined cycle. Biomass is burned to produce steam and the steam turns a turbine and drives a generator, producing electricity. Because of potential ash build-up, only certain types of biomass materials are used for direct combustion. Heat is used to thermochemically convert biomass into a pyrolysis oil. The oil, which is easier to store and transport than solid biomass material, is then burned like petroleum to generate electricity.

Biomass can be used as a primary energy source or as a secondary energy source to power gas turbines. As a secondary energy source, biomass is used to make a fuel, which can be used to fire a gas turbine. The heat produced from the electricity generating process is captured and utilized to produce domestic purposes and can be used in steam turbines to generate additional electricity. Co-generation is the simultaneous production of electricity and useful thermal energy from a single source.

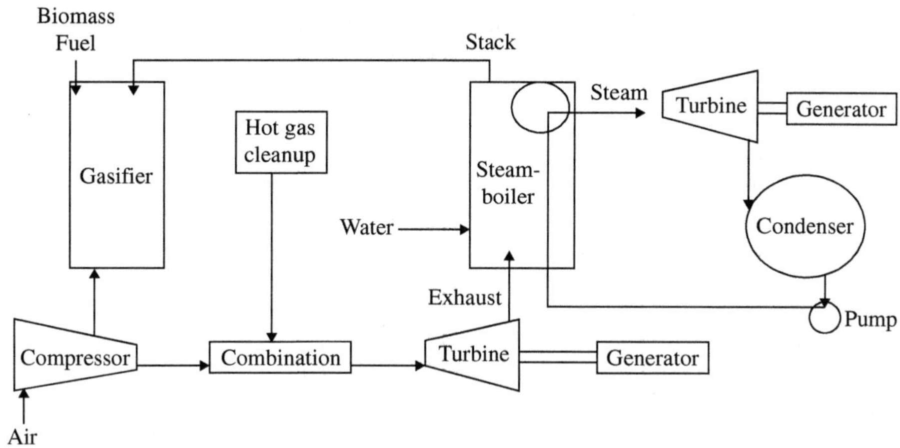

Fig. 6.12 Integrated biomass gas turbine/steam combined-cycle power plant

Practically all biomass-based electricity generation plants employ steam turbine systems. Such electricity generation is established in developed countries in order to upgrade lignocellulosic-based waste materials. Most systems are based on low-pressure boilers (about 20–25 bar) with efficiencies slightly below 20%. Modern biomass-powered high-pressure (60–100 bar) boiler turbine systems produce electricity with efficiencies approaching 32%. Figure 6.12 shows the integrated biomass gas turbine/steam combined-cycle power plant.

6.5.4 Hydrogen from Biomass via Non-Conventional Processes

Hydrogen production from gasified biomass by sponge iron reactor is also reported (Nath and Das, 2003). The sponge-iron process (or steam-iron process) offers a simple possibility to store the energy of synthesis gas. Friedrich et al. (1995) worked on the purification of nitrogen-containing reduction gas from a biomass gasifier using wood and wood waste. The process involves two steps:

1. Cleaning of gas from solid biomass or methane

$$Fe_3O_4 + 4CO \rightarrow 3Fe + 4CO_2 \tag{6.24}$$

2. Energy storage in sponge iron

$$3Fe + 4H_2O \rightarrow Fe_3O_4 + 4H_2 \tag{6.25}$$

The redox process (RP) development was eventually abandoned as other technologies such as pressure swing absorption (PSA) and cryogenic separation began to dominate. In recent years there has been a renewed interest in developing the RP.

In this process dry biomass is fed into a gasifier where it is converted to a combustible fuel gas. On exiting the gasifier the fuel gas temperature was assumed to be 1175 K. The fuel gas then passes through a cyclone to remove entrained char and ash.

The gasifier fuel gas quality is an important issue for the RP development. The effect of the fuel gas composition has been modeled. The modeling showed that the potential hydrogen production efficiency is strongly dependent on the combustion products/syngas (CP/SG) ratio $[(CO_2+H_2O)/(CO+H_2)=CP/SG]$ of the fuel gas.

A core aspect of the model is the determination of the chemical conversion of gasifier SG to hydrogen by the RP. For conditions where steam availability is not limiting, the chemical conversion relates to the difference between the initial and final combustion/fuel ratio of the fuel gas. The initial CP/SG ratio is determined by the gasifier and the biomass feedstock. The final CP/SG ratio is determined by the thermochemical properties of the metal oxide material. Ideally the difference between the initial $(CP/SG)_{initial}$ and final $(CP/SG)_{final}$ ratios should be as large as possible. In reality the availability of steam for the re-oxidation of the metal oxide is limiting for conditions where the difference in the CP/SG ratios are large.

One way to increase the difference between the $(CP/SG)_{initial}$ and $(CP/SG)_{final}$ ratios is to increase the $(CP/SG)_{final}$ ratio by modifying the metal oxide material. This may be achieved by using mixed metal oxides. A goal must be to find the optimum CP/SG ratio. The optimum metal oxide CP/SG ratio depends on the gasifier fuel gas composition.

Hydrogen was produced from organic solid wastes by gasification using solar process heat (Antal et al., 1974). A detailed review with many references of the technology describes solar gasification of carbonaceous materials to produce a syngas quality intermediate for production of hydrogen and other fuels is found in the literature (Yogev et al., 1998; Nath and Das, 2003). Shahbazov and Usupov (1994) have shown good yield of hydrogen from agricultural wastes using a parabolic mirror reflector. The use of a palladium diaphragm in this respect is reported to achieve solar-assisted hydrogen separations from the gases generated by pyrolysis of hazelnut shells at 775–975 K. During the experimental studies, mass loss of hazelnut shells, combustible gas quantities, membrane temperatures in the solar reactor and the amounts of pure hydrogen gas were determined (Midilli et al., 2000). Walcher et al. (1996) have mentioned a plan to utilize agricultural wastes in a heliothermic gasifier.

Several novel heat sources and chemistries have also been explored for hydrogen from organic biomass. Safrany (1971) has proposed the use of a thermonuclear device to vaporize waste organic materials in an underground, large-scale plasma process. In the 1980s, two novel processes for hydrogen from carbonaceous materials were presented (Milne et al., 2002).

Hydrogen can be produced chemically by using sodium hydroxide (NaOH) as a catalyst, and aluminum (Al) and its alloys can react with water to generate hydrogen gas. Although other metals such as zinc and chromium can perform the same reaction, aluminum is among the most promising materials for future development because it is safer and easier to transport than some other hydrogen storage materials like sodium borohydride.

The initial reaction (1) consumes NaOH and produces both hydrogen gas and an aluminate byproduct. Upon reaching its saturation limit, the aluminate compound decomposes (2) into sodium hydroxide and a crystalline precipitate of aluminum hydroxide. This process is similar to the reactions inside an aluminum battery.

$$\text{(1) } Al + 3H_2O + NaOH \rightarrow NaAl(OH)_4 + 1.5\ H_2 \tag{6.26}$$

$$\text{(2) } NaAl(OH)_4 \rightarrow NaOH + Al(OH)_3 \tag{6.27}$$

The overall reaction is described by reaction (3).

$$\text{(3) } Al + 3H_2O \rightarrow Al(OH)_3 + 1.5\ H_2 \tag{6.28}$$

In this process, aluminum acts as a compact hydrogen storage device because 1 kg of aluminum can theoretically produce up to 0.111 kg of hydrogen (or 11.1%).

Hydrogen also can be produced chemically by using waste hydrochloric acid (HCl) and scrap metal can react with acid to generate hydrogen gas.

$$2Al + 6HCl \rightarrow 2AlCl_3 + 3H_2 \tag{6.29}$$

$$2Fe + 6HCl \rightarrow 2FeCl_3 + 3H_2 \tag{6.30}$$

$$Zn + 2HCl \rightarrow ZnCl_2 + H_2 \tag{6.31}$$

6.5.5 Hydrogen from Biomass Gasification by Steam Reforming

Steam reforming or gasification by steam reforming is a method of producing hydrogen-rich gas from carbonaceous biorenewables. Steam reforming process provides the opportunity to convert biorenewable materials into clean fuel gases or syngas (H_2 + CO). On an industrial scale, it is the dominant method for producing hydrogen. Table 6.8 shows the range of the main operating parameters in pyrolysis and steam gasification processes (Demirbas, 2002b).

Reforming reactions between hydrocarbons and water to generate hydrogen are endothermic, and conventional steam reforming of petroleum thus depends on the combustion of additional hydrocarbons to provide the heat needed to drive the reforming reaction. In contrast, the energy required for the aqueous-phase reforming of oxygenated hydrocarbons may be produced internally, by allowing a fraction of the oxygenated compound to form alkanes through exothermic reaction pathways (Cortright et al., 2002).

In the steam reforming reaction, steam reacts with hydrocarbons in the feed to predominantly produce carbon monoxide and hydrogen, commonly called synthesis gas. Steam reforming can be applied to various solid waste materials including, municipal organic waste, waste soil, sewage sludge, paper mill sludge, black liquor, refuse-derived fuel, and agricultural waste. Steam reforming of natural gas, sometimes referred to as steam methane reforming is the most common method of producing commercial bulk hydrogen. Steam reforming of natural gas is currently

Table 6.8 Range of the main operating parameters in pyrolysis and steam gasification processes

Process	Temp (K)	Heating rate (K/s)	Residence time (s)	Water/biomass
Pyrolysis	675–875	0.1–1	600–2000	0.1–2
Fast pyrolysis	925–1175	250–300	1–3	0.2–0.6
Steam gasification	975–1225	300–500	0.5–2	0.8–2

the least expensive method of producing hydrogen, and used for about half of the world's production of hydrogen (Demirbas, 2008b).

Hydrogen production from carbonaceous solid wastes requires multiple catalytic reaction steps: For the production of high-purity hydrogen, the reforming of fuels is followed by two water–gas shift reaction steps, a final carbon monoxide purification and carbon dioxide removal. Steam reforming, partial oxidation and autothermal reforming of methane are well-developed processes for the production of hydrogen. Stepwise steam reforming of methane for production of carbon monoxide-free hydrogen has been investigated at various process conditions by Choudhary and Goodman (2000). The process consists of two steps involving the decomposition of methane to carbon monoxide-free hydrogen and surface carbon in the first step followed by steam gasification of this surface carbon in the second step. The amount of carbon monoxide-free hydrogen formed in the first step hydrogen is produced in the second step of the reaction. The mixture of gases (primarily CH_4 and CO_2) can be separated or simply methanated and returned to the first step (Choudhary and Goodman, 1999).

Steam, at high temperatures (975–1375 K) is mixed with methane gas in a reactor with a Ni-based catalyst at pressures of 3–25 bar to yield carbon monoxide (CO) and hydrogen (H_2). Steam reforming is the process by which methane and other hydrocarbons in natural gas are converted into hydrogen and carbon monoxide by reaction with steam over a nickel catalyst on a ceramic support. The hydrogen and carbon monoxide are used as initial material for other industrial processes.

$$CH_4 + H_2O \leftrightarrows CO + 3H_2 \qquad \Delta H = +251 \text{ kJ/mol} \tag{6.32}$$

It is usually followed by the shift reaction:

$$CO + H_2O \leftrightarrows CO_2 + H_2 \qquad \Delta H = -42 \text{ kJ/mol} \tag{6.33}$$

The theoretical percentage of hydrogen to water is 50%. The further chemical reactions for most hydrocarbons that take place are:

$$C_nH_m + n\, H_2O \leftrightarrows n\, CO + (m/2 + n)\, H_2 \tag{6.34}$$

It is possible to increase the efficiency to over 85% with an economic profit at higher thermal integration. There are two types of steam reformers for small-scale hydrogen production: conventional reduced-scale reformers and specially designed reformers for fuel cells.

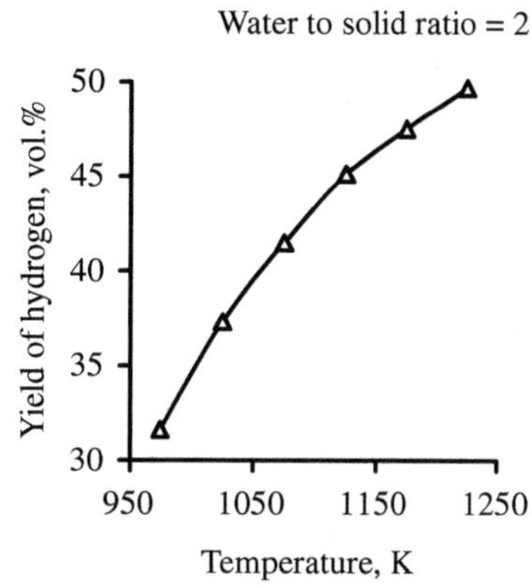

Fig. 6.13 Yield of hydrogen of beech wood by steam reforming at different temperatures

Commercial catalysts consist essentially of Ni supported on a-alumina. Mg-promoted catalysts showed a greater difficulty for Ni precursor's reduction besides different probe molecules (H_2 and CO) adsorbed states. In the conversion of cyclohexane, Mg inhibited the formation of hydrogenolysis products. Nonetheless, the presence of Ca did not influence the metallic phase. The impregnated Ni/MgO-catalyst performed better than the other types (Santos et al., 2004).

Figure 6.13 shows the yield of hydrogen of beech wood gasification by steam reforming at different temperatures (Demirbas, unpublished). The yields of hydrogen from beech wood gasification by steam reforming increase from 31.6 to 47.9% for (water/solid) = 1 and from 31.6 to 49.7% for (water/solid) = 2 (Fig. 6.13) with increasing of temperature from 975 to 1225 K.

Cox et al. (1995) portray a new approach to thermochemical gasification of biomass to hydrogen. The process is based on catalytic steam gasification of biomass with concurrent separation of hydrogen in a membrane reactor that employs a permselective membrane to separate the hydrogen as it is produced. The process is particularly well-suited for wet biomass and may be conducted at temperatures as low as 575 K.

A study of almond shell steam gasification in a fluidized bed revealed that, over the range 775 to 1075 K, smaller particle size yielded more hydrogen than did higher temperatures (Rapagna, 1996). The catalytic converter, using different steam reforming nickel catalysts and dolomite, was tested over a range of 933 to 1103 K. Fresh catalyst at the highest temperature yielded 60% by volume of hydrogen (Rapagna and Foscolo, 1998).

6.5.6 Hydrogen from Biomass via Air–Steam Gasification

The air–steam gasification would be one of the practical conversion technologies among biomass gasification (Li et al., 2004). In general, the gasification temperature is higher than that of pyrolysis and the yield of hydrogen from the gasification is higher than that of the pyrolysis.

Partial oxidation of hydrocarbons is the exothermic reaction with oxygen and steam. The amounts of oxygen and water vapor are controlled so that the reaction proceeds without the need for external energy. An example reaction for this process is:

$$2C_8H_{18} + 2H_2O + 9O_2 \rightarrow 12CO + 4CO_2 + 20H_2 \tag{6.35}$$

The carbon dioxide from the gas mixture is removed by washing with water and the carbon monoxide is removed by absorption or membrane separation.

Yields of hydrogen of hazelnut shell, tea waste and spruce wood samples by air–steam gasification at different temperatures are given in Figs. 6.13 to 6.15, respectively (Demirbas, unpublished work). As can be seen in Fig. 6.13, the percentage of hydrogen in the gaseous products from the samples of hazelnut shell increased from 43.3 to 56.9% by volume, respectively, while the final air–steam gasification temperature was increased from 900 to 1250 K. The hydrogen yields were 43.3, 47.3, 50.4, 52.5, 54.0, 55.4, 56.2 and 56.9% by volume for 900, 950, 1000, 1050, 1100, 1150, 1200 and 1250 K, respectively.

The percentage of hydrogen in gaseous products from the samples of hazelnut shell, tea waste and spruce wood increased from 43.3 to 56.9%, 41.0 to 53.9% and 40.0 to 51.5% by volume, respectively, while the final air–steam gasification temperature was increased from 900 to 1250 K (Figs. 6.14 to 6.16).

Comparison of pyrolysis and air–steam gasification shows that the yield of hydrogen from biomass is generally higher by air–steam gasification than that by pyrolysis, because with interaction of water and char from decomposition of biomass intermediate products are formed, which leads to more hydrogen-rich gas yield by the steam reforming.

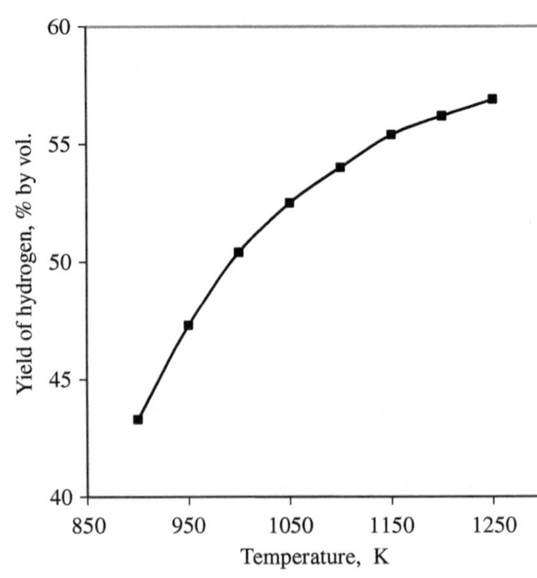

Fig. 6.14 Hydrogen yield of hazelnut shell by air–steam gasification at different temperatures

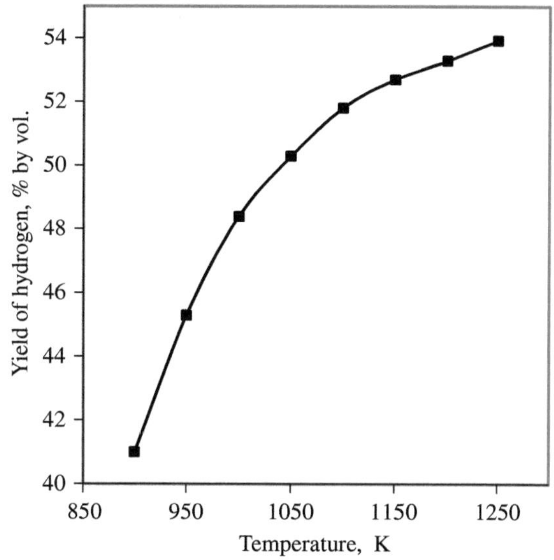

Fig. 6.15 Yield of hydrogen of tea waste by air–steam gasification at different temperatures

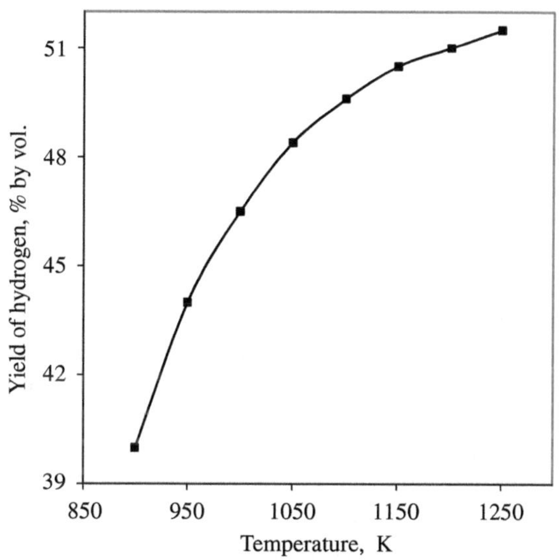

Fig. 6.16 Hydrogen yield of spruce wood by air–steam gasification at different temperatures

The effect of biomass structural components (i.e., hemicelluloses, cellulose and lignin) on air–steam gasification was considerably important by comparison with the experimental results and the structural components given in Table 6.2. From the results, the product gas composition was highly dependent on hemicelluloses and lignin contents of biomass material. The yields of hydrogen in gaseous products from biomass samples by air–steam gasification increase with increasing hemicelluloses and lignin contents.

6.5.7 Hydrogen from Biomass by Supercritical Water Gasification

The supercritical fluid extraction (SFE) is a process in which a highly compressed gas (fluid) is brought into contact with a relatively non-volatile solid or liquid at temperatures at or slightly above the critical temperature of the solvent. Under such conditions, the condensed phase will begin to volatize, which is interpreted as the supercritical fluid phase (Vayisoglu et al., 1996). The SFE is one of the best methods to obtain liquid fuels from coals. The SFE extraction is carried out in an autoclave at above the critical temperature and the pressure of the solvent. The yield of soluble material increases with increasing pressure (Paul and Wise, 1971).

Comparison with other biomass thermochemical gasification such as air gasification or steam gasification, the supercritical water gasification (SCWG) can directly deal with the wet biomass without drying, and has high gasification efficiency in relatively lower temperature. The cost of hydrogen production from supercritical water gasification of wet biomass was several times higher than the current price of hydrogen from steam methane reforming. Biomass is gasified in supercritical water at a series of temperatures and pressures during different resident times and the product gas is composed of H_2, CO_2, CO, CH_4, and a small amount of C_2H_4 and C_2H_6. Supercritical water is a promising reforming media for the direct production of hydrogen at 875–1075 K temperatures with a short reaction time (2–6 s). As the temperature is increased from 875 to 1075 K the H_2 yield increases from 53 to 73% by volume, respectively. In addition to being a high mass transfer effect, supercritical water also participates in reforming reaction. Pressure has a negligible effect on hydrogen yield above the critical pressure of water (Demirbas, 2004, 2005b).

Hydrogen production by biomass gasification in supercritical water (SCW) is a promising technology for utilizing high moisture content biomass, but reactor plugging is a critical problem for biomass gasification in the tubular reactor. Model compound (glucose) and real biomass (corn cob) were gasified under SCW conditions to generate hydrogen-rich fuel gas and a performance testing of the new SCW fluidized-bed system was conducted. The product gas composed of H_2, CH_4, CO_2, CO and small amount of C_2H_4 and C_2H_6 was obtained. The effects of solution concentration, temperature, pressure and oxidant concentration on gasification were studied. The results showed that using a fluidized-bed reactor for biomass gasification in SCW has many advantages and good prospects (Lu et al. 2008).

Biomass gasification in supercritical water is a promising technology for hydrogen production by utilizing wet biomass, having only been developed in last two decades (Lu et al., 2007). The catalyst shows a destructive effect on the organic compounds and H_2 and CO_2 form at the end of the catalytic steam reforming process. The main advantage of gasification in supercritical water is the high solid conversion, i.e., low levels of char and tars. For wet biomass containing large amounts of water up to 90%, supercritical water gasification (SCWG) appears as a useful technology (Yanik et al., 2007). Biomass energy of low quality can be converted to hydrogen energy of high quality by the SCWG process (Guo et al., 2007).

Products from biomass by supercritical water (SCW) depend on the nature and structure of the biomass. The effects of SCW on the biomass constituents should be separately studied. For example SCW affects unsaturated compounds, and unsaturated fatty and resin acids, rather than those of saturated ones under different reaction conditions (Watanabe et al., 2006). The diffusion or mass transfer rate of SCW into the individual component of biomass has been studied separately (Antal et al., 2000; Feng et al., 2004).

$$C_6H_{10}O_5 + 7H_2O \rightarrow 6CO_2 + 12H_2 \quad (6.36)$$

In this idealized, stoichiometric equation (Eq. 6.36), cellulose (represented as $C_6H_{10}O_5$) reacts with water to produce H_2 and CO_2, the commercial production of H_2 from methane by the catalytic steam reforming process (Wagner and Froment, 1992).

Recently, the supercritical fluid treatment has been considered to be an attractive alternative in science and technology as a chemical reaction field. The molecules in the supercritical fluid have high kinetic energy like the gas and high density like the liquid. Therefore, it is expected that the chemical reactivity can be high. In addition, the ionic product and dielectric constant of supercritical water are important parameters for chemical reaction. Therefore, the supercritical water can be realized from the ionic reaction field to the radical reaction field. For example, the ionic product of the supercritical water can be increased by increasing pressure, and the hydrolysis reaction field is realized. Therefore, the supercritical water is expected as a solvent for converting biomass into valuable substances (Hao et al., 2003).

Especially, the temperature in which the gasification reaction can take place depends on the category and the structure of the biomass greatly. Many reactions in supercritical water gasification have begun when the temperature and the pressure exceeds the critical point of water. In supercritical water gasification, the reaction generally takes place at the temperature over 875 K and a pressure higher than the critical point of water. With temperatures higher than 875 K, water becomes a strong oxidant, and oxygen in water can be transferred to the carbon atoms of the biomass. CO is the primary product in SCWG and it reacts with steam to produce additional H_2 and CO_2 by the water–gas shift reaction. The hydrogen atoms of water and of the biomass are set free and form H_2. The gas product consists of hydrogen, CO_2, CH_4 and CO (Antal, 1978).

A problem of general nature in SCGW is the required heat exchange between the reactor outlet and inlet streams. To achieve an acceptable thermal efficiency, it is crucial for the process that the heat of the inlet stream is utilized as far as possible to pre-heat the feedstock stream (mainly water) to reaction conditions. At the same time, heating of the biomass slurry in the inlet tube of a reactor is likely to cause fouling/plugging problems because the thermal decomposition (> 523 K) starts already far below the desired reaction temperature (> 873 K) (Kruse et al., 2000). A schematic diagram of experimental apparatus of supercritical water gasification is depicted in Fig. 6.17. Supercritical water is a promising reforming media for the direct production of hydrogen at 875–1075 K temperatures with a short reaction time between 2 and 6 s. In addition to being a high mass transfer effect, supercritical

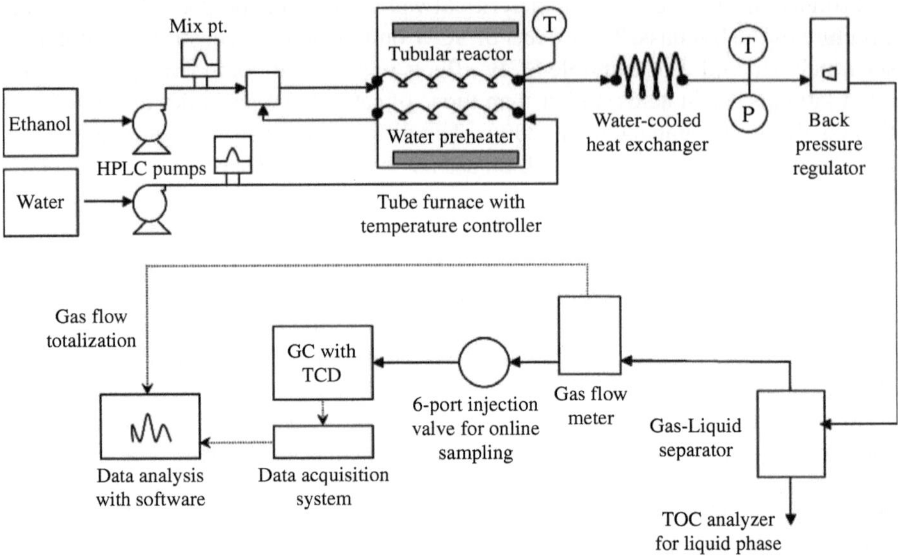

Fig. 6.17 Schematic diagram of experimental apparatus of supercritical water gasification

Table 6.9 Typical product yield (% by vol.) distributions for experiments with and without Ru/Al_2O_3 catalyst at 973 K with 1 wt.% glucose feed (15.9 g of catalyst/g mol of glucose)

	H_2	CO_2	CH_4	CO
Direct reforming	54.5	31.4	8.5	5.6
Catalytic reforming	68.9	29.8	1.3	0.1

water also participates in the reforming reaction. The molecules in the supercritical fluid have high kinetic energy like the gas and high density like the liquid. Therefore, it is expected that the chemical reactivity can be high. In addition, the ionic product and dielectric constant of supercritical water, which are important parameters for chemical reactions, can be continuously controlled by regulating pressure and temperature. Pressure has a negligible effect on the hydrogen yield above the critical pressure of water (Gadhe and Gupta, 2007). As the temperature is increased from 875 to 1075 K the H_2 yield increases from 53 to 73% by volume, respectively. Only a small amount of hydrogen is formed at low temperatures, indicating that direct reformation reaction of ethanol as a model compound in SCW is favored at high temperatures (> 975 K). With an increase in the temperature, the hydrogen and carbon dioxide yields increase, while the methane yield decreases. The water excess leads to a preference for the formation of hydrogen and carbon dioxide instead of carbon monoxide. The formed intermediate carbon monoxide reacts with water to hydrogen and carbon dioxide. The low carbon monoxide yield indicates that the water–gas shift reaction approaches completion (Byrd et al., 2007).

Thermochemical gasification of biomass has been identified as a possible system to produce renewable hydrogen with less dependence on exhaustible fossil energy resources. Compared with other biomass thermochemical gasifications, such as air gasification or steam gasification, the SCWG has high gasification efficiency at lower temperatures and can deal directly with wet biomass without drying (Yan et al., 2006).

The capillaries (1mm ID and 150 mm length tubular reactors) are heated rapidly (within 5 s) in a fluidized sand bed to the desired reaction temperature. Experimentation with the batch capillary method has revealed that, especially at low temperatures and high feed concentrations, char formation occurs. A fluidized-bed reactor might be a good alternative to solve the problems related to this char and ash formation (Kruse et al., 2000).

To study the effect of the Ru/Al_2O_3 catalyst on hydrogen yield for reforming of glucose in supercritical water, the experiments were compared to reactions with and without catalytic runs under identical conditions. Typical product distributions are shown in Table 6.9 for experiments with and without a Ru/Al_2O_3 catalyst at 973 K with 1 wt.% glucose feed (Byrd et al., 2007). There was a significant reduction in carbon monoxide and methane yields in the presence of the catalyst. The main products of the reaction were hydrogen, methane, carbon dioxide, and carbon monoxide. The low carbon monoxide yield (0.1% by vol.) indicates that the water–gas shift reaction approaches completion.

Cellulose and sawdust were gasified in supercritical water to produce hydrogen-rich gas, and Ru/C, Pd/C, CeO_2 particles, nano-CeO_2 and nano-$(CeZr)xO_2$ were selected as catalysts. The experimental results showed that the catalytic activities were Ru/C > Pd/C > nano-$(CeZr)xO_2$ > nano-CeO_2 > CeO_2 particle in turn. The 10 wt.% cellulose or sawdust with CMC can be gasified almost completely with a Ru/C catalyst to produce 2–4 g hydrogen yield and 11–15 g potential hydrogen yield per 100 g feedstock at the condition of 773 K, 27 MPa, 20 min residence time in supercritical water (Hao et al., 2005).

Catalysts for low-temperature gasification include combinations of stable metals, such as ruthenium or nickel bimetallics and stable supports, such as certain titania, zirconia, or carbon. Without catalyst the gasification is limited (Kruse et al., 2000). Sodium carbonate is effective in increasing the gasification efficiency of cellulose (Minowa et al., 1997). Likewise, homogeneous, alkali catalysts have been employed for high-temperature supercritical water gasification.

The yields of hydrogen from steam gasification increase from 29 to 45% for (water/solid) = 1 and from 29 to 47% for (water/solid) = 2 with increasing of temperature from 975 to 1225 K. Figure 6.18 shows comparisons for the yield of hydrogen from supercritical fluid extraction (SFE), pyrolysis and steam gasification (water/solid = 2) of beech wood at different temperatures. Distilled water was used in the SFE (the critical temperature of pure water is 647.7 K). From Fig. 6.18, the yield of hydrogen from SFE was considerably high (49%) at lower temperatures. The pyrolysis was carried out at moderate temperatures and steam gasification at the highest temperatures (Demirbas, 2006a).

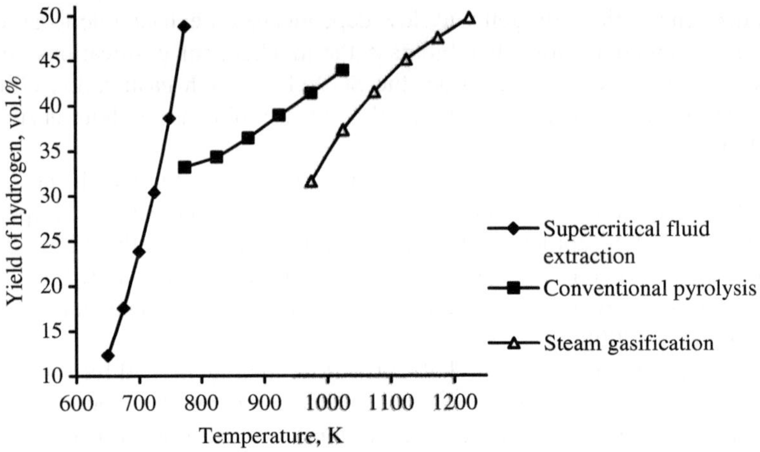

Fig. 6.18 Plots for yield of hydrogen from supercritical fluid (water) extraction, pyrolysis and steam gasification (water/solid = 2) of beech wood at different temperatures

6.5.7.1 Advantages of Supercritical Water

The reforming in supercritical water (SCW) offers several advantages over the conventional technologies because of the unusual properties of supercritical water. The density of supercritical water is higher than that of steam, which results in a high space-time yield. The higher thermal conductivity and specific heat of supercritical water is beneficial for carrying out the endothermic reforming reactions (Gadhe and Gupta, 2007). In the supercritical region, the dielectric constant of water is much lower. Further, the number of hydrogen bonds is much smaller and their strength is considerably weaker. As a result, SCW behaves as an organic solvent and exhibits extraordinary solubility toward organic compounds containing large non-polar groups and most permanent gases (Demirbas, 1998; Byrd et al., 2007). Another advantage of SCW reforming is that the H_2 is produced at a high pressure, which can be stored directly, thus avoiding the large energy expenditures associated with its compression. The SCWG process becomes economical as the compression work is reduced owing to the low compressibility of liquid feed when compared to that of gaseous H_2 (Byrd et al., 2007).

6.5.8 Hydrogen-rich Gas from Shells via Supercritical Water Extraction

Agricultural shell samples (walnut, almond, hazelnut cotton cocoon, and sunflower shells) were subjected to supercritical water extraction for producing hydrogen-rich gas mixtures (Demirbas, 2004). Table 6.10 shows structural analysis results of the shell samples. In a typical run of this study, the autoclave was loaded with a 5.0 g sample and 50–70 g of deionized water. The supercritical fluids extraction (SFE) parameters were adjusted as follows: the temperatures for supercritical water runs were 650 ± 3, 670 ± 3, 690 ± 3, 710 ± 3, 730 ± 3 and 750 ± 3 K, and pressures measured were 23.2, 25.4, 28.0, 32.5 and 38.7 MPa.

Figure 6.19 shows the plots for yields of hydrogen from supercritical water extraction of shell samples versus temperature. The yields of hydrogen from supercritical water extraction increase with increasing extraction temperatures. The yield of hydrogen from almond shell increases from 6.7 to 13.5% with the temperature increase from 650 to 750 K, respectively (Demirbas, 2004).

Figure 6.20 shows the plots for yields of hydrogen from supercritical water extraction of shell samples versus cellulose content at different temperatures. The yields of hydrogen from supercritical water extraction increase with increasing cellulose contents of the shell samples (Demirbas, 2004).

Table 6.10 Structural analyses results of shell samples (wt.% of dry, ash and extractive-free basis)

Sample of shell	Hemicellulose	Cellulose	Lignin
Almond	29.6 ± 2.8	52.0 ± 5.6	20.9 ± 2.1
Cotton cocoon	10.5 ± 1.1	33.5 ± 3.5	50.0 ± 5.3
Hazelnut	31.4 ± 3.2	27.7 ± 2.9	44.4 ± 4.7
Sunflower	35.6 ± 3.8	49.7 ± 5.2	17.5 ± 1.9
Walnut	22.7 ± 2.3	26.3 ± 2.8	53.8 ± 5.7

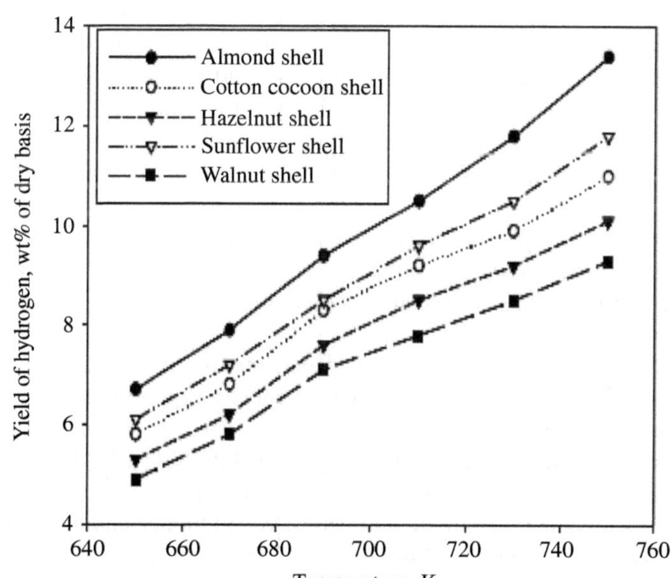

Fig. 6.19 Plots for yields of hydrogen from supercritical water extraction of shell samples versus temperature

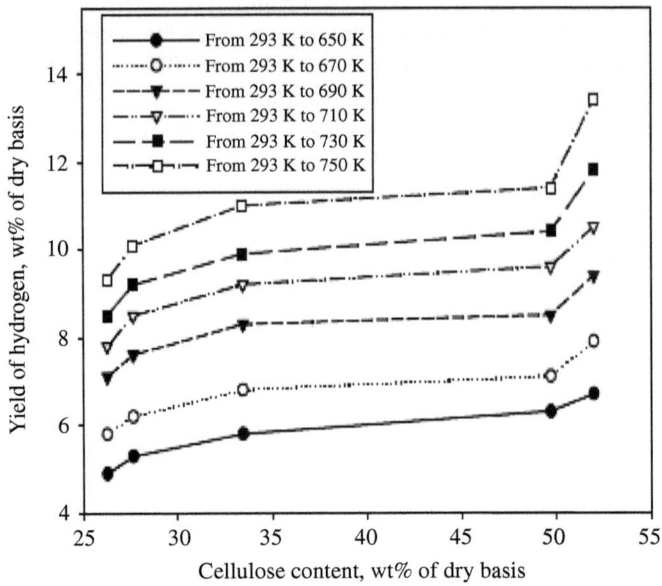

Fig. 6.20 Plots for yields of hydrogen from supercritical water extraction of shell samples versus cellulose content at different temperatures

6.5.9 Production of Hydrogen from Mosses and Algae via Pyrolysis and Steam Gasification

Most studies concerning mosses have been taxonomic and medical. Some species of Bryophytes, such as *Atricum*, *Dicranum*, *Mnium*, *Polytrichum*, and *Sphagnum*, are important for the economy. Approximately 200 million tons of peat formed from *Sphagnum* per year is used as a fuel in the world (Cetin and Yurdakulol, 1985, 1988). The chemical compositions of algae are given in Table 6.11 (Demirbas, 2007). Algae were mainly composed of proteins, lipids and water-soluble carbohydrates.

Two moss samples (*Polytrichum commune*, *Thuidium tamariscinum*), one alga sample (*Cladophora fracta*), and one microalga sample (*Chlorella protothecoides*) were subjected to pyrolysis and steam gasification for producing hydrogen-rich gas (Demirbas, unpublished work).

The temperature of the reaction vessel was measured with an iron-constantan thermocouple and controlled at ±3 K. The pyrolysis experiments were performed at temperatures of 575, 625, 675, 725, 775, 825 and 925 K. The steam gasification experiments were carried out at temperatures of 825, 875, 925, 975, 1025, 1075, 1125, 1175 and 1225 K (Demirbas, unpublished work).

Table 6.12 shows the proximate analysis data and higher heating values (HHV) of samples. The HHV (MJ/kg) of the moss and alga samples as a function of fixed carbon (FC), wt.% can be calculated from:

$$HHV = 0.322\,(FC) + 10.7123 \qquad (6.37)$$

Table 6.11 Chemical compositions of algae on a dry matter basis (%)

Species of sample	Proteins	Carbohydrates	Lipids	Nucleic acid
Scenedesmus obliquus	50–56	10–17	12–14	3–6
Scenedesmus quadricauda	47	–	1.9	–
Scenedesmus dimorphus	8–18	21–52	16–40	–
Chlamydomonas rheinhardii	48	17	21	–
Chlorella vulgaris	51–58	12–17	14–22	4–5
Chlorella pyrenoidosa	57	26	2	–
Spirogyra sp.	6–20	33–64	11–21	–
Dunaliella bioculata	49	4	8	–
Dunaliella salina	57	32	6	–
Euglena gracilis	39–61	14–18	14–20	–
Prymnesium parvum	28–45	25–33	22–38	1–2
Tetraselmis maculata	52	15	3	–
Porphyridium cruentum	28–39	40–57	9–14	–
Spirulina platensis	46–63	8–14	4–9	2–5
Spirulina maxima	60–71	13–16	6–7	3–4.5
Synechoccus sp.	63	15	11	5
Anabaena cylindrica	43–56	25–30	4–7	–

Table 6.12 Proximate analysis data and higher heating values (HHV) of samples (wt.% dry basis)

Sample	Fixed carbon	Volatile matter	Ash	HHV (MJ/kg)
Polytrichum commune	19.4	65.8	14.8	17.0
Thuidium tamarascinum	15.4	72.3	12.3	15.5
Cladophora fracta	28.1	65.6	6.3	19.8
Chlorella protothecoides	39.6	54.6	5.8	23.6

The HHVs can be calculated using Eq. 6.37 and represent high correlation obtained by means of regression analysis. The correlation coefficient r is 0.999.

The yields of bio-oil from the samples via pyrolysis are presented as a function of the temperature (K) in Fig. 6.21. The yield of bio-oil from pyrolysis of the samples increased with temperature, as expected. The yields were increased up to 750 K in order to reach the plateau values at 775 K. The maximum yields for *Polytrichum commune, Thuidium tamarascinum, Cladophora fracta* and *Chlorella protothecoides* were 31.6, 37.3, 45.0 and 50.8% of the sample at 925 K, respectively. The bio-oil yields of pyrolysis from algae were higher than those of mosses. Bio-

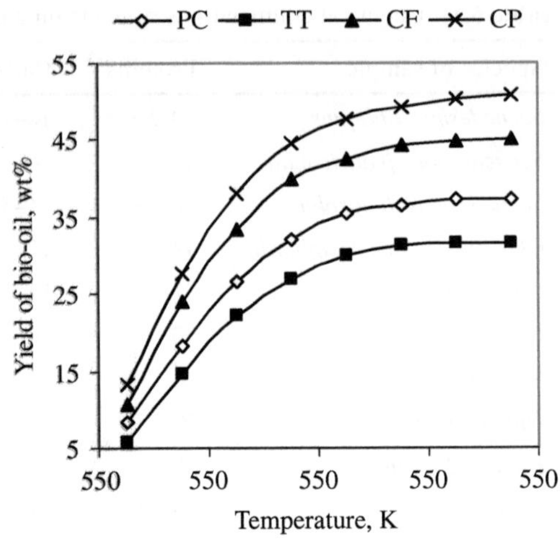

Fig. 6.21 Plots for yield of bio-oil from moss and alga samples by pyrolysis at different temperatures (K). *Polytrichum commune* (PC), *Thuidium tamarascinum* (TT), *Cladophora fracta* (CF), and *Chlorella protothecoides* (CP)

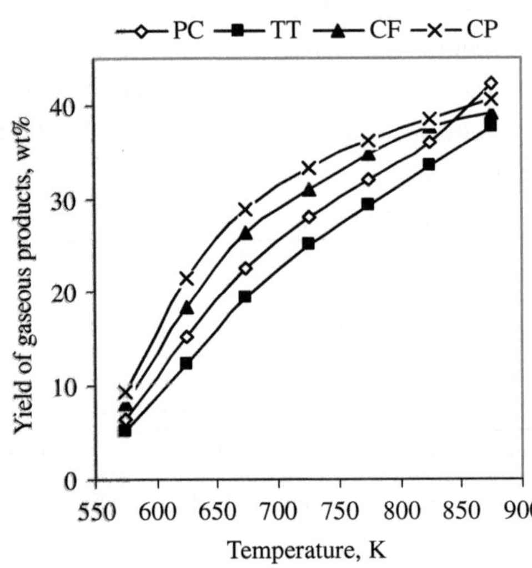

Fig. 6.22 Plots for yields of gaseous products from the samples by pyrolysis. *Polytrichum commune* (PC), *Thuidium tamarascinum* (TT), *Cladophora fracta* (CF), and *Chlorella protothecoides* (CP)

oil comparable to fossil oil was obtained from microalgae (Miao and Wu, 2004). In the pyrolysis process, the yield of charcoal decreases with increasing pyrolysis temperature. The yield of the liquid product is highly excessive at temperatures between 625 and 725 K.

The HHVs for bio-oils from mosses 21.5–24.8 MJ/kg and the HHVs for bio-oils from alga and microalga 32.5 and 39.7 MJ/kg, respectively, were obtained by pyrolysis at temperatures ranging from 775 to 825 K. In general, algae bio-oils are higher quality than bio-oils from mosses.

Figure 6.22 shows the effect of temperature on yields of gaseous products from the samples by pyrolysis. As can be seen in Fig. 6.22, the yields of gaseous products from the samples of *Polytrichum commune, Thuidium tamarascinum, Cladophora fracta and Chlorella protothecoides* increased from 5.3 to 40.6%, 6.5 to 42.2%, 8.2

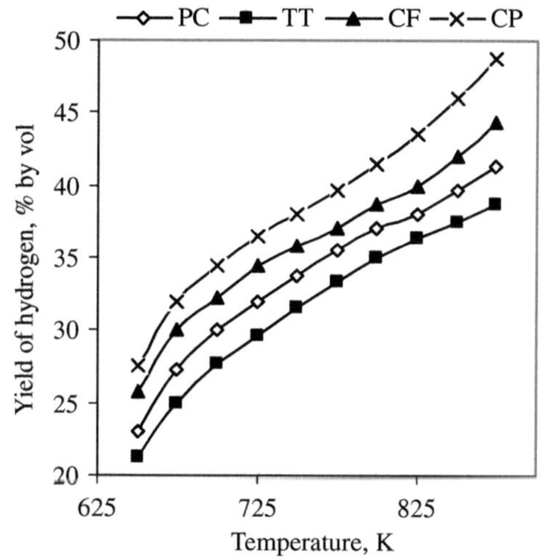

Fig. 6.23 Plots for yields of hydrogen in gaseous products from the samples by pyrolysis. *Polytrichum commune* (PC), *Thuidium tamarascinum* (TT), *Cladophora fracta* (CF), and *Chlorella protothecoides* (CP)

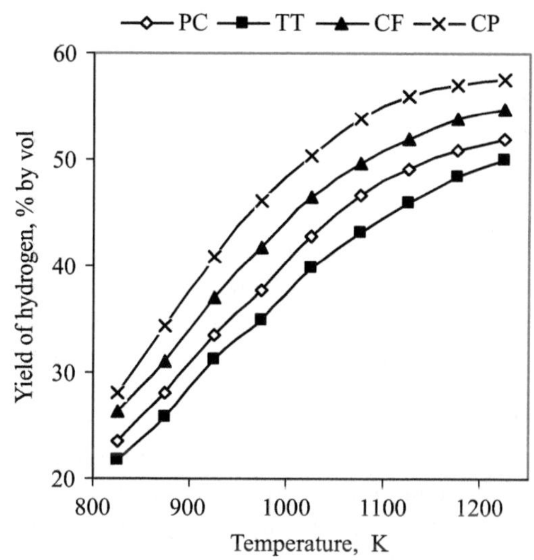

Fig. 6.24 Plots for yields of hydrogen in gaseous products from the samples by steam gasification. *Polytrichum commune* (PC), *Thuidium tamarascinum* (TT), *Cladophora fracta* (CF), and *Chlorella protothecoides* (CP)

to 39.2% and 9.5 to 40.6% by volume, respectively, while the final pyrolysis temperature was increased from 575 to 875 K.

Figure 6.23 shows the plots for yields of hydrogen in gaseous products from the samples by pyrolysis. The percentage of hydrogen in gaseous products from the samples of *Polytrichum commune*, *Thuidium tamarascinum*, *Cladophora fracta* and *Chlorella protothecoides* increased from 21.3 to 38.7%, 23.0 to 41.3% and 25.8 to 44.4% and 27.6 to 48.7% by volume, respectively, while the final pyrolysis temperature was increased from 650 to 875 K.

Figure 6.24 shows the plots for yields of hydrogen in gaseous products from the samples by steam gasification. The percentage of hydrogen in gaseous products from the above-mentioned samples increased from 21.8 to 50.0%, 23.5 to 52.0%,

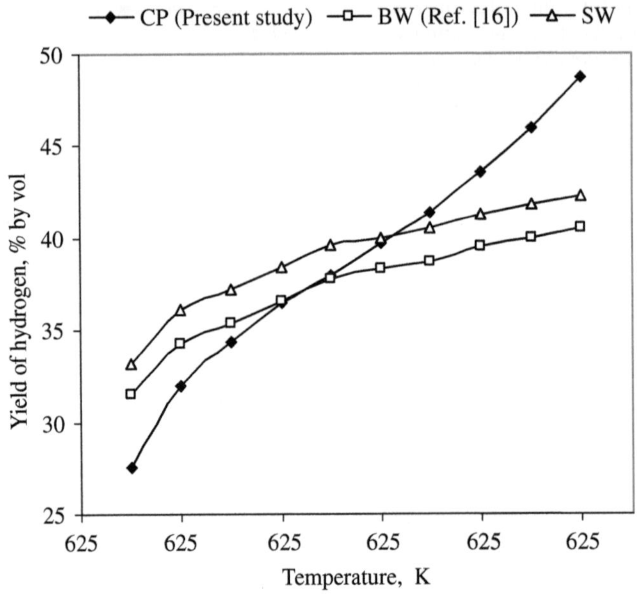

Fig. 6.25 Plots for yields of hydrogen in gaseous products from microalga and wood samples by pyrolysis. *Chlorella protothecoides* (CP), beech wood (BW) and spruce wood (SW)

26.3 to 54.7% and 28.1 to 57.6% by volume, respectively, while the final gasification temperature was increased from 825 to 1225 K.

Figure 6.25 shows the yields of hydrogen in gaseous products from microalga and wood samples by pyrolysis. The hydrogen in gaseous products from the samples of beech wood and spruce wood increased from 31.5 to 40.5% and 33.3 to 42.3% by volume, respectively, while the final pyrolysis temperature was increased from 650 to 875 K (Demirbas and Arin, 2004). Microalgae gaseous products are higher quality than gaseous products from wood (Fig. 6.25). In general, algae gaseous products are higher quality than gaseous products from mosses.

Table 6.13 shows the yields of bio-oil by pyrolysis from moss and alga samples (Demirbas, 2006b). As can be seen from Table 6.13, the bio-oil yield for *Chlorella protothecoides* (a microalga sample) rose from 12.8 to 55.3% as the temperature rose from 575 to 775 K, and then gradually the percentage decreased to 51.8%, obtained at 875 K with a heating rate of 10 K/s. The bio-oil yield for *Polytrichum commune* (a moss sample) rose from 10.3 to 39.1% as the temperature rose from 575 to 775 K, and then decreased to 36.7% obtained at 875 K with a heating rate of 10 K/s (Demirbas, 2006b). For alga, maximum bio-oil yields of between 48.2 and 46.8%, and for microalga 55.3 and 53.7% were obtained at temperatures ranging from 775 to 825 K, whereas for wood, cotton stalk, tobacco stalk and sunflower bagasse, maximum oil yields between 39.7 and 49.4% were obtained at temperatures in the range from 775 to 825 K (Putun, 2002; Gercel, 2002).

Table 6.13 Yields of bio-oil by pyrolysis from moss and alga samples at different temperatures (K)

Sample	575	625	675	725	775	825	875
Polytrichum commune	10.3	20.0	27.5	35.8	39.1	38.4	36.7
Dicranum scoparium	6.0	15.5	21.8	30.7	34.3	33.8	31.7
Thuidium tamarascinum	5.6	14.2	20.7	29.5	33.6	33.4	31.3
Sphagnum palustre	7.9	17.7	25.3	33.5	37.0	36.3	34.6
Drepanocladus revolvens	6.7	16.4	23.5	31.7	35.4	34.7	32.9
Cladophora fracta	10.5	23.5	33.2	43.4	48.2	46.8	44.6
Chlorella protothecoides	12.8	27.4	38.4	50.2	55.3	53.7	51.6

Table 6.14 Yields of gaseous product by pyrolysis from moss and alga samples at different temperatures (K)

Sample	575	625	675	725	775	825	875
Polytrichum commune	6.5	14.8	22.6	26.4	29.2	36.6	42.2
Dicranum scoparium	5.8	12.5	19.8	25.0	27.6	35.0	40.8
Thuidium tamarascinum	5.3	11.2	17.9	23.5	25.6	33.2	39.3
Sphagnum palustre	5.5	11.9	18.3	24.2	26.5	34.0	39.8
Drepanocladus revolvens	5.6	12.3	18.9	24.7	27.0	34.5	40.4
Cladophora fracta	8.2	19.7	28.2	32.6	35.7	38.0	39.7
Chlorella protothecoides	9.5	21.8	29.5	33.7	36.3	38.1	39.5

Table 6.14 shows the yields of gaseous product by pyrolysis from moss and alga samples (Demirbas, 2006b). From Table 6.14, the yields of gaseous products for *Chlorella protothecoides* increased from 9.5 to 39.5% as the temperature rose from 575 to 875 K. The char yields of pyrolysis from mosses were higher than those of algae.

With the interaction of water and char from decomposition of biomass, intermediate products are known to occur which leads to more hydrogen-rich gas yield by the steam reforming process. The pyrolysis was carried out at the moderate temperatures and steam gasification at the highest temperatures. In order to clarify the steam gasification mechanism in detail, more kinetic study is necessary. These results suggest that fundamental information obtained in the gasification of each component could possibly be used to predict the composition of product gas generated in air–steam gasification of biomass.

Summary

Biohydrogen is a renewable biofuel produced from biorenewable feedstocks by a variety of methods, including chemical, thermochemical, biological, biochemical, and biophotolytical methods. The topic of biohydrogen touches on energy security, environmental concerns, foreign exchange savings, and socioeconomic issues related to the rural sectors of all countries in the world. It is also an environmentally friendly alternative automotive fuel that can be used in an internal combustion engine.

There are three types of microorganisms that produce hydrogen: cyanobacteria, anaerobic bacteria, and fermentative bacteria. The cyanobacteria directly decompose water to hydrogen and oxygen in the presence of light by photosynthesis. Photosynthetic bacteria use organic substrates like organic acids, whereas anaerobic bacteria use organic substances as the sole source of electrons and energy, converting them into hydrogen. A promising method is the biological production of hydrogen by fermentation. Fermentative hydrogen production is the fermentative conversion of organic substrates to biohydrogen. This process is manifested by a diverse group of bacteria using multienzyme systems involving three steps similar to anaerobic conversion.

Hydrogen can be produced from biorenewable feedstocks via thermochemical conversion processes such as pyrolysis, gasification, steam gasification, steam reforming of bio-oils, and supercritical water gasification (SWG) of biomass.

Comparison with other biomass thermochemical gasification such as air gasification or steam gasification, the supercritical water gasification (SCWG) can directly deal with the wet biomass without drying, and has high gasification efficiency in relatively lower temperatures. The cost of hydrogen production from supercritical water gasification of wet biomass was several times higher than the current price of hydrogen from steam methane reforming. Biomass is gasified in supercritical water at a series of temperatures and pressures during different resident times and the product gas is composed of H_2, CO_2, CO, CH_4, and a small amount of C_2H_4 and C_2H_6.

References

Adams, M.W.W. 1990. The structure and mechanism of iron-hydrogenases. Biochim Biophys Acta 1020:115–145.

Aiba, S., Humphrey, A.E., Milis, N.F. 1973. Biochemical engineering, 2nd edn. Academic, New York.

Akkerman, I., Janssen, M., Rocha, J., Wijffels, R.H. 2002. Photobiological hydrogen production: photochemical efficiency and bioreactor design. Int J Hydrogen Energy 27:1195–1208.

Aoyama, K., Uemura, I., Miyake, J., Asada, Y. 1987. Fermentative metabolism to produce hydrogen gas and organic compounds in a cyanobacterium Spirulina platensis. J Ferment Bioeng 83:17–20.

Antal, M.J. 1978. Tower power: producing fuels from solar energy, *in* Toward a solar civilization, R.H. Williams (Ed.) MIT Press, Cambridge, MA.

Antal, M.J., Feber, R.C., Tinkle, M.C. 1974. Proceedings 1st World Hydrogen Energy Conference, Miami Beach, FL.

Antal, M.J., Allen, S., Schulman, D., Xu, X., Divilio, R. 2000. Biomass gasification in supercritical water. Ind Chem Eng Res 39:4040–4053.

Antal, T.K., Lindblad, P. 2005. Production of H_2 by sulphur-deprived cells of the unicellular cyanobacteria *Gloeocapsa alpicola* and *Synechocystis sp.* PCC 6803 during dark incubation with methane or at various extracellular pH. J Appl Microbiol 98:114–120.

Backman, R., Fredrick, W.J., Hupa, M. 1993. Basic studies on black-liquor pyrolysis and char gasification. Biores Technol 46:153–158.

Benemann, J.R. 1996. Hydrogen biotechnology: progress and prospects. Nature Biotechnol 14:1101–1103.

Benemann, J.R. 1997. Feasibility analysis of photobiological hydrogen production. Int J Hydrogen Energy 22:979–987.

Benemann, J.R. 1998. The technology of biohydrogen. Plenum, New York.

Benemann, J.R., Weare, N.M. 1974. Hydrogen evolution by nitrogen-fixing. *Anabaena cylindrica* cultures. Science 184:174–175.

Byrd, A.J., Pant, K.K., Gupta, R.B. 2007. Hydrogen production from glucose using $Ru/Al_2\ O_3$ catalyst in supercritical water. Ind Eng Chem Res 46:3574–3579.

Caglar, A., Demirbas, A. 2002a. Hydrogen-rich gas mixture from olive husk via pyrolysis. Energy Convers Manage 43:109–117.

Caglar, A., Demirbas, A. 2002b. Conversion cotton cocoon shell to hydrogen-rich gaseous products by pyrolysis. Energy Convers Manage 43:489–497.

Cetin, B., Yurdakulol, E. 1985. Gerede-Aktas (Bolu) ormanlarinin karayosunlari (Musci) florasi. Doga Bilim Der 9:29–38 (in Turkish).

Cetin, B., Yurdakulol, E. 1988. Yedigoller milli parkinin karayosunu florasi. Turkish J Bot 12:128–146 (in Turkish).

Chen, G., Spliethoff, H., Yang, L.B., Andries, J. 2003. Hydrogen production from gasification: pyrolysis of biomass. III. International Slovak Biomass Forum, Bratislava, Slovakia, 3–4 Feb 2003.

Chornet, E. 2001. Biomass to hydrogen via fast pyrolysis and catalytic steam reforming. Proceedings of the 2001 USA. DOE Hydrogen Program Review (NREL/CP 570-30535).

Claassen, P.A.M., Lier, J.B.V., Lopez, C.A.M., Niel, E.W.J., Sijitsma, L., Stams, A.J.M., Vriess, S.S.D., Weusthusis, R.A. 1999. Utilization of biomass for the supply of energy carries. Appl Microbiol Biotechnol 52:741–755.

Cortright, R.D., Davda, R.R., Dumesic, J.A. 2002. Hydrogen from catalytic reforming of biomass-derived hydrocarbons in liquid water. Nature 418:964–967.

Cox, J.L., Tonkovich, A.Y., Elliott, D.C., Baker, E.G., Hoffman, E.J. 1995. Hydrogen from biomass: a fresh approach. Proceedings of the 2nd Biomass Conference of the Americas (NREL/CP-200-8098; CONF-9508104) August 1995, Portland, Oregon.

Czernik, S., French, R., Feik, C., Chornet, E. 2001. Production of hydrogen from biomass derived liquids. Proceedings of the 2001 USA. DOE Hydrogen Program Review (NREL/CP 570-30535).

Das, D., Veziroglu, T.N. 2008. Advances in biological hydrogen production processes. Int J Hydrogen Energy 33:6046–6057.

Demirbas, A. 1998. Yields of oil products from thermochemical biomass conversion processes. Energy Convers Manage 39:685–690.

Demirbas, A. 2000. Mechanisms of liquefaction and pyrolysis reactions of biomass. Energy Convers Manage 41:633–646.

Demirbas, A. 2001. Yields of hydrogen of gaseous products via pyrolysis from selected biomass samples. Fuel 80:1885–1891.

Demirbas, A. 2002a. Gaseous products from biomass by pyrolysis and gasification: effects of catalyst on hydrogen yield. Energy Convers Manage 43:897–909.

Demirbas, A. 2002b. Hydrogen production from biomass by gasification process. Energy Sources 24:59–68.

Demirbas, A. 2004. Hydrogen-rich gas from fruit shells via supercritical water extraction. Int J Hydrogen Energy 29:1237–1243.
Demirbas, A. 2005a. Hydrogen production via pyrolytic degradation of agricultural residues. Energy Sources 27:769–775.
Demirbas, A. 2005b. Hydrogen production from biomass via supercritical water extraction. Energy Sources 27:1409–1417.
Demirbas, M.F. 2006a. Hydrogen from various biomass species via pyrolysis and steam gasification processes. Energy Sources Part A 28:245–252.
Demirbas, A. 2006b. Oily products from mosses and algae via pyrolysis. Energy Sources Part A 28:933–940.
Demirbas, A. 2007. Production of biofuels from macroalgae and microalgae. Energy Edu Sci Technol 18:59–65.
Demirbas, A. 2008a. Biohydrogen generation from organic wastes. Energy Sources Part A 30:475–482.
Demirbas, A. 2008b. Hydrogen production from carbonaceous solid wastes by steam reforming. Energy Sources Part A 30:924–931.
Demirbas, A., Arin, G. 2004. Hydrogen from biomass via pyrolysis: relationships between yield of hydrogen and temperature. Energy Sources 26:1061–1069.
Demirbas, A., Karslioglu, S., Ayas A. 1996. Hydrogen resources: conversion of black liquor to hydrogen-rich gaseous products. Fuel Sci Technol Int 14:451–465.
Demirbas A., Caglar, A. 1998. Catalytic reforming of biomass and heavy oil residues to hydrogen. Energy Edu Sci Technol 1:45–52.
Demirbas, A., Arin, G. 2004. Hydrogen from biomass via pyrolysis: relationships between yield of hydrogen and temperature. Energy Sources 26:1061–1069.
Dauvillee, D., Chochois, V., Steup, M., Haebel, S., Eckermann, N., Ritte, G., Ral, J.P., Colleoni, C., Hicks, G., Wattebled, F., Deschamps, P., D'hulst, C., Lienard, L., Cournac, L., Putaux, J.L., Dupeyre, D., Ball, S.G. 2006. Plastidial phosphorylase is required for normal starch synthesis in *Chlamydomonas reinhardtii*. Plant J 48:274–285.
Feng, W., van der Kooi, H.J., Arons, J.D.S. 2004. Biomass conversions in subcritical and supercritical water: driving force, phase equilibria, and thermodynamic analysis. Chem Eng Proc 43:1459–1467.
Friedrich, K., Kordesch, K., Simader, G., Selan, M. 1995. The process cycle sponge-iron/hydrogen/iron oxide used for fuel conditioning in fuel cells. Proceedings of International Symposium on Fuel Cell System, New York.
Gadhe, J.B., Gupta, R.B. 2007. Hydrogen production by methanol reforming in supercritical water: catalysis by in situ-generated copper nanoparticles. Int J Hydrogen Energy 32:2374–2381.
Gaffron, H. 1939. Reduction of carbon dioxide with molecular hydrogen in green algae. Am J Bot 27:273–283.
Gaffron, H. 1940. Carbon dioxide reduction with molecular hydrogen in green algae. Am J Bot 27:273–283.
Gaffron, H., Rubin, J. 1942. Fermentative and photochemical production of hydrogen in algae. J Gen Physiol 26:219–240.
Gercel, H.F. 2002. The effect of a sweeping gas flow rate on the fast pyrolysis of biomass. Energy Sources 24:633–642.
Gest, H., Kamen, M.D. 1949a. Studies on the metabolism of photosynthetic bacteria; photochemical production of molecular hydrogen by growing cultures of photosynthetic bacteria. J Bacteriol 58:239–245.
Gest, H., Kamen, M.D. 1949b. Photoproduction of molecular hydrogen by *Rhodospirillum rubrum*. Science 109:558–559.
Gest, H., Blankenship, R.E. 2004. Time line of discoveries: anoxygenic bacterial photosynthesis. Photosynthesis Res 80:59–70.
Gfeller, R.P., Gibbs, M. 1984. Fermentative metabolism of *Chlamydomonas reinhardtii*: I. Analysis of fermentative products from starch in dark-light. Plant Physiology 75:212–218.

Groenestijn, J.W.V., Hazewinkel, J.H.O., Nienoord, M., Bussmann, P.J.T. 2002. Energy aspects of biological hydrogen production in high rate bioreactors operated in the thermophilic temperature range. Int J Hydrogen Energy 27:1141–1147.

Guan, Y.F., Deng, M.C., Yu, X.J., Zhang, W. 2004. Two-stage photobiological production of hydrogen by marine green alga *Platymonas subcordiformis*. Biochem Eng J 19:69–73.

Guo, L.J., Lu, Y.J., Zhang, X.M., Ji, C.M., Guan, Y., Pei, A.X. 2007. Hydrogen production by biomass gasification in supercritical water: a systematic experimental and analytical study. Catal Today 129:275–286.

Hallenbeck, P.C., Benemann, J.R. 2002. Biological hydrogen production; fundamentals and limiting processes. Int J Hydrogen Energy 27:1185–1193.

Han, S.-K., Shin, H.-S. 2004. Biohydrogen production by anaerobic fermentation of food waste. Int J Hydrogen Energy 29:569–577.

Hao, X.H., Guo, L.J., Mao, X., Zhang, X.M., Chen, X.J. 2003. Hydrogen production from glucose used as a model compound of biomass gasified in supercritical water. Int J Hydrogen Energy 28:55–64.

Hao, X.H., Guo, L.J., Zhang, X.M., Guan, Y. 2005. Hydrogen production from catalytic gasification of cellulose in supercritical water. Chem Eng J 110:57–65.

Hawkes, F.R., Dinsdale, R.M., Hawkes, D.L., Huss, I. 2002. Sustainable fermentative hydrogen production: challenges for process optimization. Int J Hydrogen Energy 27:1339–1347.

Howarth, D.C., Codd, G.A. 1985. The uptake and production of molecular hydrogen by unicellular cyanobacteria. J Gen Microbiol 131:1561–1569.

Hussy, I., Hawkes, F.R., Dinsdale, R., Hawkes, D.L. 2005. Continuous fermentative hydrogen production from sucrose and sugarbeet. Int J Hydrogen Energy 30:471–483.

Kamen, M.D., Gest, H. 1949. Evidence for a nitrogenase system in the photosynthetic bacterium *Rhodospirillum rubum*. Science 109:560.

Klas, D. 2004. Biomass for renewable energy and fuels. Encyclopedia of energy. Elsevier, Oxford.

Kruse, A., Meier, D., Rimbrecht, P., Schacht, M. 2000. Gasification of pyrocatechol in supercritical water in the presence of potassium hydroxide. Ind Eng Chem Res 39:4842–2848.

Koskinen, P.E.P., Kaksonen, A.H., Puhakka, J.A. 2007. The relationship between instability of H_2 production and compositions of bacterial communities within a dark fermentation fluidized-bed bioreactor. Biotechnol Bioeng 97:742–758.

Kotay, S.M., Das, D. 2008. Biohydrogen as a renewable energy resource: prospects and potentials. Int J Hydrogen Energy 33:258–263.

Ledjeff-Hey, K., Kalk, T., Mahlendorf, F., Niemzig, O., Trautman, A., Roes, J. 2000. Portable PEFC generation with propane as fuel. J Power Sources 86:166–172.

Levin, D.B., Pitt, L., Love, M. 2004. Biohydrogen production: prospects and limitations to practical application. Int J Hydrogen Energy 29:173–85.

Li, S., Xu, S., Liu, S., Yang, C., Lu, Q. 2004. Fast pyrolysis of biomass in free-fall reactor for hydrogen-rich gas. Fuel Proc Technol 85:1201–1211.

Lindblad, P., Chirstensson, K., Lindberg, P., Fedorov, A., Pinto, F., Tsygankov, A. 2002. Photoproduction of H_2 by wildtype *Anabaena* PCC 7120 and a hydrogen uptake deficient mutant: from laboratory experiments to outdoor culture. Int J Hydrogen Energy 27:1271–1281.

Lower, H.H. 1963. Chemistry of coal utilization, Suppl vol. Wiley, New York, pp. 740–815.

Lu, Y., Guo, L., Zhang, X., Yan, Q. 2007. Thermodynamic modeling and analysis of biomass gasification for hydrogen production in supercritical water. Chem Eng J 131:233–244.

Lu, Y.J., Jin, H., Guo, L.J., Zhang, X.M., Cao, C.Q., Guo, X. 2008. Hydrogen production by biomass gasification in supercritical water with a fluidized bed reactor. Int J Hydrogen Energy 33:6066–6075.

Madamwar, D., Garg, N., Shah, V. 2000. Cyanobacterial hydrogen production. World J Microbiol Biotechnol 16:757–767.

Maschio, G., Lucchesi, A., Stoppato, G. 1994. Production of syngas from biomass. Biores Technol 48:119–126.

Masukawa, H., Mochimaru, M., Sakurai, H. 2002. Hydrogenases and photobiological hydrogen production utilizing nitrogenase system in cyanobacteria. Int J Hydrogen Energy 27:1471–1474.

Melis, A., Happe, T. 2001. Hydrogen production. Green algae as a source of energy. Plant Physiol 127:740–48.

Melis, A., Melnicki, M.R. 2006. Integrated biological hydrogen production. Int J Hydrogen Energy 31:1563–1573.

Mertens, R., Liese, A. 2004. Biotechnological applications of hydrogenases. Curr Opin Biotechnol 15:343–348.

Miao, X., Wu, Q. 2004. High yield bio-oil production from fast pyrolysis by metabolic controlling of *Chlorella protothecoides*. J Biotechnol 110:85–93.

Midilli, A., Rzayev, P., Hayati, O., Teoman, A. 2000. Solar hydrogen production from hazelnut shells. Int J Hydrogen Energy 25:723–732.

Milne, T.A., Elam, C.C., Evans, R.J. 2002. Hydrogen from biomass: state of the art and research challenges 1–82. National Renewable Energy Laboratory, Golden, CO.

Minowa, T., Zhen, F., Ogi, T. 1997. Cellulose decomposition in hot-compressed water with alkali or nickel catalyst. J. Supercritical Fluids 13:253–259.

Miura, Y. 1995. Hydrogen production by biophotolysis based on microalgal photosynthesis. Proc Biochem 30:1–7.

Miura, Y., Akano, T., Fukatsu, K., Miyasaka, H., Mizoguchi, T., Yagi, K., Maeda, I., Ikuta, Y., Matsumoto, H. 1997. Stably sustained hydrogen production by biophotolysis in natural day/night cycle. Energy Conver Manage 38:S533–S537.

Miura, Y., Akano, T., Fukatsu, K., Miyasaka, H., Mizoguchi, T., Yagi, K., Maeda, I., Ikuta, Y., Matsumoto, H. 1995. Hydrogen production by photosynthetic microorganisms. Energy Convers Manage 36:903–906.

Mohan, D., Pittman, C.U., Steele, P.H. 2006. Pyrolysis of wood/biomass for bio-oil: a critical review. Energy Fuels 20:848–889.

Nandi, R., Sengupta, S. 1998. Microbial production of hydrogen: an overview. Critical Rev Microbiol 24:61–84.

Nath, K., Das, D. 2003. Hydrogen from biomass. Current Sci 85:265–271.

Nath, K., Das, D. 2004. Biohydrogen production as a potential energy resource: present state-of-art. J Sci Ind Res 63:729–738.

Ormerod, J.G., Gest, H. 1962. IV. Hydrogen photosynthesis and alternative metabolic pathways in photosynthetic bacteria. Bacteriol Rev 26:51–66.

Ormerod, J.G., Ormerod, K.S., Gest. H. 1961. Light-dependent utilization of organic compounds and photoproduction of molecular hydrogen by photosynthetic bacteria; relationships with nitrogen metabolism. Arch Biochem Biophys 94:449–463.

Patil, K.Z. 1987. Hydrogen. Chem Age India 38:519–527.

Paul, P.F.M., Wise, W.S. 1971. The principle of gas extraction. Mills and Boon, London.

Prince, R.C., Kheshgi, H.S. 2005. The photobiological production of hydrogen: potential efficiency and effectiveness as a renewable fuel. Crit Rev Microbiol 31:19–31.

Putun, A.E. 2002. Biomass to bio-oil via fast pyrolysis of cotton straw and stalk. Energy Sources 24:275–285.

Rapagna, F.P.U. 1996. Hydrogen from biomass by steam gasification. Hydrogen Energy Progress XI, Proceedings of the 11th World Hydrogen Energy Conference, Stuttgart, Germany, 23–28 June 1996, 1:907–912.

Rapagna, S., Foscolo, P.U. 1998. Catalytic gasification of biomass to produce hydrogen-rich gas. Int J Hydrogen Energy 23:551–557.

Safrany, D.R. 1971. A potential large scale plasma process: synthesis of inexpensive hydrogen by using a thermonuclear device to vaporize waste organic materials. Chem Eng Progress Symp Ser 67:103–108.

Salmenoja, K. 1993. Black-liquor gasification: theoretical and experimental studies. Biores Technol 46:167–171.

Sasikala, K., Ramana, C.V., Rao, P.R., Kovacs, K.L. 1993. Anoxygenic phototrophic bacteria: physiology and advances in hydrogen technology. Adv Appl Microbio 38:211–295.

Serebryakova, L.T., Sheremetieva, M., Tsygankov, A. 1998. FEMS Microbiology Lett 166:89.
Shahbazov, Sh.J., Usupov, I. 1994. Non-trading sources of energy for hydrogen. Int J Hydrogen Energy 19:863–864.
Stevens, D.J. 2001. Hot gas conditioning: Recent progress with larger-scale biomass gasification systems. National Renewable Energy Laboratory, NREL/SR-510-29952, Golden, CO.
Sung, S. 2004. Biohydrogen production from renewable organic wastes. Final Technical Report. Iowa State University, Ames, IA.
Susta, M.R., Luby, P., Mat, S.B. 2003. Biomass energy utilization and environment protection commercial reality and outlook, Power Gen Asia. http://www.powergeneration.siemens.com/download/pool/industrialheatpower_02.pdf. Accessed 2009.
Tran, D.Q., Charanjit, R. 1978. A kinetic model for pyrolysis of Douglas fir bark. Fuel 57:293–298.
Tsygankov, A.A., Kosourov, S.N., Tolstygina, I.V., Ghirardi, M.L., Seibert, M. 2006. Hydrogen production by sulfur-deprived *Chlamydomonas reinhardtii* under photoautrophic conditions. Int J Hydrogen Energy 31:1574–1585.
Van Ginkel S., Sung, S., Lay, J.J. 2001. Biohydrogen production as a function of pH and substrate concentration. Environ Sci Technol 35:4726–4730.
Van Ginkel, S.W., Logan, B. 2005. Increased biological hydrogen production with reduced organic loading. Water Res 39:3819–3826.
Vayisoglu, E.S., Bartle, K.D., Erbatur, N.G., Frere, B., Snape, C.E., Erbatur, O. 1996. Chemical composition of SFE extracts obtained from coal and maceral concentrates. Fuel Process Technol 46:99–115.
Vijayaraghavan, K., Soom, M.A.M. 2006. Trends in biohydrogen generation: a review. Environ Sci 3:255–271.
Viswanathan, B. 2006. An introduction to energy sources. Indian Institute of Technology, Madras, India.
Wagner, E.S., Froment, G.F. 1992. Steam reforming analyzed. Hydrocarbon Proc July:69–77.
Walcher, G., Girges, S., Weingartner, S. 1996. Hydrogen Energy Progress XI, Proceedings of the 11th World Hydrogen Conference, 1996, Stuttgart.
Wang, D., Czernik, S., Montane, D., Mann, M., Chornet, E. 1997. Biomass to hydrogen via fast pyrolysis and catalytic steam reforming of the pyrolysis oil or its fractions. Ind Eng Chem Res 36:1507–1518.
Wang, D., Czernik, S., Chornet, E. 1998. Production of hydrogen from biomass by catalytic steam reforming of fast pyrolysis oils. Energy Fuels 12:19–24.
Warnecke, R. 2000. Gasification of biomass: comparison of fixed bed and fluidized bed gasifier. Biomass Bioenergy 18:489–497.
Watanabe, M., Iida, T., Inomata, H. 2006. Decomposition of a long chain saturated fatty acid with some additives in hot compressed water. Energy Convers Manage 47:3344–3350.
Weissman, J.C., Benemann, J.R. 1977. Hydrogen production by nitrogen-starved cultures of *Anabaena cylindrica*. Appl Environ Microbiol 33:123.
White, A.L., Melis, A. 2006. Biochemistry of hydrogen metabolism in *Chlamydomonas reinhardtii* wild type and a Rubisko-less mutant. Int J Hydrogen Energy 31:455–465.
Yan, Q., Guo, L., Lu, Y. 2006. Thermodynamic analysis of hydrogen production from biomass gasification in supercritical water. Energy Convers Manage 47:1515–1528.
Yanik, J., Ebale, S., Kruse, A., Saglam, M., Yuksel, M. 2007. Biomass gasification in supercritical water: Part 1. Effect of the nature of biomass. Fuel 86:2410–2415.
Yogev, A., Kribus, A., Epstein, M., Kogan, A. 1998. Solar tower reflector systems: a new approach for high-temperature solar plants. Int J Hydrogen Energy 23:239–245.

Chapter 7
Fuel Cells

7.1 Introduction

During the past decade, fuel cells have received an enormous amount of attention all over the world as novel electrical energy conversion systems. The higher efficiencies and lower emissions make the fuel cells a valuable contribution to the power generation facilities.

As a clean energy source, hydrogen gas (H_2) has potential if used in an electricity generating fuel cell (Caglar, 2003). H_2 production by reforming of HC-based fuels in suitable fuel processors has become more and more important, in particular for both mobile and residential fuel cells applications (Specchia et al., 2005).

The major factor that has influenced the development of fuel cells over the last few years is the worldwide concern of the environmental consequences of the use of fossil fuels in the production of electricity and for the propulsion of vehicles. Fuel cells seem to be the best solution for clean, efficient and non-hazardous energy.

Fuel cells are electrochemical devices that convert the chemical energy of the fuels directly into electrical energy, and are considered to be the key technology for power generation in stationary, automotive, portable and even microscale systems. Among all kinds of fuel cells, direct methanol fuel cells have really exhibited the potential to replace current portable power sources and micropower sources in the market (Yao et al., 2006).

A fuel cell produces electricity directly from the electrochemical reaction of hydrogen, from a hydrogen-containing fuel, and oxygen from the air. Hydrogen is industrially produced by steam reformation of naphtha oil, methane and methanol. High-purity hydrogen has been mainly used as a fuel for low-temperature fuel cells such as polymer or alkaline electrolyte fuel cells (Lin and Rei, 2000).

7.1.1 History of Fuel Cells

In 1800, British scientists Sir William Nicholson (1753–1815) and Sir Anthony Carlisle (1768–1840) described the process of using electricity to decompose water into hydrogen and oxygen. Sir Humphrey Davy created in 1802 a simple fuel cell that allowed him to give himself a feeble electric shock. However, this result was not well-documented. The principle of the fuel cell was discovered by Christian Friedrich Schönbein (1799–1868) in 1838, who published his article about the hydrogen–oxygen fuel cell in the *Philosophical Magazine* in January 1839. Based on this work, the first fuel cell was demonstrated by Sir William Robert Grove in the February 1839 edition of the *Philosophical Magazine and Journal of Science* and later sketched, in 1842, in the same journal (Grove, 1839, 1842).

Grove is known as the father of the fuel cell. At the London Institution, where he was professor of physics (1840–1847), he used his platinum–zinc batteries to produce electric light for one of his lectures (Chaurasia, 2000; Aravindhababu et al., 1999). Grove, however took this idea one step further or, more accurately, one step in reverse in 1839. He discovered that by arranging two platinum electrodes with one end of each immersed in a container of sulfuric acid and the other ends separately sealed in containers of oxygen and hydrogen, a constant current would flow between the electrodes. The sealed containers held water as well as the gases, and he noted that the water level rose in both tubes as the current flowed. By combining several sets of these electrodes in a series circuit, he created what he called a "gas battery": the first fuel cell. Grove had speculated that the action in his gas battery occurred at the point of contact between electrode, gas, and electrolyte, but was at a loss to explain further. He also made fuel cells with ethylene and carbon monoxide.

Friedrich Wilhelm Ostwald (1853–1932), a founder of the field of physical chemistry, provided much of the theoretical understanding of how fuel cells operate. In 1893, he experimentally determined the interconnected roles of the various components of the fuel cell: electrodes, electrolyte, oxidizing and reducing agents, anions, and cations. His exploration of the underlying chemistry of fuel cells laid the groundwork for later fuel cell researchers.

In 1896, Antonio Henri Becquerel (1852–1908) made a new type of battery and used a carbon rod. Ludwig Mond and Carl Langer produced a gas-powered battery and called their system a fuel cell in 1889. In 1889, Ludwig Mond (1838–1909) and assistant Carl Langer described their experiments with a hydrogen-oxygen fuel cell that attained 6 A (ampere) per square foot at 0.73 V. Mond and Langer's cell used electrodes of thin, perforated platinum.

Dr. William W. Jacques further explored the carbon approach in 1896. His fuel cells had a carbon rod central anode in the electrolyte of molten potassium hydroxide. He made a fuel cell system of 100 cylindrical cells, which produced as much as 1500 W. Francis T. Bacon worked on fuel cells to produce alkaline systems that did not use noble metal catalysts in the 1930s. He developed and built a 6 kW alkaline hydrogen–oxygen system in 1959. In the same year, Dr. Harry Ihrig introduced

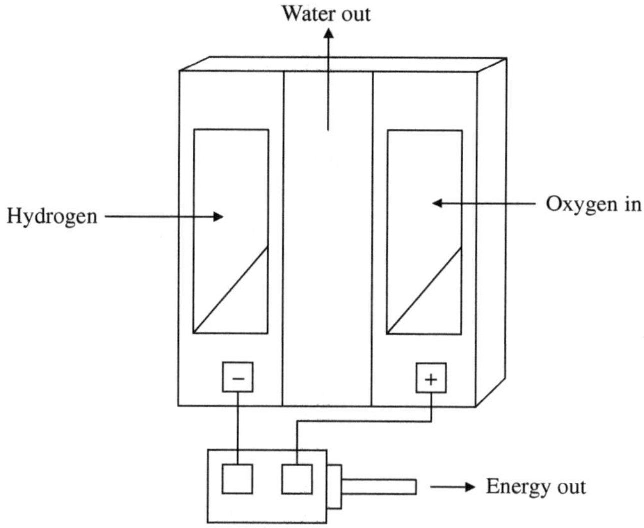

Fig. 7.1 Schematic of a hydrogen fuel cell

the first fuel cell tractor. Then a one-man submarine powered by a hydrazine and compressed oxygen fuel cell was produced. Currently, fuel cells are used to supply power and drinking water in space. The first stationary fuel cell power plant, PC-25, was developed by the International Fuel Cell Corporation (Ayhan, 2002).

Francis T. Bacon (1904–1992) began researching alkali electrolyte fuel cells in the late 1930s. In 1939, he built a cell that used nickel gauze electrodes and operated under pressure as high as 3000 psi. During World War II, Bacon worked on developing a fuel cell that could be used in Royal Navy submarines, and in 1958 demonstrated an alkali cell using a stack of 10-in. diameter electrodes for Britain's National Research Development Corporation. Though expensive, Bacon's fuel cells proved reliable enough to attract the attention of the company Pratt & Whitney. The company licensed Bacon's work for the Apollo spacecraft fuel cells.

Beginning in the late 1950s and early 1960s, there was renewed interest in the fuel cell. NASA was looking for a way to power a series of upcoming manned space flights. Using batteries for power had already been ruled out due to weight considerations. Solar energy was too expensive at the time and nuclear power was determined to be too risky. In NASA's search for an alternative, the fuel cell was thought to be a possible solution. NASA sponsored efforts to develop practical working fuel cells that could be used during these space flights. These efforts led to the development of the first proton exchange membrane fuel cell (PEMFC). In 1955, while NASA was conducting research, a scientist working at General Electric (GE) modified the original fuel cell design. In the early 1960s, aircraft engine manufacturer Pratt & Whitney licensed the Bacon patents for the alkaline fuel cell (AFC), and won a contract from NASA to supply these fuel cells to the Apollo spacecraft. Figure 7.1 shows a schematic of hydrogen fuel cell.

During the 1970s, fuel cell technology was developed for systems on Earth. The oil embargos of 1973 and 1979 helped to push along the research effort of the fuel cell as the US government was looking for a way to become less dependent on petroleum imports. A number of companies and government organizations began serious research into overcoming the obstacles to widespread commercialization of the fuel cell. Throughout the 1970s and 1980s, a large research effort was dedicated to developing the materials needed, identifying the optimum fuel source and drastically reducing the cost of this technology. During the 1980s, fuel cell technology began to be tested by utilities and automobile manufacturers. Technical breakthroughs during the decade included the development of the first marketable fuel cell-powered vehicle in 1993 by the Canadian company, Ballard.

7.2 Fundamentals of a Fuel Cell

A fuel cell is a device in which the energy of a fuel is converted directly into electricity direct current (DC) by an electrochemical reaction without resorting to a burning process, rather than being converted to heat by a combustion reaction (Berger, 1968; Bockris and Srinivasan, 1969; McAuliffe, 1980; Kordesh, 1998; Larmine and Dicks, 1999; Chaurasia et al., 2003). The energy chemically stored in the fuels is converted into electric current by means of an electrochemical process in the fuel cell. More precisely, it is an electrochemical cell in which, at a uniform temperature, the free-energy change of a fuel/oxidizer reaction is converted directly and continuously to electric energy for as long as the reactants are supplied and the products are removed at an appropriate rate (Veziroglu, 1975).

A fuel cell is an electrochemical conversion device. It produces electricity from fuel and an oxidant, which react in the presence of an electrolyte. The reactants flow into the cell, and the reaction products flow out of it, while the electrolyte remains within it. Fuel cells are different from electrochemical cell batteries in that they consume reactant, which must be replenished, whereas batteries store electrical energy chemically in a closed system. The chemical energy of the fuel is released in the form of an electrical energy instead of heat when the fuel is oxidized in an ideal electrochemical cell. Energy conversion by a fuel cell depends largely

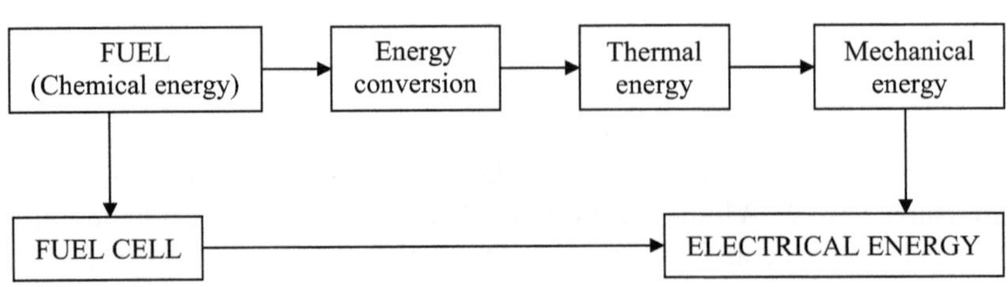

Fig. 7.2 Direct energy conversion with fuel cells compared to conventional indirect technologies

upon catalytic electrodes, which accomplish the electrochemical reaction to convert fuel into electric energy without involving the burning process. Efficiencies of fuel cells (40–85%) are considerably high when compared to heat engines. Catalysts are so expensive that electricity from most fuel cells costs about a thousand times more than the same amount derived from conventional sources. The need is to develop the catalysts from the different precursors to succeed in the necessary chemical reactions in an effective way (Demirbas, 2007a). Figure 7.2 shows the comparison of direct energy conversion with fuel cells to conventional indirect technologies.

A fuel cell is an electrochemical energy conversion device similar to a battery, but differing from the latter in that it is designed for continuous replenishment of the reactants consumed, i.e., it produces electricity from an external supply of fuel and oxygen as opposed to the limited internal energy storage capacity of a battery. Additionally, the electrodes within a battery react and change as a battery is charged, or discharged, whereas a fuel cell's electrodes are catalytic and relatively stable. Figure 7.3 shows the schematic of a fuel cell.

Typical reactants used in a fuel cell are hydrogen on the anode side and oxygen on the cathode side (a hydrogen cell). Usually, reactants flow in and reaction products flow out. Virtually continuous long-term operation is feasible as long as these flows are maintained.

The chemical energy of the fuel is released in the form of an electrical wave instead of heat when the fuel is oxidized in an ideal electrochemical cell. Many investigators believe that a stream of electrons produces the electricity current; however, it should not be forgotten that an electron always has a particle-wave character, which when transferring through an electrical conductor is a wave not a particle, i.e., an electron, during electrical current (Demirbas, 2002).

Thus, energy conversion by a fuel cell depends largely upon catalytic electrodes, which accomplish the chemical reaction to convert fuel into electric energy without involving the burning process. The only basic difference between a fuel cell and an electrochemical cell is that, in a fuel cell, the combustible chemicals that generate the electrical current are stored separately and supplied to the electrodes on demand. In principle, therefore, a fuel cell can generate electricity by burning a fuel to electrical current.

Like electrochemical cells, all fuel cells consist of a pair of electrodes, i.e., cathode and anode, and an electrolyte, plus an external circuit for electrical current and

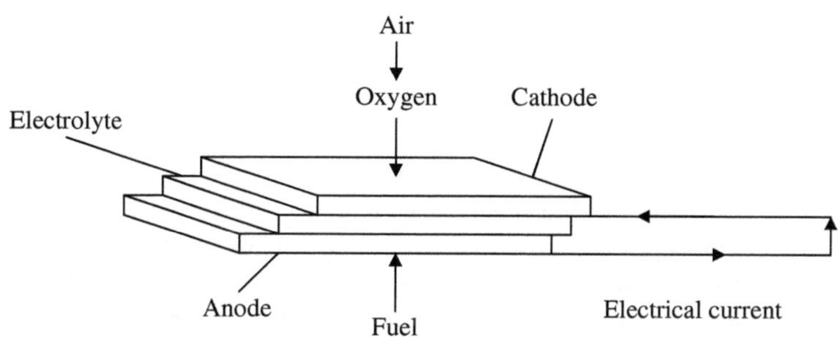

Fig. 7.3 Schematic of a fuel cell

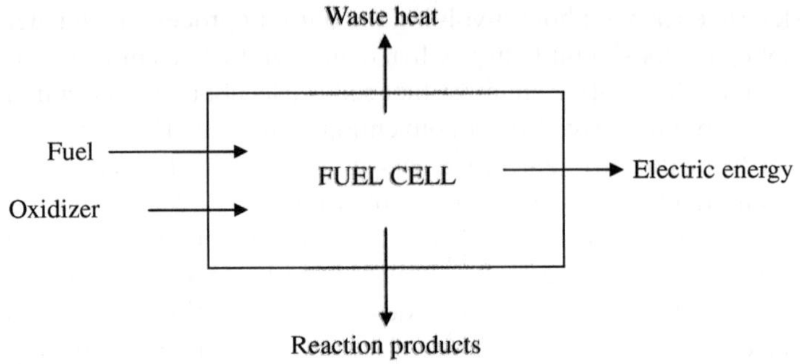

Fig. 7.4 A simplified fuel cell device

internal mechanism for allowing ion migration to complete the circuit. The output of electricity, which is always low-potential DC, is controlled by electrical potential or current regulators. In addition, they must have the associated hardware necessary for the controlled delivery of fuel and oxidizer to the electrodes, and each input and output is controlled in a suitable manner. A fuel cell can be simplified as in Fig. 7.4.

7.3 Different Types of Fuel Cells

Various types of fuel cells have been developed to generate power according to the applications and load requirements (Chaurasia, 2000). There are several types of electrolyte, which plays a key role in the different types of fuel cells. It must permit only the appropriate ions to pass between the anode and cathode. The main electrolyte types are alkali, molten carbonate, phosphoric acid, proton exchange membrane (PEM), and solid oxide. The first three are liquid electrolytes, the last two are solids.

There are several types of fuel cells with different characteristics and uses. Fuel cells are usually classified according to the electrolyte that is used. Table 7.1 shows some common fuel cell types and their uses.

7.3.1 Proton Exchange Membrane Fuel Cells (PEMFCs)

Proton exchange membrane fuel cells (PEMFCs) work with a polymer electrolyte in the form of a thin, permeable sheet. The PEMFCs, otherwise known as polymer electrolyte fuel cells (PEFC), are of particular importance for the use in mobile and small/medium-sized stationary applications (Pehnt, 2001). The PEM fuel cells are considered to be the most promising fuel cell for power generation (Kazim, 2000).

Table 7.1 Some common fuel cell types and their uses

Type of fuel cell	Area of use	Electrolyte	Temperature (K)
Proton exchange membrane (PEM)	Transportation, space travel, small CHP, mobile equipment	Polymer	325–355
Alkaline (AFC)	Transportation, space travel	Alkaline	325–475
Direct methanol (DMFC)	Transport, mobile equipment	Polymer	355–475
Phosphoric acid (PAFC)	CHP, power plants	Phosphoric acid	465–485
Molten carbonate (MCFC)	CHP, power plants	Molten carbonate	875–925
Solid oxide (SOFC)	CHP, power plants	Solid oxide	875–1275

CHP: Combined heat and power

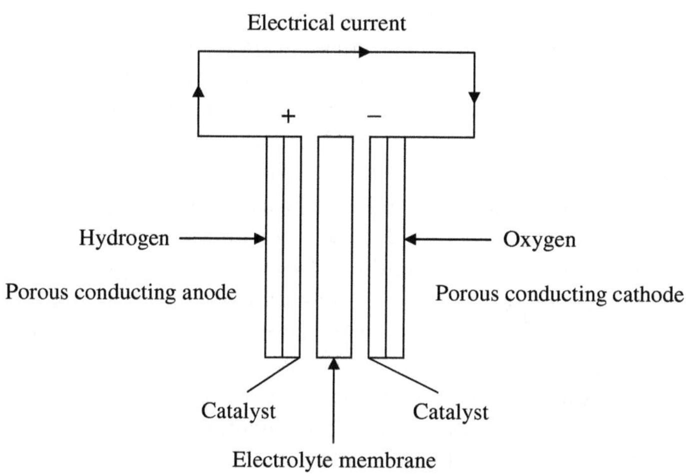

Fig. 7.5 Polymer electrolyte membrane fuel cell (PEMFC)

The advantages of PEMFCs over the other fuel cell systems are as follows (Chaurasia et al., 2003):

1. Quick-start capability
2. High power density
3. Relatively simple design
4. Light-weight fuel cell
5. Ability to adjust against variable power
6. Spontaneous off facility

Figure 7.5 shows the polymer electrolyte membrane fuel cell (PEMFC). There are two porous metal plates connected in a circuit with a membrane between them. In addition, the side of the metal plates facing the electrolyte membrane is coated with platinum as a catalyst. This catalyst is important in order for the fuel cell to work. The gases, H_2 and O_2, will pass through the porous plates and hit the catalyst.

The efficiency of PEMFCs ranges from about 40 to 50%, and operating temperature is about 255 K. The PEMFCs and direct methanol fuel cells (DMFCs) are considered to be promising power sources, especially for transportation applications. The PEMFCs with potentially much higher efficiencies and almost zero emissions offer an attractive alternative to the internal combustion engines for automotive applications. This fuel cell has many important attributes such as high efficiency, clean, quiet, low-temperature operation, capable of quick start-up, no liquid electrolyte and simple cell design (Hu et al., 2004).

7.3.2 Direct Use of Methanol in Fuel Cells

Methanol can be used blended with conventional fuels without engine modification or pure as a fuel. The direct methanol fuel cell (DMFC) is an extremely promising power source for portable applications due to its simple handling and processing of fuel. A DMFC converts chemical energy into electricity via oxidization of methanol. Research into the electrocatalysis of methanol has been stimulated by intense interest in advancing fuel cells (Gao et al., 2004). Main advantages of direct use of methanol in fuel cells are: (a) methanol is fed into fuel cell, (b) complicated catalytic reforming is not needed, and (c) the actual power densities of a DMFC are clearly lower than those of a conventional hydrogen-fed polymer electrolyte fuel cell. Methanol oxidation reaction at Pt and Pt alloy electrodes has attracted considerable attention during the last few decades in relation to direct methanol fuel cells (Seiler et al., 2004).

Methanol can be used as one possible replacement for conventional motor fuels. The use of methanol as a motor fuel received attention during the oil crises of the 1970s due to its availability and low cost. Methanol is an attractive fuel: it is liquid at room temperature, it has limited toxicity, high energy density, and is an inexpensive energy source (Shukla et al., 1998). Methanol is a favored fuel as it has twice the energy density of liquid hydrogen and since it is a liquid at normal ambient temperatures and it can thus be stored and transported easily and inexpensively (Collins, 2001).

Since methanol can be used as a blend with conventional fuels, without engine modification, or even pure as a fuel, it can also be used in traditional combustion engines and indirect methanol fuel cells. Methanol is also being viewed as an excellent base fuel for fuel cell vehicles. It can also be used as a base product for making biodiesel from vegetable oils (Demirbas, 2006). Methanol has been seen as a possible large-volume motor fuel substitute at various times during gasoline shortages. Methanol as a fuel can be used in two different ways: directly in the direct methanol fuel cell or indirectly in the indirect methanol fuel cell. The direct method liquid methanol is oxidized to carbon dioxide (CO_2) without initial reforming. The indirect method transforms methanol into hydrogen via a reforming step while the direct way feeds methanol straight to the fuel cell.

The direct methanol fuel cell (DMFC) is an extremely promising power source for portable applications due to its simple handling and processing of fuel. The direct use of methanol in fuel cells is considerably attractive from the point of view of simplicity of system design and hence cost (Shukla et al., 2002). The performance of DMFCs has improved markedly in the past 5 years (He et al., 2004). The mutual effects of the oxygen reduction reaction and methanol oxidation reaction occurring simultaneously on a Pt/Vulcan fuel cell catalyst in O_2-saturated 0.5 M H_2SO_4 solution containing 0.1 or 0.01 M methanol and the formation of a mixed potential were studied using differential electrochemical mass spectrometry (Jusys and Behm, 2004).

A relatively new member of the fuel cell family, the DMFC is similar to the PEM cell in that it uses a polymer membrane as an electrolyte. The DMFC is a special form of low-temperature fuel cell. It can be operated at 355–475 K temperatures depending on the fuel feed and type of electrolyte used. In a DMFC, methanol is fed directly into the fuel cell without the intermediate step of reforming the alcohol into hydrogen (Collins, 2001).

The DMFC, based on a polymer electrolyte fuel cell (PEFC), uses methanol directly for electric power generation and promises technical advantages for power trains. The fuel can be delivered to the fuel cell in a gaseous or liquid form. The actual power densities of a DMFC are clearly lower than those of a conventional hydrogen-fed polymer electrolyte fuel cell. In addition, methanol permeates through the electrolyte and oxidizes at the cathode. This results in a mixed potential at the cathode (Hohlein et al., 2000).

Methanol is one of the few alcohols that can be fed directly into a fuel cell and can be converted electrochemically at the anode. The DMFC can be fed with a gaseous or liquid fuel feed. The liquid DMFC generally uses a diluted methanol in water mixture (typically 1–2 molar) and only a fraction of the methanol is used at the anode (Collins, 2001). The DMFC, like an ordinary battery, provides DC electricity according to the following half reactions.

Methanol is oxidized to CO_2 with half reaction at the anode:

$$CH_3OH + H_2O \rightarrow CO_2 + 6H^+ + 6e \tag{7.1}$$

and a cathode at which oxygen is reduced to water

$$1.5O_2 + 6H^+ + 6e \rightarrow 3H_2O \tag{7.2}$$

Accordingly, the overall cell reaction is:

$$CH_3OH + 1.5O_2 \rightarrow CO_2 + 2H_2O \tag{7.3}$$

The DMFC is similar to the PEMFC in that the electrolyte is a polymer and the charge carrier is the hydrogen ion (proton). However, the liquid methanol (CH_3OH) is oxidized in the presence of water at the anode generating CO_2, hydrogen ions and the hydrogen ions travel through the electrolyte and react with oxygen from the air. Because water is consumed at the anode in the oxidization half reaction, pure methanol cannot be used without provision of water via either passive transport such as back diffusion, or active transport such as pumping. Since the fuel cell operates isothermally, all the free energy associated with this reaction should in principle be

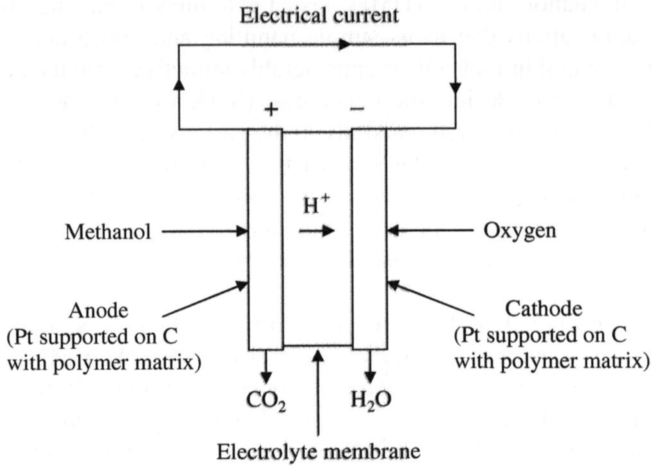

Fig. 7.6 Diagram of the operation of the DMFC membrane electrode assembly

converted to electrical energy. Figure 7.6 shows a schematic diagram of an operating DMFC system. In the direct methanol fuel cell, methanol is directly oxidized at the surface of the Pt/Ru (platinum/ruthenium) catalytic electrodes on the anode side to generate electricity without the need for any intermediate fuel processing and fuel reforming steps (Yao et al., 2006).

7.3.3 Characteristics of Main Types of Fuel Cells

The characteristics of the main types of fuel cells are shown in Table 7.2. Alkali fuel cells (AFCs) operate on compressed hydrogen and oxygen. Alkaline electrolytes are generally less corrosive than acidic ones, and permit higher power output at ambient temperature and pressure. They are therefore preferred for fuel calls. The H_2/O_2 fuel cell operating with an alkaline electrolyte is the most highly developed and reliable of all fuel cells. High-purity hydrogen has been mainly used as a fuel for low-temperature fuel cells such as polymer or alkaline electrolyte fuel cells (Lin and Rei, 2000). They generally use an aqueous KOH solution as their electrolyte in AFCs. Efficiency of an AFC is about 70%, and operating temperature is about 470 K. Molten carbonate fuel cells (MCFCs) use high-temperature compounds of salt carbonates (such as Na_2CO_3, $MgCO_3$) as the electrolyte. Efficiency of MCFCs ranges from 60 to 80%, and operating temperature is about 925 K. Phosphoric acid fuel cells (PAFCs) use phosphoric acid as the electrolyte. Efficiency of PAFCs ranges from 40 to 80%, and operating temperature is between 425 and 475 K. Solid oxide fuel cells (SOFCs) use a hard, ceramic compound of metal oxides (such as CaO, NiO, ZrO_2, $ZrO_2(Y_2O_3)$, Ni–$ZrO_2(Y_2O_3)$, $LaMnO_3$) as electrolyte. The SOFCs are an emerging technology including higher efficiencies and reduced emissions

Table 7.2 Characteristics of the main types of fuel cells

Type of fuel cell	Operating temperature, K	Efficiency, %	Cell output, kW
Alkali	425–2475	60–70	0.3–5
Molten carbonate	925	60–80	0.1
Phosphoric acid	425–475	40–80	50–200
Proton exchange membrane	255	40–50	50–200
Solid oxide	1275	50–60	50–100

(Karakoussis et al., 2001). Efficiency of a SOFC is about 60%, and operating temperatures are about 1275 K (Table 7.2).

A fuel cell is not a heat engine, because is operates at a uniform temperature between ambient temperature (~293 K) to 1300 K. A heat engine is any device for transforming heat into mechanical work. The generation of electricity by burning fossil fuels using a boiler causes serious air pollution. All heat engines operate between two temperatures: the higher temperature T_1 is that of the heat source, and the lower temperature T_2 is that of the heat sink. According to the second law of thermodynamics the thermal efficiency (η_T) of a heat engine can be expressed as:

$$\eta_T = \frac{T_1 - T_2}{T_1} \tag{7.4}$$

where T_1 and T_2 are absolute temperatures. The theoretical efficiency of a Carnot engine is about 64%. The maximum η_T of most modern steam turbines is about 40%, but most other heat engines, such as internal combustion engines and reciprocating steam engines, are fortunate to achieve 30% efficiency. Clearly, the η_T of a heat engine can be 100% only if the temperature T_2 is absolute zero, which is impossible. As a result, efficiencies of fuel cells are high compared to heat engines.

The general advantages of fuel cells are:

1. Operates cleanly
2. Efficient (40–85%)
3. No moving parts
4. Reliable
5. Quiet start
6. Waste heat generated can be captured for other uses

7.4 Catalysts Used in Fuel Cells

The fuel cell-powered cars will operate on conventional fuels. There has been much work reported in the literature by different researchers on the catalysts used in preparation of the catalytic electrodes of fuel cells (Ito et al., 1991; Yamashita et al.,

Table 7.3 Catalysts prepared from different precursors and used in present study of fuel cells

Precursor	Catalyst	Composition	Loading ratio (wt.%)
$H_2PtCl_6 . 6H_2O$	Pt	Single	5.0
$RuCl_3 . 3H_2O + H_2PtCl_6 6H_2O$.	Ru	Composite	2.5
	Pt		2.5
K_2PtCl_4	Pt	Single	5.0
$RuCl_3 . 3H_2O + K_2PtCl_4$	Ru	Composite	2.5
	Pt		2.5

1991; Ning et al., 1999; Parker et al., 2004; Cao and Bergens, 2004; He et al., 2004; Seiler et al., 2004; Jusys and Behm, 2004). The new catalysts will be used in the on-board fuel reformer of an automotive fuel cell system to convert gasoline, or other hydrocarbon fuels, into hydrogen-rich gas.

Carbon-supported platinum (Pt) and platinum–ruthenium (Pt–Ru) alloy are one of the most popular electrocatalysts in polymer electrolyte fuel cells (PEFCs). Pt supported on electrically conducting carbons, preferably carbon black, is being increasingly used as an electrocatalyst in fuel cell applications (Parker et al., 2004). Carbon-supported Pt could be prepared at loadings as high as 70 wt.% without a noticeable increase of particle size. Unsupported and carbon-supported nanoparticle $Pt–Ru_{adatom}$ ($Pt–Ru_{ad}$) catalysts prepared using the surface reductive deposition technique were evaluated as anode catalysts in liquid feed PEM-DMFCs. It was found that the surface composition of unsupported $Pt–Ru_{ad}$ nanoparticles has a significant influence on their activities as anode catalysts in direct methanol fuel cells (Cao and Bergens, 2004). Carbon-supported $Pt–Ru_{ad}$ catalysts display higher mass activities than unsupported $Pt–Ru_{ad}$. The electrochemical deposition of Pt–Ru nanoparticles on carbon nanotube electrodes and their electrocatalytic properties have been investigated by He et al. (2004). Table 7.3 shows the catalysts Ru and Pt, prepared from various precursors serving as electrolytic electrodes in the fuel cells (Chaurasia et al., 2003).

Preferential oxidation of CO in a simulated reformed gas to CO_2 by using selective CO oxidation catalysts was investigated. The effects of the preparation method, O_2, water vapor, and CO_2 concentration in the feed stream on the selective CO oxidation over Au/CeO_2 catalysts were investigated in the temperature range of 323–463 K (Luengnaruemitchai et al., 2004). The activity of the Au catalyst depends very strongly upon the preparation method, with co-precipitation prepared Au/CeO_2 catalysts exhibiting the highest activities.

The process by which the hydrogen is burned in the presence of oxygen is:

$$2H_2 + O_2 \rightarrow 2H_2O + 494 \text{ kJ (heat)} \quad (7.5)$$

The process for fuels cells is very similar, except that this time we get electricity instead of heat:

$$2H_2 + O_2 \rightarrow 2H_2O + 494 \text{ kJ (electricity)} \tag{7.6}$$

The bond energy of the H–O bond is 460 kJ. The bond energy of H–H bonds is 431 kJ. The bond energy of the O=O bonds is 486 kJ.

$$4(H) + 2(O) \rightarrow 2H_2O + 1348 \text{ kJ} \tag{7.7}$$

The formation of 2 mol of water from (H) and (O) atoms yields 1348 kJ.

The decomposition of 2 molecules of water requires breaking 4 H–O bonds and thus the input of 1840 kJ.

$$4(H) + 2(O) \rightarrow 4(H\text{–}O) + 1840 \text{ kJ} \tag{7.8}$$

The formation of 4 H–O bonds from (H) and (O) atoms yields 1840 kJ. The maximum energy, 2 mol H_2 and 1 mol O_2, available from a PEMFC (with 50% efficiency) is 920 kJ.

7.5 Use of Alternative Fuel in Fuel Cells

Alternative fuels can be used to power a fuel cell such as hydrogen, methane, natural gas, methanol, ethanol, liquefied petroleum gas and landfill gas, which can be produced from renewable energy sources such as biomass and wind.

Fuel cells have several important benefits over conventional electrical energy generation from sources such as coal, as they are more efficient at converting fuel sources to end-use energy. The fuel cell vehicle will have no harmful emissions such as nitrogen oxide (NO_x), sulfur oxide (SO_x), or particulates. Fuel cells provide waste heat utilization in co-generation, units which serve to raise the overall energy efficiency (65–85%).

7.5.1 Biodiesel-Fueled Cells

In the past decade, biodiesel has been gaining worldwide popularity as an alternative energy source because of its many benefits: this environment friendly fuel reduces tailpipe emissions, visible smoke and noxious odors (Ma and Hanna, 1999; Demirbas, 2003). Biodiesel is the only alternative fuel that produces basically no emissions during manufacture (Van Gerpen, 2005). Biofuels such as bioethanol, biomethanol, biohydrogen and biodiesel generally result in lower emissions than those of fossil-based engine fuels (Demirbas, 2007b). Many studies on the performances and emissions of compression ignition engines, fueled with pure biodiesel and blends with diesel oil, have been performed and are reported in the literature (Laforgia et al., 1994; Cardona et al., 1998). Biodiesel is being introduced to help reduce emissions from vehicles (Dicks et al., 2004). Figure 7.7 shows a schematic diagram of an operating gasoline, liquefied petroleum gas, methanol, ethanol and natural gas to electricity via fuel cell system.

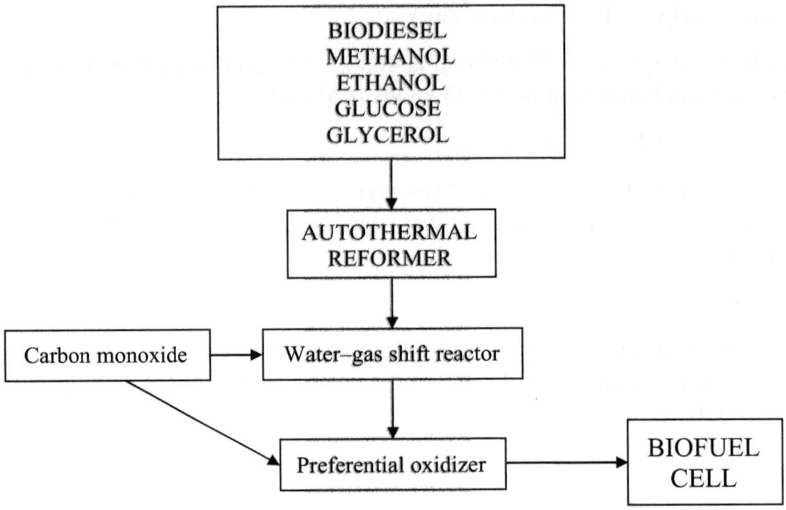

Fig. 7.7 Biodiesel, methanol, ethanol, glucose and glycerol to electricity via biofuel cell

The European Project BIOFEAT (Biodiesel Fuel Processor for a Fuel Cell Auxiliary Power Unit for a Vehicle) started in 2003. The overall aims of the use of biofuels were to reduce engine emissions and increase fuel economy (Specchia et al., 2005). An integrated auxiliary power unit (APU) based on a 10 kW_e integrated biodiesel fuel processor has been designed and is being developed. Autothermal reforming (ATR) and thermal cracking (TC) were considered for converting the fuel into a hydrogen-rich gas suitable for PEM fuel cells (Sgroi et al., 2005). Specific programs were dedicated to modeling, project coordination and management. ATR and TC processes have been evaluated as part of the investigation into the primary fuel processor. The fuel processor also includes a gas clean up system that will reduce the carbon monoxide in the primary processor exit gas to below 10 ppm via a new heat-integrated CO clean-up unit, based on the assembly of catalytic heat exchange plates, so as to meet the operational requirements of a PEMFC stack (Sgroi et al., 2005).

The following reactions take place in the ATR; a part of the feed is oxidized partially as follows (Specchia et al., 2005):

$$2C_{19}H_{36}O_2 + 17O_2 \rightarrow 38CO + 38H_2 \tag{7.9}$$

$$2H_2 + O_2 \rightarrow 2H_2O \tag{7.10}$$

$$2CO + O_2 \rightarrow 2CO_2 \tag{7.11}$$

The remaining feed is catalytically reformed via the steam reforming and water–gas shift reactions:

$$C_{19}H_{36}O_2 + 17H_2O \rightarrow 19CO + 35H_2 \tag{7.12}$$

$$CO + H_2O \rightarrow CO_2 + H_2 \tag{7.13}$$

The main unit is the catalytic primary process reactor for H_2 gross production, based on the ATR of biodiesel. After the primary step, secondary units for both the CO clean-up process and the simultaneous increase of the H_2 concentration are employed: the H_2 content from the reformated gas can be increased through the water–gas shift (WGS) reaction by converting the CO with steam to CO_2 and H_2. The high thermal shift (HTS) reactor is operating at 575–625 K followed by a low thermal shift (LTS) reactor operating at 475–535 K (Ruettinger et al., 2003). A preferential oxidation (PROX) step is required to completely remove the CO by oxidation to CO_2 on a noble metal catalyst. The PROX reaction is assumed to take place in an isothermal bed reactor at 425 K after the last shift step (Rosso et al., 2004).

The TC biodiesel fuel processor instead consists of a two-reactor system, in which one reactor, the cracker, is used for H_2 production, while the other one is being regenerated by gasification of the deposited solid C with steam yielding H_2, CO_2, CO and CH_4 (Ledjeff-Hey et al., 2000). The product gas streams of the TC and the gasification unit are combined, cooled down to the shift inlet temperature and then fed to the shift reactor and the CO-purification step. First, the cracking of biodiesel takes place for the production of H_2:

$$C_{19}H_{36}O_2 \rightarrow 17C + 2CO + 19H_2 \tag{7.14}$$

The regeneration process of the deposited C is then carried out in parallel, in a second reactor, via endothermic gasification:

$$C + H_2O \rightarrow CO + H_2 \tag{7.15}$$

The cracking and gasification reactions are both carried out at an operating temperature of 1175 K.

7.6 Hybrid Electric Vehicles and Supercapacitors

Plug-in hybrids are outfitted with a battery pack sufficient to power the vehicle for 35 miles or more on battery power alone. A plug-in hybrid electric vehicle (PHEV) is a hybrid vehicle with batteries that can be recharged by connecting a plug to an electric power source. It shares the characteristics of both conventional hybrid electric vehicles, having an electric motor and a backup internal combustion engine (ICE) for power, and of battery electric vehicles, also having a plug to connect to the electric grid. Most PHEVs on the road today are passenger cars, but there are also PHEV versions of commercial passenger vans, utility trucks, school buses, motorcycles, scooters, and military vehicles. Electric vehicles are clean, quiet, largely independent of imported petroleum, and can be powered using wind generation or another source of renewable electricity. Figure 7.8 shows the transportation fuels and petroleum-fueled, biorenewable-fueled, and renewable electricity-powered vehicles.

Supercapacitors are electrical storage devices that can deliver a higher amount of energy in a short time. Hybrid electric and fuel cell-powered vehicles need such a surge of energy to start, more than can be provided by regular batteries. Superca-

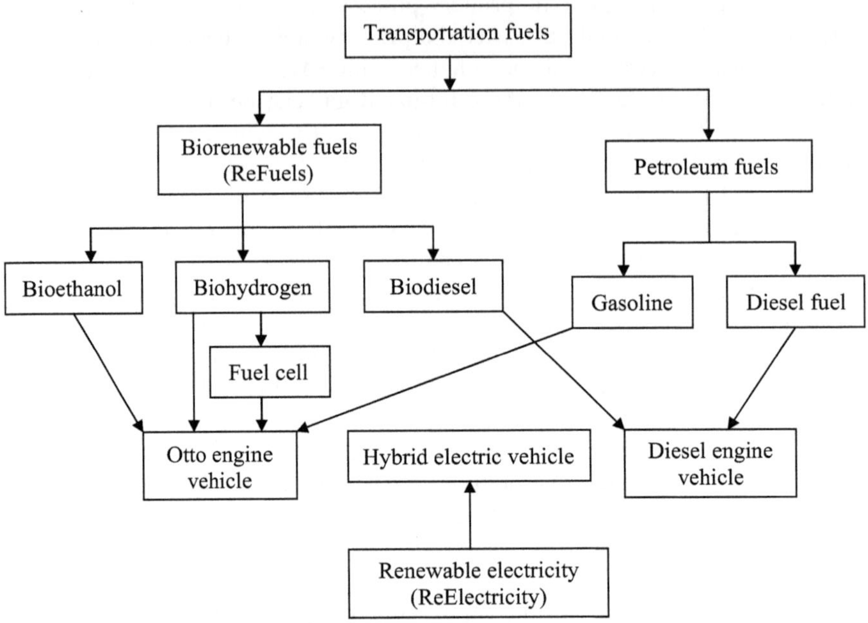

Fig. 7.8 Transportation fuels and petroleum-fueled, biorenewable-fueled, and renewable electricity-powered vehicles

pacitors are also needed in a wide range of electronic and engineering applications, wherever a large, rapid pulse of energy is required. The most important difference between a supercapacitor and a battery is the principle of electrochemical energy storage. Electrochemical energy can be stored in two fundamentally different ways. Capacitors store the energy within the electrochemical double layer at the electrode/electrolyte interface (Viswanathan, 2006).

Electrochemical capacitors have been studied for many years. The first patents date back to 1957, where a capacitor based on high surface area carbon was described by Becker. Later in 1969 first attempts to market such devices were undertaken by Standard Oil Company of Ohio (SOHIO). However, only in the 1990s did electrochemical capacitors become famous in the context of hybrid electric vehicles. The electrochemical capacitor (EC) was supposed to boost the battery or the fuel cell in the hybrid electric vehicle to provide the necessary power for acceleration, and additionally allow for recuperation of brake energy (Viswanathan, 2006).

Stand-alone photovoltaic systems are designed to operate independently from the electric utility grid, and are generally designed and sized to supply certain DC and/or AC electrical loads. These types of systems may be powered by a photovoltaic array only, or may use wind, an engine-generator or utility power as an auxiliary power source in what is called a photovoltaic hybrid system. Figure 7.9 shows how a typical photovoltaic hybrid system might be configured (Viswanathan, 2006).

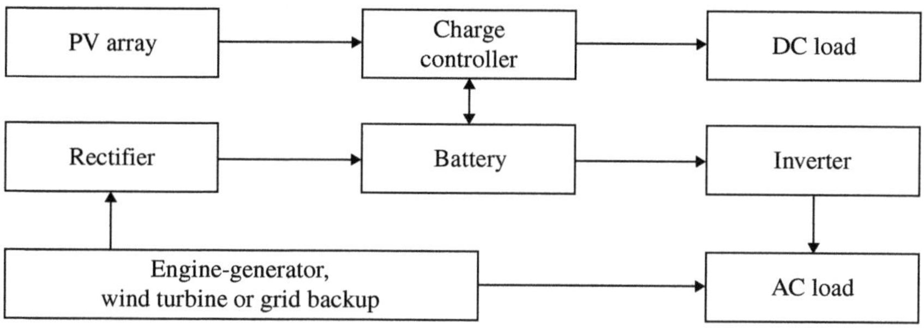

Fig. 7.9 Diagram of a photovoltaic hybrid system

Summary

In recent years, fuel cells have received considerable research attention worldwide. Fuel cells are electrochemical devices that convert the chemical energy of the fuels directly into electrical energy, and are considered to be the key technology for power generation in stationary, automotive, portable and even microscale systems. A fuel cell produces electricity directly from the electrochemical reaction of hydrogen, from a hydrogen-containing fuel, and oxygen from the air. Hydrogen is industrially produced by steam reformation of naphtha oil, methane and methanol.

Many investigators believe that a stream of electrons results in the electrical current, however, it should not be forgotten that an electron always has a particle-wave character and only a wave is transferred through an electrical conductor, i.e., an electron is not passed through the conductor, in an electrical current. The main electrolyte types are alkali, molten carbonate, phosphoric acid, proton exchange membrane (PEM), and solid oxide. Proton exchange membrane fuel cells (PEMFCs) work with a polymer electrolyte in the form of a thin, permeable sheet. The PEMFCs, otherwise known as polymer electrolyte fuel cells (PEFC), are of particular importance for the use in mobile and small/medium-sized stationary applications (Pehnt, 2001). The PEM type fuel cells are considered to be the most promising fuel cell for power generation.

Methanol can be used in a blend with conventional fuels without engine modification or pure as a fuel. The direct methanol fuel cell (DMFC) is an extremely promising power source for portable applications due to its simple handling and processing of fuel. A DMFC converts chemical energy into electricity via oxidization of methanol. Research into the electro-catalysis of methanol has been stimulated by intense interest in advancing fuel cells. Main advantages of direct use of methanol in fuel cells are: (a) methanol is fed into a fuel cell; (b) complicated catalytic reforming is not needed; (c) the actual power densities of a DMFC are clearly lower than those of a conventional hydrogen-fed polymer electrolyte fuel cell.

Fuel cells have several important benefits over conventional electrical energy generation from sources such as coal. They are more efficient at converting fuel sources to end-use energy. The fuel cell vehicle will have no harmful emissions such as nitrogen oxide (NO_x), sulfur oxide (SO_x), or particulates. Fuel cells provide waste heat utilization in co-generation, units which serve to raise the overall energy efficiency (65–85%).

References

Aravindhababu, P., Mohan, G., Sasikala, J. 1999. Emerging energy conversion techniques for next millennium, *in* Proceedings National Solar Energy Convention 1999, Indore, 20–22 Dec 1999. Pushpkunj, Indore.

Ayhan, A.F. 2002. Design of a piezoelectrically actuated microvalve for flow control in fuel cells. M.S. Thesis, University of Pittsburgh, School of Engineering, Pittsburgh, PA.

Berger, C. 1968. Handbook of fuel cell technology. Prentice-Hall, Englewood Cliffs, NJ.

Bockris, O.M.J., Srinivasan, E. 1969. Fuel cells: their electrochemistry. McGraw-Hill, New York.

Caglar, A. 2003. Gaseous products from solid wastes. Energy Edu Sci Technol 10:107–110.

Cao, D., Bergens, S.H. 2004. $Pt–Ru_{adatom}$ nanoparticles as anode catalysts for direct methanol fuel cells. J Power Sources 134:170–180.

Chaurasia, P.B.L. 2000. Solar energy utilization technology through chemical reactions: a report. Indian Council of Agricultural Research, New Delhi.

Chaurasia, P.B.L., Ando, Y., Tanaka, T. 2003. Regenerative fuel cell with chemical reactions. Energy Convers Manage 44:611–628.

Collins, J.A. 2001. Development of electrocatalyst materials for direct methanol fuel cells. Energy Marie Curie Fellowship Conference, Profactor GmbH, Steyr, Austria, 16–19 May 2001.

Demirbas, A. 2002. Similarities and differences in electricity and light concepts: convenient dimensions theory. Energy Edu Sci Technol 9:69–74.

Demirbas, A. 2003. Biodiesel fuels from vegetable oils via catalytic and non-catalytic supercritical alcohol transesterifications and other methods: a survey. Energy Convers Manage 44:2093–109.

Demirbas, A. 2007a. Fuel cells as clean energy converters. Energy Sources Part A 29:185–191.

Demirbas, A. 2007b. Progress and recent trends in biofuels. Prog Energy Combus Sci 33:1–18.

Dicks, A.L., Diniz da Costa, J.C., Simpson, A., McLellan, B. 2004. Fuel cells, hydrogen and energy supply in Australia. J Power Sources 131:1–12.

Gao, L., Huang, H., Korzeniewski, C. 2004. The efficiency of methanol conversion to CO_2 on thin films of Pt and PtRu fuel cell catalysts. Electrochim Acta 49:1281–1287.

Grove, W.R. 1839. On voltaic series and the combination of gases by platinum. Phil Magazine J Sci XIV:127–130.

Grove, W.R. 1842. On a gaseous voltaic battery. Philosophical Magazine J Sci XXI:417–420.

He, Z., Chen, J., Liu, D., Zhou, H., Kuang, Y. 2004. Electrodeposition of Pt–Ru nanoparticles on carbon nanotubes and their electrocatalytic properties for methanol electrooxidation. Diamond Related Mater 13:1764–1770.

Hohlein, B., von Andrian, S., Grube, Th., Menzer, R. 2000. Critical assessment of power trains with fuel-cell systems and different fuels. J Power Sources 86:243–249.

Hu, G., Fan, J., Chen, S., Liu, Y., Cen, K. 2004. Three-dimensional numerical analysis of proton exchange membrane fuel cells (PEMFCs) with conventional and interdigitated flow fields. J Power Sources 136:1–9.

Ito, E., Yamashita, M., Saito, Y. 1991. A composite Ru–Pt catalyst for 2-propanol dehydrogenation adaptable to the chemical heat pump system. Chem Soc Jpn Chem Lett 1:351–354.

Jusys, Z., Behm, R.J. 2004. Simultaneous oxygen reduction and methanol oxidation on a carbon-supported Pt catalyst and mixed potential formation-revisited. Electrochimica Acta 49:3891–3900.

Karakoussis, V., Brandon, N.P., Leach, M., van der Vorst, R. 2001. The environmental impact of manufacturing planar and tubular solid oxide fuel cells. J. Power Sources 101:10–26.

Kazim, M. 2000. Economical and environment assessments of proton exchange membrane fuel cell in public undertakings. Energy Convers Manage 42:763–72.

Kordesh, K. 1998. Fuel cell and their applications. VCH, Weinheim.

Laforgia, D., Ardito, V. 1994. Biodiesel fueled IDI engines: performances, emissions and heat release investigation. Biores Technol 51:53–59.

Larmine, J., Dicks, A. 1999. Fuel cell system explained. Wiley, New York.

Lin, Y.-M., Rei, M.-H. 2000. Process development for generating high purity hydrogen by using supported palladium membrane reactor as steam reformer. Int J Hydrogen Energy 25:211–219.

Luengnaruemitchai, A., Osuwan, S., Gulari, E. 2004. Selective catalytic oxidation of CO in the presence of H_2 over gold catalyst. Int J Hydrogen Energy 29:429–435.

Ma, F., Hanna, M.A. 1999. Biodiesel production: a review. Biores Technol 70:1–15.

McAuliffe, C.A. 1980. Hydrogen and energy. Macmillan, London, pp. 73–77.

Ning, M., Ando, Y., Tanaka, T., Takashima, T. 1999. Study of photocatalytic 2-propanol dehydrogenation for solar thermal cell. American Institute of Aeronautics and Astronautics, AIAA-2000-2863 (35th IECEC-2000; 7.24–7.28).

Parker, S.F., Taylor, J.W., Albers, P., Lopez, M., Sextl, G., Lennon, D., McInroy, A.R., Sutherland, I.W. 2004. Inelastic neutron scattering studies of hydrogen on fuel cell catalysts. Vibration Spectrosc 35:179–182.

Pehnt, M. 2001. Life-cycle assessment of fuel cell stacks. Int J Hydrogen Energy 26:91–101.

Rosso, I., Galletti, C., Saracco, G., Garrone, E., Specchia, V. 2004. Development of A zeolites-supported noble-metal catalysts for CO preferential oxidation: H_2 gas purification for fuel cell. Appl Catal B 48:195–203.

Ruettinger, W., Ilinich, O., Farrauto, R.J. 2003. A new generation of water gas shift catalysts for fuel cell applications. J Power Sources 118:61–65.

Seiler, T., Savinova, E.R., Friedrich, K.A., Stimming, U. 2004. Poisoning of PtRu/C catalysts in the anode of a direct methanol fuel cell: a DEMS study. Electrochimica Acta 49:3927–3936.

Sgroi, M., Bollito, G., Saracco, G., Specchia, S. 2005. BIOFEAT: Biodiesel fuel processor for a vehicle fuel cell auxiliary power unit: Study of the feed system. J Power Sources 149:8–14.

Shukla, A.K., Christensen, P.A., Dickinson, A.J., Hamnett, A. 1998. A liquid-feed solid polymer electrolyte direct methanol fuel cell operating at near-ambient conditions. J Power Sources 76:54–59.

Shukla, A.K., Jackson, C.L., Scott, K., Murgia, G. 2002. A solid-polymer electrolyte direct methanol fuel cell with a mixed reactant and air anode. J Power Sources 111:43–51.

Specchia, S., Tillemans, F.W.A., van den Oosterkamp, P.F., Saracco, G. 2005. BIOFEAT: Conceptual design and selection of a biodiesel fuel processor for a vehicle fuel cell auxiliary power unit. J Power Sources 145:683–690.

Van Gerpen, J. 2005. Biodiesel processing and production. Fuel Process Technol 86:1097–1107.

Veziroglu, T.N. 1975. Hydrogen energy: Part B. Plenum, New York.

Viswanathan, B. 2006. An introduction to energy sources. Indian Institute of Technology, Madras, India.

Yamashita, M., Kawamura, T., Suzuki, M., Saito, Y. 1991. Characteristics of suspended Ru/carbon catalyst for 2-propanol dehydrogenation applicable to chemical heat pump. Bull Chem Soc Jpn 64:272–278.

Yao, S.C., Tang, X., Hsieh, C.C., Alyousef, Y., Vladimer, M., Fedder, G.K., Amon, C.H., 2006. Micro-electro-mechanical systems (MEMS)-based micro-scale direct methanol fuel cell development. Energy 31:636–649.

Chapter 8
The Hydrogen Economy

8.1 Introduction

The hydrogen economy is a vision for a future in which hydrogen replaces fossil fuels. There are a variety of methods for generating, storing and delivering hydrogen since no single method has yet proven superior. Economically, the wasteful hydrogen process translates to electricity from hydrogen and fuel cells costing at least four times as much as electricity from the grid. In fact, electricity would be much more efficiently used if it were sent directly to the appliances instead. Low density and extremely low boiling point of hydrogen increases the energy cost of compression or liquefaction and the investment costs of storage. The low energy content of solar irradiation dictates that photosynthetic processes operate at high conversion efficiencies and places severe restrictions on photobioreactor economics (Hallenbeck and Benemann, 2002). Looking forward, towards the end of the century, solar-hydrogen type of fuels could be an option for a global energy economy.

The hydrogen economy is a proposed method of deriving the energy needed for motive power (cars, boats, airplanes), buildings or portable electronics. Hydrogen production plays a very important role in the development of a hydrogen economy. One of the promising hydrogen production approaches is conversion from biomass, which is abundant, clean and renewable. Alternative thermochemical (pyrolysis and gasification) and biological (biophotolysis, water–gas shift reaction and fermentation) processes can be practically applied to produce hydrogen (Ni et al., 2006).

The transition to a hydrogen economy will require a huge investment in new infrastructure to produce, store and deliver hydrogen to end-users. It will also require establishing hydrogen stationary systems, as well as developing and manufacturing fuel cells. In the long-term, a hydrogen-based economy will have an impact on all these sectors. The transition to the hydrogen-powered system could take several decades, because of the slow turnover of the existing stock of capital. The transition to hydrogen is likely to begin later in most developing economies than in the industrialized countries. The earlier industrialized countries/economies begin the transition to hydrogen, the faster they can achieve their energy sustainability (UNEP, 2006).

The local availability of biomass, solar and wind resources could provide the basis for the production of hydrogen in those countries where fossil fuel resources are scarce. Biomass, in particular, could be a low-cost option for some countries. Biomass, as a product of photosynthesis, is the most versatile non-petroleum renewable resource that can be utilized for sustainable production of hydrogen. Therefore, a cost-effective energy production process could be achieved in which agricultural wastes and various other biomasses are recycled to produce hydrogen economically (Nat and Das, 2003). Fortunately, hurdles in biomass gasification have been economic rather than technical. Until recently, biomass gasification has been employed to produce low-value products like electricity or heat, which rarely justify the capital and operating costs. But the increasing demand for hydrogen promises to make biomass gasification economically viable in the near future.

A hydrogen economy would be based on two electrolytic processes both associated with heavy energy losses: electrolysis and fuel cells. Furthermore, between the conversion of electricity into hydrogen by electrolysis and the reconversion of hydrogen to electricity by fuel cells, the energy carrier gas has to be packaged by compression or liquefaction. It has to be distributed by road vehicles or pipelines, stored and transferred. Fuel cells convert fuel energy to electricity. They can contribute to a secure energy future only as long as they are used to generate electricity from original fossil source energy that otherwise would have been converted to electricity with conventional heat engines. Compared to thermal power plants, internal combustion engines and gas turbines fuel cells promise to be more efficient, cleaner, more convenient and less expensive. These arguments speak for fuel cells operated on natural gas, gasoline, diesel, methanol, ethanol, etc. (Bossel et al., 2003).

The National Renewable Energy Laboratory (NREL), USA has explored the economic feasibility of producing hydrogen from biomass through various thermochemical routes. One study has been carried out to incorporate recent experimental advances in addition to plausible changes in direction from previous analysis performed by NREL. An assessment of hydrogen production technologies concludes that biomass gasification is the most economical process for renewable hydrogen production (McKinley et al., 1990), whereas other analyses have shown that biomass gasification/shift conversion is economically unfavorable compared to natural gas–steam reforming, except for very low-cost biomass and potential environmental incentives. Pyrolysis with a valuable co-product approach yields hydrogen in the price range of 6–8 US $/GJ, which is promising application in the near-term (Nath and Das, 2003).

The pyrolysis-based technology, in particular, because of the co-products opportunity, has the most favorable economics. An added advantage of biomass as a renewable feedstock is that it is not intermittent, but can be used to produce hydrogen as and when required. With scientific and engineering advancements, biomass can be viewed as a key and economically viable component to a renewable-based hydrogen economy. Economic viability of different types of energy generation processes is summarized in Table 8.1 (Bockris, 1981; Tanisho, 1996; Benemann, 1997).

Table 8.1 Unit cost of energy obtained by different processes

Type of energy	Conversion efficiency (%)	Unit cost of energy of fuel (US $/MBTU)
Photobiological hydrogen	~ 10	~ 10
Fermentative hydrogen	~ 10	~ 40
Hydrogen from coal, biomass	–	4
Hydrogen from advanced electrolysis	–	10
Hydrogen from photochemical	–	21
Fermentative ethanol	15–30	~ 31.5

8.2 Cost of Hydrogen

Current research predicts that current oil and gas reserves will only last a few more decades. The twentieth century was the century of the petrochemical economy. However, the petroleum is a finite source for fuel that is rapidly becoming scarcer and more expensive. Biorefineries will not eliminate the need for petrochemicals in the twenty-first century, but they will play a key role in reducing our level of dependence on imported petroleum and making the twenty-first century one of an increasingly sustainable, domestic, and environmentally responsible biofuel economy (bioeconomy). Oil price increases have also increased the level of interest in bioenergy. Production of bioenergy, biofuel, and bioindustrial products derived from agriculture is accelerating rapidly. A strong bioeconomy may provide significant environmental benefits.

Most of the biofuel production processes developed to date are immature and have never been implemented on an industrial scale. Projections of the amount of biofuel depend on the development of the bioeconomy and society. Biofuels are generally considered as offering many priorities, including sustainability, reduction of greenhouse gas emissions, regional development, social structure and agriculture, security of supply (Reijnders, 2006).

There are several reasons for biofuels to be considered as relevant technologies by both developing and industrialized countries. These include energy security, environmental concerns, foreign exchange savings, and socioeconomic issues, mainly related to the rural sector.

The estimation for the cost of biohydrogen is affected by a range of drivers that could change in direction and importance over time. These include:

1. Supply cost, market price and demand
2. Storage cost
3. Distribution cost
4. Production cost
5. Competing, non-energy markets for biomass
6. Access to market

The calculation of biohydrogen prices should be designed to maintain in the future the equilibrium between demand and supply, taking into account the costs of planned investments. They should also take into account the rest of the economy and the environment. Two very important characteristics of energy prices are equity and affordability. Biohydrogen prices must reflect the cost, imposed by the specific consumer category on the economy. Since energy prices based on apparent long run marginal costs may not be sufficient to finance the development of the energy sector, the prices should be adjusted, so that the energy sector can be financed without subsidies to enhance its autonomy. In competitive markets this form of adjustment may not be possible.

The marginal biohydrogen is the most expensive fuel that would be used where a target was met with the least cost. Fuels would be used in turn in order of price from lowest to highest. Our concern is with the most expensive fuel that would be used to meet the target, as that is the one that would be displaced by a lower-cost fuel. There are three cost factors that affect the biohydrogen costs:

1. Operating costs
2. Distribution and blending costs
3. Capital costs

The operating costs are defined here as all costs that are not capital or feedstock costs. The literature contains a large variation in the estimation of operating costs. Where there is a range of costs reported in different sources, a mid-point has been used.

The transition to a hydrogen-based energy economy, where the main chemical energy carrier is hydrogen and the main non-chemical energy form is electricity, is being made gradually, and is likely to continue to the middle or end of the twenty-first century (Midilli et al., 2005). Hydrogen opens access to a broad range of primary energy sources, including fossil fuels, nuclear energy and, increasingly, renewable energy sources (e.g., wind, solar, ocean, and biomass), as they become more widely available. Thus, the availability and price of hydrogen as a carrier should be more stable than any single energy source.

Current world hydrogen production is approximately 50 million tons per year (45 billion kg per year), which represents 2% of world energy demand (Raman, 2003). However, current total annual worldwide hydrogen consumption is in the range of 400–500 billion Nm^3. Of this quantity, approximately 97% is represented by captive or internal production and only about 3% is provided from merchant sources (MEDI, 2003).

Hydrogen can be produced from a variety of domestic, renewable sources of energy. On-site hydrogen production can be done by steam gasification of natural gas (or, alternatively, of hydrogen carriers such as methanol or ammonia) or by electrolysis. Steam gasification is generally much cheaper than electrolysis at industrial scales because of the low price of natural gas compared to electricity. Today, virtually all hydrogen is produced by steam gasification (Berry et al., 1996; Kaarstad and Audus, 1997). Costs of liquid hydrogen from different methods com-

Table 8.2 Costs and contents of thermochemical energies for selected energy carriers

	KWh/l	US $/kWh	US $/l gasoline-equivalent
Bio-liquid hydrogen	2.64	0.123	1.00
Diesel oil	10.00	0.050	0.45
Fossil-liquid hydrogen	2.64	0.073	0.58
Gasoline	9.10	0.060	0.55
Heavy fuel oil	11.70[a]	0.015	0.14
Hydro-liquid hydrogen	2.64	0.136	1.18
Natural gas	10.20[a]	0.016	0.15
Wind-liquid hydrogen	2.64	0.268	2.36
Solarthermic-liquid hydrogen	2.64	0.359	3.16
Photovoltaic-liquid hydrogen	2.64	1.27	11.18

[a] Units in kWh/kg

pared to fossil fuels and fossil liquid hydrogen are listed in Table 8.2. Hydrogen can be produced using wind and solar electricity.

The strategy is based on producing hydrogen from biomass pyrolysis using a co-product strategy to reduce the cost of hydrogen. It was concluded that only this strategy could compete with the cost of the commercial hydrocarbon-based technologies (Wang et al., 1998).

This strategy will demonstrate how hydrogen and biofuel are economically feasible and can foster the development of rural areas when practiced on a larger scale. The process of biomass to activated carbon is an alternative route to hydrogen with a valuable co-product that is practiced commercially. Due in part to the advantages mentioned in Chap. 5, the production of hydrogen by SCWG from biomass provides for reducing the cost of production.

Hydrogen is produced commercially in almost a dozen processes. Most of them involve the extraction of the "hydro" part from hydrocarbons. The most widely used, least costly process is "steam reforming," in which natural gas is made to react with steam, releasing hydrogen (Hohhmann, 2002). While using steam to reform natural gas has proven the cheapest way to produce commercial hydrogen, natural gas is still a hydrocarbon and emits CO_2 in the conversion process (Rifkin, 2002; Arni, 2004).

Hydrogen is a synthetic energy carrier. The synthesis of hydrogen requires energy. Production, packaging, storage, transfer and delivery of the hydrogen gas, in essence all key components of an economy, are so energy-consuming that alternatives should be considered. The production technology would be site-specific and include steam reforming of methane and electrolysis in hydropower-rich countries.

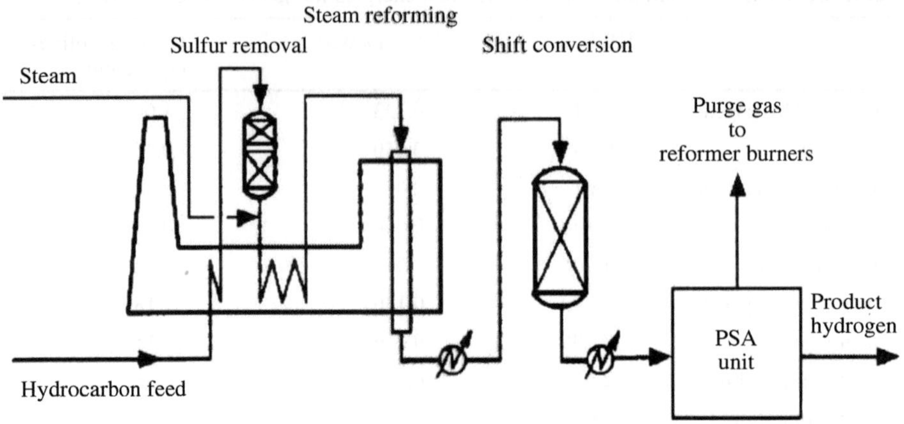

Fig. 8.1 A schematic diagram of hydrogen from hydrocarbon feed by steam reforming

Hydrogen may be the only link between physical energy from renewable sources and chemical energy. It is also the ideal fuel for modern clean energy conversion devices like fuel cells or even hydrogen engines.

Hydrogen for fleet vehicles would probably dominate in the transportation sector. To produce hydrogen via electrolysis and the transportation of liquefied hydrogen to rural areas with pipelines would be expensive. The cost of hydrogen distribution and refueling is very specific. However, there are some barriers to the development of a hydrogen economy. They are technological, economic, supply, storage, safety and policy barriers. Reducing these barriers is one of the driving factors in the government's involvement in hydrogen and fuel cell research and development.

Today the least expensive way of making hydrogen is through steam gasification of natural gas. Figure 8.1 shows a schematic diagram of hydrogen from hydrocarbon feed by steam reforming. Steam methane reforming (SMR), methanol reforming, and gasoline reforming are based on the same fundamental principles with modified operating conditions depending on the hydrogen-to-carbon ratio of the feedstock. Typically the feedstock is pre-treated to remove sulfur, a poison which deactivates nickel-reforming catalysts.

The resulting cost of hydrogen is typically about 5–6 US $/GJ when made from natural gas at a price of 3 US $/GJ (Kaarstad and Audus, 1997). The conversion of natural gas to hydrogen almost doubles the price per unit of energy from 3 to 5–6 US $/GJ. The hydrogen has to show one or more substantial advantage over natural gas if it is to compete in the energy market. One such advantage would be the lack of CO_2 and other greenhouse gas emissions if hydrogen is produced at end-use, and it can sometimes be that the technology for using hydrogen offers higher efficiency (i.e., fuel cells) or other favorable effects (e.g., low weight for space fuel application). At the present state of technologies for utilizing hydrogen it seems that the technology factor alone cannot justify the change to hydrogen as an energy carrier.

Table 8.3 The economics of decarbonization conversion to hydrogen

	Production cost US $/GJ
Today's oil and natural gas price	3.0
Hydrogen from natural gas, CO_2 to atmosphere	5.6
Hydrogen from natural gas with CO_2 removed and stored underground	6.9
Hydrogen from coal with CO_2 removed and stored underground	13.2
Hydrogen from water electrolysis based on an electricity price of 3.6 ¢/kWh	20.0

Conventionally produced hydrogen gas (with CO_2 emitted to atmosphere) costs about twice that of natural gas or oil and about three times more than coal. At present only the space industry seems to be willing to pay the high cost of hydrogen energy (Demirbas and Caglar, 1998). The relative competitiveness of the various carbon-containing and non-carbon energy carriers are given in Table 8.3 (Surmen and Demirbas, 2005). It is obvious that hydrogen made from coal, oil, or natural gas will be more costly than these primary energy sources themselves. It is also obvious that catching and storing CO_2 is more expensive than emitting the CO_2 directly into the atmosphere.

In general terms, it is expected that decarbonizing hydrogen made from natural gas or coal will increase the production cost by between 50 and 75%. Hydrogen gas that is produced conventionally (with CO_2 emitted to atmosphere) costs about twice as much as natural gas or oil and about three times more than coal.

Biomass residues are the cheapest feedstocks. Because of the low sulfur content of biomass, a sulfur removal system is not likely to be required. The current selling price of H_2 in industry is generally between 5 US $/GJ and 14 US $/GJ, depending on the size of the production facility. This range is for hydrogen as produced by the plant: purified but not compressed or stored. For a conceptual process, the cost of hydrogen has been estimated to be 7.70 US $/GJ for the base case (production capacity: 35.5 tons of hydrogen per day) (Wang et al., 1997). The preliminary economic analysis of hydrogen from biomass by steam gasification indicates that the necessary selling price of hydrogen is well within current market values.

The running cost of the generated hydrogen was carried out by processing 10 tons of the biomass. It was transported by truck for $ 16. The cost of 1 kg of purified hydrogen is only $ 0.68 (Rabah and Eldighidy, 1989).

Hydrogen will play an important role in a future energy economy mainly as a storage and transportation medium for renewable energy sources. Renewable shares of 69% on the total energy demand will lead to hydrogen shares of 34% in 2050 (Rohland et al., 1992). The ability to store hydrogen more easily than electricity can help with load leveling and in balancing the intermittent nature of renewable energy sources. The share of individual primary energy sources in meeting final energy needs are given in Table 8.4 (Demirbas and Caglar, 1998). Hydrogen is also one of

Table 8.4 Share of individual primary energy sources in meeting final energy needs (%)

Source of energy	1998	2025	2050
Fossil fuels	88	62	29
Nuclear energy	10	2	2
Hydrogen from solar energy	–	7	31
Electricity from solar energy	–	11	16
Heat from solar energy	–	18	22
Energy from solar energy	2	25	35
Hydrogen	–	11	34

Table 8.5 Economics of decarbonization to electricity and hydrogen, US \$/GJ

Oil and natural gas price	3
Electricity from coal with CO_2 removed and stored underground	21.9
Electricity from natural gas with CO_2 removed and stored underground	18.8
Coal-fired plant electricity price	12.5
Hydrogen from H_2O electrolysis based on an electricity price: 3.6 ¢/kWh	20.0
Hydrogen from coal with CO_2 removed and stored underground	13.2
Hydrogen from natural gas with CO_2 removed and stored underground	6.9
Hydrogen from natural gas, CO_2 to atmosphere	5.6

Table 8.6 The cost of hydrogen versus other fuels

Fuel	Excluding efficiency differences	Including efficiency differences
Gasoline	0.21	0.21
Diesel fuel, No. 2	0.17	0.09
Electrolytic hydrogen	0.33–1.22	0.22–0.81
Electrolytic hydrogen (diesel operation)	0.33–2.22	0.18–0.67
Coal hydrogen (diesel operation)	0.24	0.13

Table 8.7 Better utilization of energy resources

1 ton coal converted to gasoline fuel-runs	704 km
1 ton coal converted to electricity fuel-runs	768 km
1 ton coal converted to methanol fuel-runs bus	832 km
1 ton coal converted to hydrogen fuel-runs bus	1024 km

the few energy carriers that enable renewable energy sources to be introduced into transport systems (Rohland et al., 1992).

The economics of decarbonization to electricity and hydrogen is shown in Table 8.5 (Kaarstad and Audus, 1997). Table 8.6 shows what the cost of hydrogen versus other fuels is. Better utilization of energy resources are given in Table 8.7.

Hydrogen can be transported by two systems: road delivery system (cryogenic liquid trucks, compressed tube trailers) and pipeline delivery system. Hydrogen pipeline costs could be reduced by placing the pipelines in sewers, securing utility status, or converting existing natural gas pipelines to carry a mixture of hydrogen/ natural gas (town gas). A hydrogen economy also involves hydrogen transport by trucks and ships. There are other options for hydrogen distribution, but road transport will always play a role, be it to serve remote locations or to provide back-up fuel to filling stations at times of peak demand. Hydrogen pipelines exist, but they are used to transport a chemical commodity from one production site to another. The energy required to move the gas is of secondary importance, because energy consumption is part of the production and energy expenditures are one part of the overall production cost. This is not so for hydrogen energy transport through pipelines. Normally, pumps are installed at regular intervals to keep the gas moving. These pumps are energized by energy taken from the delivery stream (UNEP, 2006). The infrastructure investment for fueling stations could reach 60% of the total capital costs.

Delivery by cryogenic liquid hydrogen tankers is the most economical pathway for medium market penetration. They could transport relatively large amounts of hydrogen and reach markets located throughout large geographic areas. Tube trailers are better suited for relatively small market demand and the higher costs of delivery could compensate for losses due to liquid boil-off during storage. Pipelines are most effective for handling large flows. Operating costs for pipelines are relatively low. In the long run, when hydrogen is a very common energy carrier, distribution with pipeline is probably the preferred option. Hydrogen generated from rooftop solar electricity and stored at low pressure in stationary tanks may be a viable solution for private buildings.

Compressed gas and liquid storage are the most commercially viable options today, but completely cost-effective storage systems have yet to be developed. The safety aspects with all storage options, particularly the novel hydride storage options, must not be underestimated (Schlapbach and Züttel, 2001).

Summary

The hydrogen economy is a vision for a future in which hydrogen replaces fossil fuels. Economically, the wasteful hydrogen process translates to electricity from hydrogen and fuel cells costing at least four times as much as electricity from the grid. In fact, electricity would be much more efficiently used if it were sent directly to the appliances instead. The transition to a hydrogen economy would require a substantial investment in new infrastructure to produce, store and deliver hydrogen to end-users, establishing hydrogen stationary systems, as well as to develop and manufacture fuel cells. The transition to the hydrogen-powered system could take several decades, because of the slow turnover of the existing stock of capital. The transition to a hydrogen economy is likely to begin later in most developing economies than in the industrialized countries.

A hydrogen economy would be based on two electrolytic processes both associated with heavy energy losses: electrolysis and fuel cells. Compared to thermal power plants, internal combustion engines and gas turbines fuel cells promise to be more efficient, cleaner, more convenient and less expensive. The calculation of biohydrogen prices should be designed to maintain in the future the equilibrium between supply and demand, taking into account the costs of planned investments. They should also take into account the rest of the economy and the environment. Two very important characteristics of energy prices are equity and affordability.

Hydrogen is a synthetic energy carrier. The synthesis of hydrogen requires energy. Production, packaging, storage, transfer and delivery of the hydrogen gas, in essence all key component of an economy, are so energy-consuming that alternatives should be considered. The production technology would be site-specific and include steam reforming of methane and electrolysis in hydropower-rich countries. Conventionally produced hydrogen gas costs about twice that of natural gas or oil and about three times more than coal. At present only the space industry seems to be willing to pay the high cost of hydrogen energy.

There are two hydrogen transport systems: road delivery system (cryogenic liquid trucks, compressed tube trailers) and pipeline delivery system. Hydrogen pipeline costs could be reduced by placing the pipelines in sewers, securing utility status, or converting existing natural gas pipelines to carry a mixture of hydrogen/natural gas.

References

Arni, S. 2004. Hydrogen-rich gas production from biomass via thermochemical pathways. Energy Edu Sci Technol 13:47–54.

Benemann, J.R. 1997. Feasibility analysis of photobiological hydrogen production. Int J Hydrogen Energy 22:979–987.

Berry, G.D., Pasternak, A.D., Rambach, G.D., Smith, J.R., Schock, R.N. 1996. Hydrogen as a future transportation fuel. Energy 21:289–303.

Bockris, J.O'M. 1981. The economics of hydrogen as a fuel. Int J Hydrogen Energy 6:223–241.

Bossel, U., Eliasson, B., Taylor, G. 2003. The future of the hydrogen economy: bright or bleak? Fuel Cell Seminar, 3–7 Nov 2003, Miami Beach, FL.

Demirbas, A., Caglar, A. 1998. Catalytic reforming of biomass and heavy oil residues to hydrogen. Energy Edu Sci Technol 1:45–52.

Hallenbeck, P.C., Benemann, J.R. 2002. Biological hydrogen production; fundamentals and limiting processes. Int J Hydrogen Energy 27:1185–1193.

Hohhmann, P. 2002. Tomorrow's energy: hydrogen, fuel cells, and the prospects for a cleaner planet. MIT Press, Cambridge, MA.

Kaarstad, O., Audus, H. 1997. Hydrogen and electricity from decarbonised fossil fuels. Energy Convers Manage 38:S431–S436.

McKinley, K.R., Browne, S.H., Neill, D.R., Seki, A., Takahashi, P.K. 1990. Hydrogen fuel from renewable resources. Energy Sources 12:105–110.

Midilli, A., Ay, M., Dincer, I., Rosen, M.A. 2005. On hydrogen and hydrogen energy strategies I: current status and needs. Renew Sust Energy Rev 9:255–271.

MEDI (Manitoba Energy Development Initiative). 2003. Department of Energy, Science and Technology. Preliminary hydrogen opportunities report. http://www.manitobaenergy.com. Accessed April 2003.

Ni, M., Leung, D.Y.C., Leung, M.K.H., Sumalty, K. 2006. An overview of hydrogen production from biomass. Fuel Proc Technol 87:461–472.

Rabah, M.A., Eldighidy, M.S. 1989. Low cost hydrogen production from waste. Int J Hydrogen Energy 14:221–227.

Raman, V. 2003. Hydrogen infrastructure-market development. Air Products and Chemicals Inc., Allentown, PA. http://www.cleanair.org. Accessed 2009.

Rifkin, J. 2002. The hydrogen economy. Jeremy P. Tarcher/Penguin, New York.

Rohland, B., Nitsch, J., Wendt, H. 1992. Hydrogen and fuel cells: the clean energy system. J Power Sources 37:271–277.

Schlapbach, L., Züttel, A. 2001. Hydrogen-storage materials for mobile applications. Nature 414:353–357.

Surmen, Y., Demirbas, A. 2002. Thermochemical conversion of residual biomass to hydrogen for Turkey. Energy Sources 24:403–411.

Tanisho, S. 1996. Feasibility study of biological hydrogen-production from sugar cane by fermentation. Hydrogen Energy Progress XI. Proceedings of 11th WHEC, Stuttgart, 3:2601–2606.

UNEP (United Nations Environment Program). 2006. The hydrogen economy: a non-technical review. UNEP, Paris.

Wang, D., Czernik, S., Chornet, E. 1998. Production of hydrogen from biomass by catalytic steam reforming of fast pyrolysis oils. Energy Fuels 12:19–24.

Wang, D., Czernik, S., Montane, D., Mann, M., Chornet, E. 1997. Biomass to hydrogen via fast pyrolysis and catalytic steam reforming of the pyrolysis oil or its fractions. Ind Eng Chem 36:1507–1518.

Chapter 9
Hydrogen Policy

9.1 Introduction

Current energy policies address environmental issues including environmentally friendly technologies to increase energy supplies and encourage cleaner, more efficient energy use, air pollution, greenhouse effect (mainly reducing carbon dioxide emissions), global warming and climate change (Demirbas, 2008). In general, an energy policy includes issues of energy production, distribution and consumption. It is also the manner in which a given entity has decided to address these issues. The attributes of energy policy may include international treaties, legislation on commercial energy activities (trading, transport, storage, etc.), incentives to investment, guidelines for energy production, conversion and use (efficiency and emission standards), taxation and other public policy techniques, energy-related research and development, energy economy, general international trade agreements and marketing, energy diversity, and risk factors contrary to possible energy crisis.

There is considerable uncertainty in the economic analysis, associated with the costs of feedstocks and oil prices, and also because of the rapid development of the technologies. Any policy to support and encourage the supply of biofuel should provide incentives for both biodiesel and ethanol supplies. Policy options include incentive payments or tax breaks.Due to rising prices for fossil fuels (especially oil, but also natural gas and to a lesser extent coal) the competitiveness of biomass use has improved considerably over time. Biomass and bioenergy are now a key option in energy policies. Security of supply, an alternative for mineral oil and reduced carbon emissions are key reasons. Targets and expectations for bioenergy in many national policies are ambitious, reaching 20–30% of total energy demand in various countries. Similarly, long-term energy scenarios also contain challenging targets.

Hydrogen could be a peaceful energy carrier for all countries. Hydrogen might be the next great fuel, as it is available worldwide and water is its only byproduct. Hydrogen has received increased attention as a renewable and environmentally friendly option to help meet today's energy needs.

Policy-makers will need to pay more attention to the implications for a transition to a hydrogen economy. The advantages of hydrogen praised by journalists (non-toxic, burns to water, abundance of hydrogen in the Universe, etc.) are misleading, because the production of hydrogen depends on the availability of energy and water, both of which are increasingly rare and may become political issues, as much as oil and natural gas are today.

Due to some technological and economical consequences, practical uses of hydrogen energy do not have wide applications in the richest countries or in the poorest countries at present. For the developed countries, active involvement in hydrogen research and development, especially through collaborative international programs, could facilitate the introduction of new hydrogen technologies as they become competitive. A major dilemma now faced by the developing countries is how to invest in hydrogen research and development for the transition to hydrogen economy. Most developing countries will probably invest in hydrogen technology rather than be developers of cutting-edge technologies. Developing countries have at least as much to gain from a move towards the hydrogen economy as industrialized ones, since they generally suffer more from urban pollution and their economies tend to be more energy intensive. International organizations have an important role to play in assisting countries in creating a market-based policy relating to hydrogen and other clean energy systems (UNEP, 2006). International organizations should support the developing countries for the transition to a hydrogen economy as well as hydrogen production and distribution, to provide capital both national and foreign.

Hydrogen's share in the energy market is increasing with the implementation of fuel cell systems for sustainable energy supply. The concept of sustainable development embodies the idea of the interlinkage and the balance between economic, social and environmental concerns. Energy is an essential input for social development and economic growth. Globally, the demand for energy is increasing in consonance with socioeconomic development. Commercial energy consumption is increasing much faster in the developing countries. Energy use in developing countries has risen more than fourfold over the past three decades and is expected to continue increasing rapidly in the future. The increase in the services that energy provides is necessary for both economic and social development in the developing countries and to provide for increased human populations. Population growth is more rapid in developing countries. It is estimated to double by 2040, while the population of the industrial world will increase by only 15% over the same period.

9.2 Political, Environmental, and Economical Impacts of Hydrogen

Energy is an essential input driving economic development. Therefore, in developed economies energy policies constitute an important component of overall regulatory frameworks shaping the improving overall competitiveness and market integration of the private business sector. Overall competitiveness includes liberalization of the

Table 9.1 Shares of alternative fuels in total automotive fuel consumption in the EU under the optimistic development scenario of the European Commission. Adapted from Demirbas (2008)

Year	Biofuel	Natural gas	Hydrogen	Total
2010	6	2	–	8
2015	7	5	2	14
2020	8	10	5	23

electricity and gas markets as well as by separation of energy production, transportation, and distribution activities.

Current European Union (EU) policies on alternative motor fuels focus on the promotion of biofuels. The definition of the marginal producer depends on the policy stance on biofuels. Biofuel pricing policy should not be employed as an anti-inflationary instrument. It should be applied in such a way, that it does not create cross subsidies between classes of consumers. In a proposed biofuels directive the introduction of a mandatory share scheme for biofuels, including as from 2009 minimum blending shares. Table 9.1 shows the shares of alternative fuels compared to the total automotive fuel consumption in the EU under the optimistic development scenario of the European Commission. The EU has set the goal of obtaining 5.75% of transportation fuel needs from biofuels by 2010 in all member states in February 2006. In the commission's view mandating the use of biofuels will (a) improve energy supply security, (b) reduce greenhouse gas (GHG) emissions, and (c) boost rural incomes and employment (Jansen, 2003; Hansen et al., 2005). The European Union accounted for nearly 89% of all biodiesel production worldwide in 2005. By 2010, the United States is expected to become the world's largest single biodiesel market, accounting for roughly 18% of world biodiesel consumption, followed by Germany.

The general EU policy objectives considered most relevant to the design of energy policy are (Jansen, 2003):

1. Competitiveness of the EU economy
2. Security of energy supply
3. Environmental protection

Producing and using biofuels for transportation offers alternatives to fossil fuels that can help provide solutions to many environmental problems. Using biofuels in motor vehicles helps reduce green house gas (GHG) emissions. Biodiesel and ethanol provide significant reductions in GHG emissions compared to gasoline and diesel fuel. Due to the low or zero content of pollutants such as sulfur in biofuels, the pollutant emission of biofuels is much lower than the emission of conventional fuels. Numerous low-emission scenarios have demonstrated that the Kyoto Protocol cannot be achieved without establishing a large role for biofuel in the global energy economy by 2050. Low-emission scenarios imply 50–70 EJ of biofuel raw material in 2050. Well-designed biofuel projects would have very significant sustainable

Table 9.2 Major benefits of biohydrogen

Economic impacts	Sustainability and fuel diversity
	Increased number of rural manufacturing jobs
	Increased investments in plant and equipment
	International competitiveness
	Reducing the dependency on imported petroleum
Environmental impacts	Reducing of greenhouse gas emissions
	Reducing of air pollution
	Non-carbon fuel
Energy security	Peaceful and domestic targets
	Reducing use of fossil fuels
	Ready availability
	Domestic distribution
	Renewability

development benefits for rural areas, including creation of rural employment, rural electricity supply, soil conservation and environmental benefits (Demirbas, 2006).

The main biofuel opportunities where suitable land is available are in developing countries. The issue of energy security has been accorded the top-most priority. Every effort needs to be made to enhance the indigenous content of energy in a time-bound and planned manner. The additional benefit of biofuel development is creation of new employment opportunities in manufacturing, construction, plant operation and servicing, and fuel supply. Rural jobs are created in fuel harvesting, transport and maintenance of processing areas (Demirbas, 2006).

It is believed that in the future biomass can become an important sustainable source of hydrogen. The future prospects for hydrogen economy or economic hydrogen production are the basic point of many articles (Veziroglu, 1998; Goltsov and Veziroglu, 2002). At the end of the twenty-first century, the world will depend on the predominant use of carbon-free and carbon-neutral sources of energy, such as flow energy, solar energy, energy from modern biomass and safe nuclear energy (Quakernaat, 1995).

Biomass has an advantage of having low environmental impact compared to that for fossil fuels. However, the price of hydrogen obtained by direct gasification of lignocellulosic biomass is about three times higher than that for hydrogen produced by steam reforming of natural gas (Spath et al., 2000).

The strategy is based on producing hydrogen from biomass pyrolysis using a co-product strategy to reduce the cost of hydrogen and concluded that only this strategy could compete with the cost of the commercial hydrocarbon-based technologies (Wang et al., 1998). This strategy will demonstrate how hydrogen and biofuel are economically feasible and can foster the development of rural areas when practiced on a larger scale. The process of biomass to activated carbon is an alternative route to hydrogen with a valuable co-product that is practiced commercially. Table 9.2 shows the major benefits of biohydrogen.

During the first steps slow pyrolysis is used to maximize the yield of charcoal using densified biomass. The slow pyrolysis is used to maximize the yield of charcoal-producing organic vapors in 40% yield. The amount of hydrogen that can be produced from biomass is relatively low at 12–14%, which is based on the biomass weight. In the proposed second process, fast pyrolysis of biomass is used to generate bio-oil and catalytic steam reforming of the bio-oil to hydrogen and carbon dioxide.

9.3 Global Biofuel Projections

Projections are important tools for long-term planning and policy settings. Renewable energy sources that use indigenous resources have the potential to provide energy services with zero or almost zero emissions of both air pollutants and greenhouse gases. Renewable energy is a promising alternative solution because it is clean and environmentally safe. Currently, renewable energy sources supply 14% of the total world energy demand. Approximately, half of the global energy supply from renewables in 2040. Photovoltaic (PV) systems and wind energy will be able to play an important role in the energy scenarios of the future. The most significant developments observed in renewable energy production have been with photovoltaics (from 0.2 to 784 Mtoe) and wind energy (from 4.7 to 688 Mtoe) between 2001 and 2040.

Various scenarios have resulted in high estimates of biofuel in the future energy system. The availability of the resources is an important factor if high shares of biofuel penetrate the electricity, heat or liquid fuel market. The rationale is to facilitate the transition from the hydrocarbon economy to the carbohydrate economy by using biomass to produce bioethanol and biomethanol as replacements for traditional oil-based fuels and feedstocks. The biofuel scenario produced equivalent rates of growth in GDP and per capita affluence, reduced fossil energy intensities of GDP, reduced oil imports and gave an energy ratio. Each scenario has advantages whether it is rates of growth in GDP, reductions in carbon dioxide emissions, the energy ratio of the production process, the direct generation of jobs, or the area of plantation biomass required to make the production system feasible (Demirbas, 2006).

Renewable resources are more evenly distributed than fossil and nuclear resources, and energy flows from renewable resources are more than three orders of magnitude higher than current global energy use. Today's energy system is unsustainable because of equity issues as well as environmental, economic, and geopolitical concerns that have implications far into the future (UNDP, 2000).

According to International Energy Agency (IEA), scenarios developed for the USA and the EU indicate that near-term targets of up to 6% displacement of petroleum fuels with biofuels appear feasible using conventional biofuels, given available cropland. A 5% displacement of gasoline in the EU requires about 5%

of available cropland to produce ethanol while in the USA 8% is required. A 5% displacement of diesel requires 13% of USA cropland, 15% in the EU. The recent commitment by the USA government to increase bio-energy threefold in 10 years has added impetus to the search for viable biofuels (IEA, 2006).

The dwindling fossil fuel sources and the increasing dependency of the USA on imported crude oil have led to a major interest in expanding the use of bioenergy. The EU has also adopted a proposal for a directive on the promotion of the use of biofuels with measures ensuring that biofuels account for at least 2% of the market for gasoline and diesel sold as transport fuel by the end of 2005, increasing in stages to a minimum of 5.75% by the end of 2010 (Hansen et al., 2005).

Figure 9.1 shows the shares of alternative fuels compared to the total automotive fuel consumption in the world as a futuristic view. Hydrogen is currently more expensive than conventional energy sources. There are different technologies presently being practiced to produce hydrogen economically from biomass. Biohydrogen technology will play a major role in the future because it can utilize the renewable sources of energy (Nath and Das, 2003).

Biofuels are expected to reduce dependence on imported petroleum with associated political and economic vulnerability, reduce greenhouse gas emissions and other pollutants, and revitalize the economy by increasing demand and prices for agricultural products. Although most attention focuses on ethanol, interest in biodiesel is also increasing. Rapeseed is the primary oil used to make European biodiesel. Currently, biodiesel use is particularly strong in Germany. Biodiesel is primarily produced from soybeans in the US. The European Union has chosen biodiesel as its main renewable liquid fuel. Fuel use of ethanol in the European Union is much less important. Low European corn production and a high proportion of diesel engines compared to the United States make biodiesel a more attractive alternative in the European Union.

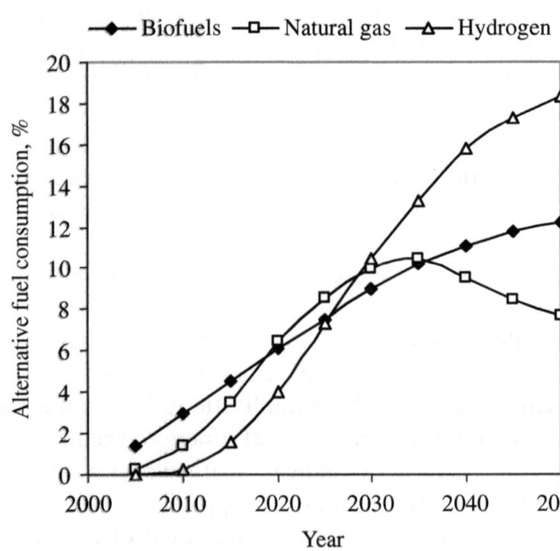

Fig. 9.1 Shares of alternative fuels compared to the total automotive fuel consumption in the world. Adapted from Demirbas (2006)

For fuels produced from biomass, various conversion routes are available that follow from the different types of biomass feedstocks. These routes include direct conversion processes such as extraction of vegetable oils followed by esterification (biodiesel), fermentation of sugar-rich crops (ethanol), pyrolysis of wood (pyrolysis oil derived diesel equivalent), and hydrothermal upgrading (HTU) of wet biomass (HTU oil derived diesel equivalent). Another possibility is to produce liquid biofuels (methanol, DME, Fischer–Tropsch liquids) from synthesis gas, which results from gasification of biomass. At present, the biofuel-producing countries in the European Union only have a small share in global production of biofuels, namely a little less than 6%. Most of the global biofuel production consists of ethanol. The main ethanol producers are the US and Brazil, whereas the share of Europe is rather small. However, Europe is the most important producer of biodiesel on the global market.

Biomass is the most used renewable energy source now and in the future. The potential of sustainable large hydropower is quite limited to some regions in the world. The potential for small hydropower (< 10 MW) is still significant and will become more significant in the future. PV systems and wind energy are technologies with annual growth rates of more than 30% during the last years that will also become more significant in the future, as well as geothermal and solar thermal sources. PV will then be the largest renewable electricity source with a production of 25.1% of global power generation in 2040.

The global and regional availability has been estimated for wind energy, solar energy especially photovoltaic (PV) cell, and biomass, together with a description of factors influencing this availability. In the long-term, countries with surplus biomass potential could develop into exporters of bioenergy.

Renewable energy sources or renewables contributed 2% of the world's energy consumption in 1998, including 7 exajoules (EJ) from modern biomass and 2 EJ for all other renewables (UNDP, 2000). The renewables are clean or inexhaustible and primary energy resources. Renewable technologies like water and wind power probably would not have provided the same fast increase in industrial productivity as fossil fuels (Edinger and Kaul, 2000).

Biomass provides a number of local environmental gains. Energy forestry crops have a much greater diversity of wildlife and flora than the alternative land use, which is arable or pasture land. In industrialized countries, the main biomass processes utilized in the future are expected to be direct combustion of residues and wastes for electricity generation, bioethanol and biodiesel as liquid fuels and combined heat and power production from energy crops. The future of biomass electricity generation lies in biomass-integrated gasification/gas turbine technology, which offers high energy conversion efficiencies. Biomass will compete favorably with fossil mass for niches in the chemical feedstock industry. Biomass is a renewable, flexible and adaptable resource. Crops can be grown to satisfy changing end-use needs.

In the future, biomass has the potential to provide a cost-effective and sustainable supply of energy, while at the same time aiding countries in meeting their greenhouse gas reduction targets. By the year 2050, it is estimated that 90% of the world population will live in developing countries.

Summary

Current energy policies also address environmental issues including environmentally friendly technologies to increase energy supplies and encourage cleaner, more efficient energy use, air pollution, greenhouse effect, global warming and climate change. Any policy to support and encourage the supply of biofuel should provide incentives for both biodiesel and ethanol supplies. Policy options include incentive payments or tax breaks.

Hydrogen could be a peaceful energy carrier for all countries. Hydrogen might be the next great fuel, as it is available worldwide and water is its only byproduct. Hydrogen has received increased attention as a renewable and environmentally friendly option to help meet today's energy needs.

Policy makers will need to pay more attention to the implications for the transition to a hydrogen economy. A major dilemma now faced by the developing countries is how to invest in hydrogen research and development for the transition to the hydrogen economy. Hydrogen's share in the energy market is increasing with the implementation of fuel cell systems for sustainable energy supply. The concept of sustainable development embodies the idea of the interlinkage and the balance between economic, social and environmental concerns.

The strategy is based on producing hydrogen from biomass pyrolysis using a co-product strategy to reduce the cost of hydrogen and concluded that only this strategy could compete with the cost of the commercial hydrocarbon-based technologies.

Hydrogen is currently more expensive than conventional energy sources. There are different technologies presently being practiced to produce hydrogen economically from biomass. Biohydrogen technology will play a major role in the future because it can utilize the renewable sources of energy.

References

Demirbas, A. 2006. Global biofuel strategies. Energy Edu Sci Technol 17:27–63.

Demirbas A. 2008. Biodiesel: a realistic fuel alternative for diesel engines. Springer, London.

Edinger, R., Kaul, S. 2000. Humankind's detour toward sustainability: past, present, and future of renewable energies and electric power generation. Renew Sustain Energy Rev 4:295–313.

Goltsov, V.A., Veziroglu, T.N. 2002. A step on the road to hydrogen civilization. Int J Hydrogen Energy 27:719–723.

Hansen, A.C., Zhang, Q., Lyne, P.W.L. 2005. Ethanol-diesel fuel blends: a review. Biores Technol 96:277–285.

IEA (International Energy Agency). 2006. Reference scenario projections. IEA, Paris.

Jansen, J.C. 2003. Policy support for renewable energy in the European Union. Energy Research Centre the Netherlands. http://www.ecn.nl/docs/library/report/2003/C03113. Accessed 2009.

Nath, K., Das, D. 2003. Hydrogen from biomass. Current Sci 85:265–271.

Quakernaat, J. 1995. Hydrogen in a global long-term perspective. Int J Hydrogen Energy 20:485–492.

Spath, P., Lane, J., Mann, M., Amos, W. 2000. Update of hydrogen from biomass: determination of the delivered cost of hydrogen. NREL Milestone Report.

UNDP (United Nations Development Programme). 2000. World Energy Assessment. Energy and the challenge of sustainability. UNDP, New York.
UNEP (United Nations Environment Program). 2006. The hydrogen economy: a non-technical review. Paris.
Veziroglu, T.N. 1998. Dawn of the hydrogen age. Int J Hydrogen Energy 23:1077–1078.
Wang, D., Czernik, S., Chornet, E. 1998. Production of hydrogen from biomass by catalytic steam reforming of fast pyrolysis oils. Energy Fuels 12:19–24.

Chapter 10
Environmental Impacts of Hydrogen

10.1 Introduction

Today the world is facing three critical problems: (1) high fuel prices, (2) climatic changes, and (3) air pollution. Research suggests that current oil and gas reserves are sufficient for only a few more decades. It is well-known that transport is almost totally dependent on fossil fuels, particularly petroleum-based fuels such as gasoline, diesel fuel, liquefied petroleum gas, and compressed natural gas. Petroleum-based fuels are well-established products that have served industry and consumers for more than one hundred years. For the foreseeable future automotive fuels will still be largely based on liquid biorenewables and gaseous biohydrogen. However the time is running out and petroleum, once considered inexhaustible, is now being depleted at a rapid rate. As the amount of available petroleum decreases, the need increases for alternate technologies to produce liquid biorenewables and gaseous biohydrogen fuels that could potentially help prolong the liquid fuels culture and mitigate the forthcoming effects of the shortage of transportation fuels.

Thermochemical conversion processes have met resistance from the environmental community and the public. An essential difference between direct combustion, pyrolysis, and gasification is that the latter two are intermediate processes for producing gaseous, liquid, and solid products that can be used in a wide variety of applications. Pyrolysis processes can be optimized for the production of biocrude oils. In the case of chemical and fuel production, the emissions from a direct process effluent can be avoided, although consideration must be given to emissions from the ultimate use of these products as they are used or combusted downstream.

While exhaust gas cleanup of non-combustion thermochemical conversion processes may be easier than that associated with direct combustion, proper design of the process and emissions control systems is necessary to ensure that health and safety requirements are met. The output products of pyrolysis and gasification reactors can contain a variety of potential process and air pollutants that must be controlled prior to discharge into the ambient air. These include particulate matter

(PM), aerosols or tars, oxides of nitrogen (NO_x), oxides of sulfur (SO_x), dioxins and furans, hydrocarbon (HC) gases, multiple metals, and carbon monoxide (CO). There are many strategies for controlling emissions from thermochemical conversion processes, and they are highly dependent on the process requirements of each individual facility.

Subsidies and incentives are provided independently from the environmental impact that ethanol may have during its entire life cycle, therefore, supporting biofuel production in the US. In 2001, the European Commission launched a policy to promote the use of biofuels for transport in order to reduce greenhouse gas emissions and the environmental impact of transport, as well as to increase security of supply, technological innovation and agricultural diversification (UN, 2006).

Environmental concerns have been raised in recent years dealing with greenhouse gases produced from the transportation industry. A contributing cause of these emissions is the combustion of fossil fuels such as diesel, gasoline and oil. A strong environmental initiative has pushed for the development of alternative fuels such as ethanol and biodiesel in pure and blended forms (Demirbas, 2008).

Increasingly, engineering practitioners and managers need to know how to respond to challenges of integrating environmentally conscious technologies, techniques, strategies, and objectives into their daily work, and, thereby, find opportunities to lower costs and increase profits while managing to limit environmental impacts (Demirbas, 2008).

The biofuels include bioethanol, biobutanol, fatty acid (m)ethylesters (biodiesel), vegetable oils, biomethanol, pyrolysis oils, fuels from Fischer–Tropsch synthesis, biogas, and biohydrogen. There are several reasons for biofuels to be considered as relevant technologies by both developing and industrialized countries. They include energy security reasons, environmental concerns, foreign exchange savings, and socioeconomic issues related to the rural sector. Due to its environmental merits, the share of biofuel in the automotive fuel market will grow fast in the next decade (Kim and Dale, 2005; Demirbas and Balat, 2006). Advantages of biofuels are the following (Puppan, 2002):

1. Biofuels are easily available from common biomass sources
2. They represent a carbon dioxide cycle in combustion
3. Biofuels have a considerable environmentally friendly potential
4. There are many benefits to the environment, economy and consumers in using biofuels
5. They are biodegradable and contribute to sustainability

There is a consensus amongst scientists that biofuels used in a sustainable manner result in no net increase in atmospheric carbon dioxide (CO_2). Some would even go as far as to declare that sustainable use of biomass will result in a net decrease in atmospheric CO_2 (Jefferson et al., 1991). This is based on the assumption that all the CO_2 given off by the use of biomass fuels was recently taken in from the atmosphere by photosynthesis. Increased substitution of fossil fuels with biomass-based fuels would therefore help reduce the potential for global warming, caused by increased atmospheric concentrations of CO_2.

10.2 Environmental Impacts of Hydrogen

Hydrogen can be burned in such a way as to produce no harmful emissions. The fundamental attraction of hydrogen is its environmental advantages over fossil fuels. But hydrogen is only as clean as the technologies used to produce and use it. If hydrogen is produced without emitting any carbon dioxide or other climate-destabilizing greenhouse gases, it could form the basis of a truly sustainable energy system.

Hydrogen can be produced from carbon-free or carbon-neutral energy sources or from fossil fuels with CO_2 capture and storage (sequestration). Thus, the use of hydrogen could eventually eliminate greenhouse gas emissions from the energy sector.

Vehicles and stationary power generation fueled by hydrogen are zero-emission devices at the point of use, with consequential local air quality benefits. Hydrogen-powered fuel cells could contribute to reducing or eliminating emissions of carbon dioxide and other greenhouse gases from road transportation vehicles. There is increasing interest in the role that hydrogen-based energy systems may play in the future, especially in the transport sector. They appear to be an attractive alternative to current fossil fuel-based energy systems in the future, since these have been proven to affect climate due to greenhouse gas emissions. Nevertheless, any future hydrogen-based economy would need to assess the possible global environmental impacts of such an alternative energy production.

The scientists have estimated that the potential effects on climate from hydrogen-based energy systems would be much lower than those from fossil fuel-based energy systems. However, such impacts will depend on the rate of hydrogen leakage during its synthesis, storage and use. The researchers have calculated that a global hydrogen economy with a leakage rate of 1% of the produced hydrogen would produce a climate impact of 0.6% of the fossil fuel system it replaces. If the leakage rate was 10%, then the climate impact would be 6% of that of the fossil fuel system.

A great attraction of hydrogen is pollution-free combustion while a host of undesirable compounds are emitted from gasoline and diesel fuel vehicles, or formed from their emissions. The main combustion product of hydrogen is water:

$$H_2 + O_2 \rightarrow 0.5H_2O \tag{10.1}$$

The use of fossil fuels has a harmful impact on the environment. During the extraction, transportation, refinement, and storage of petroleum and petroleum products, spills and leakages occur, which causes water and air pollution. Most of the fossil fuel environmental impact occurs during the end-use combustion when tremendous amounts of various gases (CO_2, CO, SO_x, NO_x, CH), soot and ash are produced and released into the atmosphere (Plass et al., 1990). Hydrogen vehicles would not produce, either directly or indirectly, significant amount of CO, HCs, particulates, SO_x, sulfur-acid deposition, ozone and other oxidants, benzene and other carcinogenic aromatic compounds, formaldehyde and other aldehydes, lead and other toxic metals, smoke or CO_2 and other greenhouse gases. The only pollutant of concern would be NO_x. If hydrogen is made from water using a clean power source, then hydrogen production and distribution will be pollution-free.

The production of hydrogen electrolytically, using clean solar power or other forms of renewable energy is essentially pollution-free. The feedstock, water, is composed of hydrogen and oxygen. Hydrogen production or distribution would produce no CO_2.

Of the three fossil fuel feedstocks for hydrogen—oil, coal and natural gas—only coal will be available at relatively cheap prices over the next several decades. However, the use of coal to produce hydrogen would not be a desirable long-term option from an environmental standpoint (Demirbas and Caglar, 1998).

An internal combustion engine fueled by hydrogen can be adjusted so that the emission of NO_x is 200 times less than in present vehicles (Williamson and Edeskuty, 1986). Emissions of NO_x increase with the temperature combustion temperature, the length of the high-temperature combustion period, and the availability of hydrogen, up to a point. There are several ways to control NO_x in a hydrogen engine: run the engine very lean, which lowers the temperature, or very rich, which reduces the oxygen supplies, and decreases the burn time or lower the engine. The results of several hydrogen engine and vehicle emissions tests are shown in Table 10.1 (Deluchi, 1989).

Table 10.1 Emissions of hydrogen and same model gasoline vehicles

Vehicle or engine	Fuel	Emission-control	Emissions, g/km				
			NO_x	CO	CH	CO_2	Reference
Dodge D-SO Pick-up	CH_2	PCV	1.19	0.44	0.26	8.66	CARB, 1986
Dodge D Pick-ups	Gasoline	CC, EGR	1.80	2.73	0.28	3.70	EPA, 1987
Musashi-2	LH_2	No CC	2.50	0.18	0.05	–	Stewart, 1986
Ford 4-cl	Gasoline	No CC	1.42	21.9	1.41	–	Watson, 1984
BMW 745i, 6 cyl	LH_2	No CC	0.35	–	–	–	Strobl and Peschka, 1986
BMW 745i, 6 cyl	LH_2	No CC	0.27	–	–	–	Peschka, 1986
BMW 745i, 6 cyl	Gasoline	No CC	2.50	–	–	–	Peschka, 1986
MB 280 TE Wagon	H_2/Gasoline	No CC	0.52	1.67	2.17	–	May, 1984

LH_2: Liquid hydrogen
PVC: Positive crankcase ventilation
CC: Catalytic converter
EGR: Exhaust gas re-circulation

Hydrogen offers the prospect of plentiful supplies of clean transportation energy. Analysis of hydrogen in efficient passenger vehicles was given a comprehensive review (Deluchi, 1989). That study addressed hydrogen production and distribution, on-board storage technology, re-fueling, vehicle performance and safety, the environmental impacts of hydrogen use, and life-cycle costs. The environmental impact analysis focuses on NO_x emissions from vehicles, the environmental impacts of making hydrogen from coal, and the contribution to the greenhouse effect of CO_2 emissions from the use of coal-based hydrogen. The life-cycle cost model compares hydrogen vehicles to similar baseline gasoline vehicles, and includes estimates of the total retail cost of liquid and gaseous hydrogen from coal and solar-powered electrolysis, the cost of fuel storage equipment, the cost of pollution control equipment, maintenance costs, other ownership and operating costs, and the external costs of hydrogen use.

While there is considerable uncertainty about several important parameters, I conclude that under certain plausible (though not necessarily likely) conditions—low hydrogen fuel and vehicle costs, high non-market costs of gasoline use, and moderately strong efforts to avoid a greenhouse warming—hydrogen vehicles could be the socially preferred highway transportation option.

Starch, cellulose or hemicellulose content of wastes, carbohydrate-rich food industry effluents or waste biological sludge can be further processed to convert the carbohydrates to organic acids and then to hydrogen gas by using proper bioprocessing technologies. Figure 10.1 shows schematic diagram for biohydrogen production from food industry wastewaters and agricultural wastes by two-stage, anaerobic dark and photofermentations (Karapinar Kapdan and Kargi, 2006).

Hydrogen is one of the major trace gases in the lower atmosphere or troposphere. The mean global mixing ratio of hydrogen is currently about 510 ppb, 500 ppb in the northern hemisphere and 520 ppb in the southern hemisphere (Simmonds et al., 2000). Hydrogen is somewhat unusual among trace gases in that, although its life cycle has been heavily influenced by human activities, its mixing ratios in

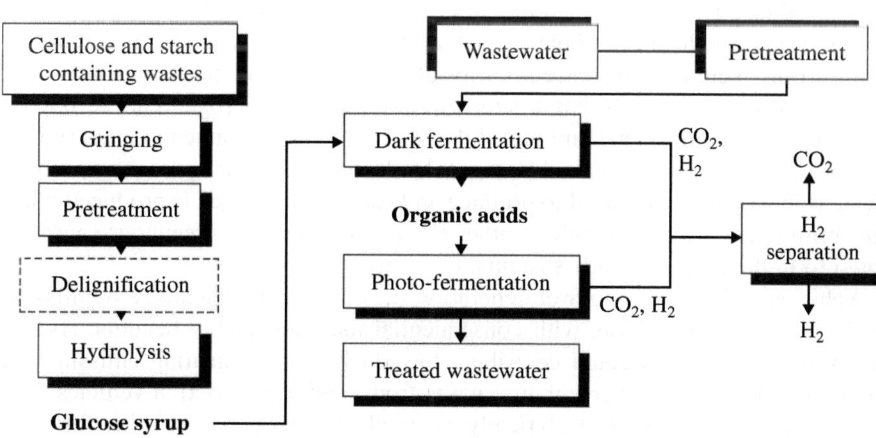

Fig. 10.1 A schematic diagram for biohydrogen production from cellulose/starch containing agricultural wastes and food industry wastewaters

the northern hemisphere are lower than those in the southern hemisphere. Because hydrogen reacts with tropospheric hydroxyl radicals, emissions of hydrogen to the atmosphere perturb the distributions of methane and ozone, the second and third most important greenhouse gases after carbon dioxide.

Emissions of hydrogen can be compared to the six criteria pollutants (CO, SO_x, NO_2, PM, ozone, and lead) and GHGs from near and long-term methods of generating hydrogen for vehicles and stationary power systems, since it is important to evaluate effects of emissions on climate, human health, ecosystem and structures. The importance of hydrogen as a greenhouse gas was quantified (Derwent et al., 2001) using the global Lagrangian chemistry transport model STOCHEM. The model was started from an initial set of trace gas concentrations in October 1994, using analyzed wind fields to run the model through to 1 January 1995. At that point, two model experiments were initiated. The first model experiment continued on without change until 31 December 1998 and this formed the base case. In the second model experiment, the transient case, the hydrogen emission source strength was increased so that a pulse containing an additional 40 Tg of hydrogen was emitted into the model by 31 January 1995. At this point, the hydrogen emission was reset to the base case value and the model experiment was continued until 31 December 1998. The impacts of the additional hydrogen on the composition of the model troposphere were followed by taking differences between the base and transient cases. These differences in composition between the two experiments were termed "excess" concentrations. There was no particular significance to the size chosen for the emission pulse and it was given the same spatial distribution as that given to the man-made sources.

Summary

Environmental concerns have been raised in recent years dealing with greenhouse gases produced from the transportation industry. The use of biohydrogen for transport in order to reduce greenhouse gas emissions and the environmental impact of transport, as well as to increase security of supply, technological innovation and agricultural diversification, has a strategic importance. Hydrogen can be burned in such a way as to produce no harmful emissions. The fundamental attraction of hydrogen is its environmental advantages over fossil fuels. But hydrogen is only as clean as the technologies used to produce and use it. If hydrogen is produced without emitting any carbon dioxide or other climate-destabilizing greenhouse gases, it could form the basis of a truly sustainable energy system.

Vehicles and stationary power generation fueled by hydrogen are zero-emission devices at the point of use, with consequential local air quality benefits. Hydrogen-powered fuel cells could contribute to reducing or eliminating emissions of carbon dioxide and other greenhouse gases from road transportation vehicles. The production of hydrogen electrolytically, using clean solar power or other forms of renewable energy is essentially pollution-free. The feedstock, water, is composed of hydrogen and oxygen. Hydrogen production or distribution would produce no CO_2. An internal combustion engine fueled by hydrogen can be adjusted so that the emission of NO_x is 200 times less than in present vehicles. Emissions of NO_x increase with the temperature combustion temperature, the length of the high-temperature combustion period, and the availability of hydrogen, up to a point.

References

CARB (Californian Air Resources Board). 1986. Mobile Source Division. Exhaust emission tests of a hydrogen fueled vehicle, MW-86-006, The Board El Monte, California, Protect 2F86EI.

Deluchi, M.A. 1989. Hydrogen vehicles: an evaluation of fuel storage, performance, safety, environmental impacts, and cost. Int J Hydrogen Energy 14:81–130.

Demirbas, A. 2008. Economic and environmental impacts of the liquid biofuels. Energy Edu Sci Technol 22:37–58.

Demirbas, A., Caglar, A. 1998. Catalytic reforming of biomass and heavy oil residues to hydrogen. Energy Edu Sci Technol 1:45–52.

Demirbas, M.F., Balat, M. 2006. Recent advances on the production and utilization trends of biofuels: a global perspective. Energy Convers Manage 47:2371–2381.

Derwent, R.G., Collins, W.J., Johnson, C.E., Stevenson, D.S. 2001. Transient behavior of tropospheric ozone precursors in a global 3-D CTM and their indirect greenhouse effects. Climatic Change 49:463–487.

EPA (Environmental Protection Agency). 1987. Test car list: passenger cars. EPA, Ann Arbor, MI.

Jefferson, W.T, Wood, D.O., Ferrari, N.A. 1991. Energy and the Environment in the 21st century. MIT Press, Cambridge, MA.

Karapinar Kapdan, I., Kargi. 2006. Biohydrogen production from waste materials. Enzyme Microbial Technol 38:569–582.

Kim, S., Dale, B.E. 2005. Life cycle assessment of various cropping systems utilized for producing biofuels: bioethanol and biodiesel. Biomass Bioenergy 29:426–439.

May, H. 1984. New results of engine driving tests of a 6-cylinder Otto-Engine in mixed hydrogen gasoline operation. 5th World Hydrogen Energy Conference, Toronto, Canada, 1984.

Peschka, W. 1986. Liquid hydrogen fueled automotive vehicles in Germany-status and development. Int J Hydrogen Energy 11:721–728.

Plass, H.J., Barbur, F., Miller, H.P., Veziroglu, T.N. 1990. Economics of hydrogen as a fuel for surface transportation. Int J Hydrogen Energy 15:663–668.

Puppan, D. 2002. Environmental evaluation of biofuels. Periodica Polytechnica Ser Soc Man Sci 10:95–116.

Simmonds, P.G., Derwent, R.G., O'Doherty, S., Ryall, D.B., Steele, L.P., Langenfelds, R.L., Salameh, P., Wang, H.J., Dimmer, C.H., Hudson, L.E. 2000. Continuous high-frequency observations of hydrogen at the Mace Head baseline atmospheric monitoring station over the 1994–1998 period. J Geophys Res 105:105–121.

Stewart, W.F. 1986. Hydrogen as a vehicular fuel, *in* Recent developments in hydrogen technology, Vol. II, K.D. Williamson, F. Edeskuty (Eds.) CRC, Boca Raton, FL.

Strobl, W., Peschka, W. 1986. Liquid hydrogen as a fuel of the future for individual transport, *in* T.N. Veziroglu, N. Getoff, P. Weinzier (Eds.), Hydrogen Energy Progress VI, Proceedings of 6th World Hydrogen Energy Conference, Vienna, Austria. Pergamon, New York.

UN (United Nations). 2006. The emerging biofuels market: regulatory, trade and development implications. United Nations Conference on Trade and Development, New York and Geneva.

Watson, H.C. 1984. An Australian hydrogen car, *in* T.N. Veziroglu, J.B. Taylor (Eds.), Hydrogen Energy Progress V, Proceedings of World Hydrogen Energy Conference, Toronto, Canada, Pergamon, New York.

Williamson, K., Edeskuty, F. 1986. Recent developments in hydrogen technology, Vol. II, CRC, Boca Raton, FL.

Index

MIX
Papier aus verantwortungsvollen Quellen
Paper from responsible sources
FSC® C105338

If you have any concerns about our products,
you can contact us on
ProductSafety@springernature.com

In case Publisher is established outside the EU,
the EU authorized representative is:
Springer Nature Customer Service Center GmbH
Europaplatz 3, 69115 Heidelberg, Germany

Printed by Libri Plureos GmbH
in Hamburg, Germany